BioNanoFluidic MEMS

MEMS Reference Shelf

Series Editor: Stephen D. Senturia
Professor of Electrical Engineering, Emeritus
Massachusetts Institute of Technology
Cambridge, Massachusetts

BioNanoFluidic MEMS
Peter Hesketh, ed.
ISBN 978-0-387-46281-3

Microfluidic Technologies for Miniaturized Analysis Systems
Edited by Steffen Hardt and Friedhelm Schöenfeld, eds.
ISBN 978-0-387-28597-9

Forthcoming Titles

Self-assembly from Nano to Milli Scales
Karl F. Böhringer
ISBN 978-0-387-30062-7

Photonic Microsystems
Olav Solgaard
ISBN 978-0-387-29022-5

Micro Electro Mechanical Systems: A Design Approach
Kanakasabapathi Subramanian
ISBN 978-0-387-32476-0

Experimental Characterization Techniques for Micro-Nanoscale Devices
Kimberly L. Turner and Peter G. Hartwell
ISBN 978-0-387-30862-3

Microelectroacoustics: Sensing and Actuation
Mark Sheplak and Peter V. Loeppert
ISBN 978-0-387-32471-5

Inertial Microsensors
Andrei M. Shkel
ISBN 978-0-387-35540-5

Peter J. Hesketh
Editor

BioNanoFluidic MEMS

Editor
Peter J. Hesketh
George W. Woodruff
School of Mechanical Engineering
Georgia Institute of Technology
Atlanta, GA 30332-0405

ISBN: 978-0-387-46281-3 e-ISBN: 978-0-387-46283-7

Library of Congress Control Number: 2007932882

Printed on acid-free paper

9 8 7 6 5 4 3 2 1

springer.com

Preface

This collaboration evolved from contributions by faculty members who participated in workshops on NanoBioFluidic Micro Electro-Mechanical Systems (MEMS) at the Georgia Institute of Technology, Atlanta, Georgia, in November, 2005 and June, 2006. The objective of these workshops was to bring together researchers, engineers, faculty, and students to review the interdisciplinary topics related to miniaturization and to nanomaterials processing, with a particular emphasis on the development of sensors and microfluidic systems. The workshops were events attended by participants from industry and academia, with lectures, hands-on laboratory sessions, student poster sessions, and panel discussions.

These chapters cover current research topics pertinent to the field, including: materials synthesis, nanofabrication methods, nanoscale structures' properties, nanopores, nanomaterial-based chemical sensors, biomedical applications, and nanodevice packaging. The emphasis has been placed on a review of fundamental principles, thereby providing an introduction to nanodevice fabrication methods. Supporting this background are discussions of recent developments and a selection of practical applications.

It should be noted that NanoBioFluidic MEMS is an enormously broad field of study, and any survey must of necessity be selective. Taken individually, topics chosen for inclusion in this volume may of be most benefit to those working within the corresponding area. Nevertheless, the aggregate of specific topic selections within this compilation should provide an effective overview of this vast, highly interdisciplinary subject, and hopefully, a glimpse into the magnitude of possibilities at the nanoscale.

The enormity of the potential for nanodevices and miniature systems cannot be overstated. An understanding of these possibilities is the first step toward the realization of practical applications and solutions to important problems in health care, agriculture, manufacturing, and the pharmaceuticals industry, among many others. The evolution of these applications will bring about such advancements as novel sensor technologies capable of contributing to such vital undertakings as the reduction of pollution and its inherent impact on global warming, and to any number of comparably imperative enterprises that promise to bring to bear new approaches to solving significant problems and raising the standard of living for people worldwide.

Chapter 1 sets the stage by surveying the past and present of core microelectronic nanotechnology, and addresses its likely future directions. It addresses a central question: is the most appropriate method for integration based upon traditional top down methods, or are bottom up methods more appropriate for manufacturing?

Chapter 2 examines the high temperature growth of a range of metal oxide nanostructures that form nanobelts, nanowires, and nanorods. These materials exhibit notably unique properties of special relevance because they become evident at the nanoscale size. These materials represent an example of a broad class of nanomaterials that promise suitability for integration with microelectronics.

Chapter 3 discusses direct write lithography methods and their processing advantages and limitations.

Chapter 4 presents an introduction to and an overview of nanofabrication methods.

Chapter 5 examines emerging nanoimprinting methods.

Chapter 6 describes methods for nondestructive nanoscale material characterization.

Chapter 7 addresses the use of micro stereo-lithography. Micro- and nanodevices need to be connected to the outside world, and this highly versatile method provides customized coupling either to individual dies or to arrays, and even to wafer-scale integrated packaging.

Chapters 8 through 10 survey nanobiofluidic system applications, including case studies for chemical sensors, nanopores-to-DNA sequencing, and biomaterial cell-surface interfaces.

Chapter 11 concludes the discussion with an exploration into integration methods for fine-pitch electrical connections to nanobiosensors.

I would very much like to thank all of the contributing authors for the timely submission of their manuscripts and for assisting in reviews of their co-authors' chapters. Thanks to Philip Duris for editorial suggestions, in particular a detailed editing of Chapter 4.

It has been a great pleasure to have been a participant in the preparation of this book, principally because of the involvement of such a knowledgeable group of faculty and researchers. The interdisciplinary nature of this important, dynamic, and challenging area of research necessitated the contributions of all involved, to whom I am deeply grateful.

Peter J. Hesketh

Contents

Contributors

Amir G. Ahmadi
School of Chemical & Biomolecular Engineering
Georgia Institute of Technology
Atlanta GA 30332-0100

Thomas Beechem
The George W. Woodruff
School of Mechanical Engineering
Georgia Institute of Technology
Atlanta, GA 30332-0405

Joseph L. Charest
The George W. Woodruff
School of Mechanical Engineering
Georgia Institute of Technology
Atlanta, GA 30332-0405

Ravi Doraiswami
The George W. Woodruff
School of Mechanical Engineering
Georgia Institute of Technology
Atlanta, GA 30332-0405

Samuel Graham
The George W. Woodruff
School of Mechanical Engineering
Georgia Institute of Technology
Atlanta, GA 30332-0405

Peter J. Hesketh
The George W. Woodruff
School of Mechanical Engineering
Georgia Institute of Technology
Atlanta, GA 30332
USA, (404)385-1358

Chenguo Hu
School of Materials
Science and Engineering
Georgia Institute of Technology
Atlanta, GA 30332-0245
USA;
Department of Applied Physics
Chongqing University
Chongqing 400044
China

Gary W. Hunter
NASA Glenn Research
Center at Lewis Field
Cleveland, OH 44135

William P. King
Department of Mechanical
Science and Engineering
University of Illinois
Urbana-Champaign
Urbana, IL 61801, USA

Hong Liu
School of Materials
Science and Engineering
Georgia Institute of Technology
Atlanta, GA 30332-0245
USA;
State Key Laboratory
of Crystal Materials
Shandong University
Jinan 250100
China

Darby B. Makel
Makel Engineering, Inc.,
1585 Marauder St.
Chico, CA 95973

James D. Meindl
School of Electrical and
Computer Engineering
Georgia Institute of Technology
Atlanta, GA 30332
USA

Raghunath Murali
School of Electrical and
Computer Engineering
Georgia Institute of Technology
Atlanta, GA 30332
USA

Sankar Nair
School of Chemical & Biomolecular
Engineering
Georgia Institute of Technology
Atlanta GA 30332-0100

Harry D. Rowland
The George W. Woodruff
School of Mechanical Engineering
Georgia Institute of Technology
Atlanta, GA 30332
USA

David W. Rosen
The George W. Woodruff
School of Mechanical Engineering
Georgia Institute of Technology
Atlanta, GA 30332

Zhong Lin Wang
School of Materials
Science and Engineering
Georgia Institute of Technology
Atlanta, GA 30332-0245
USA

Jennifer C. Xu
NASA Glenn
Research Center at Lewis Field
Cleveland, OH 44135

Chapter 1
Nanotechnology: Retrospect and Prospect

James D. Meindl

Abstract The predominant economic event of the 20th century was the information revolution. The most powerful engine driving this revolution was the silicon microchip. During the period from 1960 through 2000, the productivity of semiconductor or silicon microchip technology advanced by a factor of approximately 100 million. Concurrently, the performance of the technology advanced by a factor greater than 1000. These sustained simultaneous advances were fueled primarily by sequentially scaling down the minimum feature size of the transistors and interconnects of a microchip thereby both reducing cost and enhancing performance. In 2005 minimum feature sizes of 80 nanometers clearly indicate that microchip technology has entered the 1–100 nanometer domain of *nanotechnology* through use of a "top-down" approach. Moreover, it is revealing to recognize that the 300-millimeter diameter silicon wafers, which facilitate microchip manufacturing, are sliced from a 1–2 meter long single crystal ingot of hyper-pure silicon. This silicon ingot is produced by a "self–assembly" process that represents the essence of the "bottom-up" approach to nanotechnology. Consequently, modern silicon microchips containing over one billion transistors are enabled by a quintessential *fusion* of top-down and bottom-up nanotechnology.

Due to factors such as transistor leakage currents and short-channel effects, critical dimension control tolerances, increasing interconnect latency and switching energy dissipation relative to transistors, escalating chip power dissipation and heat removal demands as well as design, verification and testing complexity, it appears that the rate of advance of silicon microchip technology may decline drastically within the next 1–2 decades. Nanotechnology presents a generic opportunity to overcome the formidable barriers to maintaining the historical rapid rate of advance of microchip technology and consequently the information revolution itself. The breakthroughs that are needed are unlikely without a concerted global effort on the part of industries, universities and governments. Nurturing such an effort

J. D. Meindl
School of Electrical and Computer Engineering, Georgia Institute of Technology, Atlanta, GA 30332, USA

P. J. Hesketh (ed.), *BioNanoFluidic MEMS*.

is profoundly motivated by the ensuing prospect of enhancing to unprecedented levels the quality of life of all people of the world.

1.1 Introduction

Beginning about 10,000 years ago in the Middle East, the agricultural revolution was a crucial development in human history. This revolution enabled the accumulation of surplus food supplies, which gave rise to large settlements and the emergence of Western civilization itself.

The industrial revolution that began in the 18th century in Europe was the most far-reaching, influential transformation of human culture following the agricultural revolution. The consequences of the industrial revolution have changed irrevocably human labor, consumption and family structure; it has caused profound social changes, as Europe moved from a primarily agricultural and rural economy to a capitalist and urban economy. Society changed rapidly from a family-based economy to an industry-based economy.

The information revolution was *the* predominant economic event of the 20th century and promises to continue well into the 21st century and beyond. It has given us the personal computer, the multi-media cell phone, the Internet and countless other electronic marvels that influence our daily lives. The explosive emergence of the Internet and its potential to create a global information infrastructure, a global educational system and a global economy provide a unique opportunity to improve the quality of life of all people to unprecedented levels.

1.2 In Retrospect

Perhaps the three most prominent inventions that collectively launched the information revolution were the transistor in 1947 [1], the stored program digital computer in 1945 [2] and the silicon monolithic integrated circuit or "microchip" in 1958 [3]. The single most powerful engine driving the information revolution has been the silicon microchip for two compelling reasons, productivity and performance. For example, from 1960 through 2000, the productivity of silicon technology improved by a factor or more than 100 million [4, 5]. This is evident from the fact that the number of transistors contained within a microchip increased from a handful in 1960 to several hundred million in 2000, while the cost of a microchip remained virtually constant. Concurrently, the performance of a microchip improved by a factor of more than 1,000 [6]. These simultaneous sustained exponential rates of improvement in both productivity and performance are unprecedented in technological history.

The most revealing microchip productivity metric, the number of transistors per microchip, N, can be quantified by a simple mathematical expression: $N = F^{-2} \bullet D^2 \bullet PE$ where F is the minimum feature size of a transistor, D^2 is the area of the microchip and PE is the transistor packing efficiency in units of transistors per minimum feature square or [tr/F^2] [7]. One can graph $\log_2$ vs. calendar year, Y, and then take the derivative of the plot, $d(\log_2 N)/dY$, to observe that N doubled every 12 months in the early decades of the microchip [4, 8] and every 18 months in more recent decades [8]. This incisive observation is now quite widely known as Moore's Law [9].

The minimum feature size of a transistor, F, has been reduced at a rapid rate throughout the entire history of the microchip [9, 10] and is projected to continue to decrease for at least another decade [11]. Chip area, D^2, increased less rapidly than F^{-2} in the early decades of the microchip [9, 10] and maximum chip area is projected to saturate for future generations of technology [11]. Packing efficiency, PE, has increased monotonically throughout the entire history of the microchip but at a considerably smaller rate than F^{-2} [9–11]. The key observation regarding F, D and PE is that reducing the minimum feature size of a transistor, F, or "scaling" has been the most effective means of increasing the number of transistors per chip, N, and consequently improving the productivity of microchip technology.

The most appropriate metric for gauging the performance of a microchip depends greatly on its particular product application. For a microprocessor, the number of instructions per second, IPS, executed by the chip is a commonly used performance metric [12]. A useful mathematical relationship for this metric is: $IPS = IPC \bullet f_C$ where IPC is number of instructions per cycle and f_C is the number of cycles per second or clock frequency of the chip. The IPC executed by a microprocessor depends strongly on both the hardware microarchitecture of the chip and its software instruction set architecture. Throughout the history of the microprocessor its microarchitecture has been influenced significantly by the capabilities and limitations of silicon monolithic microchip technology [12]. This has become quite evident with the recent advent of the chip multiprocessor (or cell microprocessor), CMP, [13, 14], which consists of a (growing) number of complex cells each of which is effectively a microprocessor. The principal purpose of the CMP is to increase the number of instructions per cycle, IPC, executed by the chip. The microarchitecture of a chip multiprocessor is particularly enabled by the cost and latency reductions resulting directly from reduced feature size or scaling of transistors. Consequently, it is clear that scaling effectively enables increases in IPC.

Moreover, the more than 1,000 times increase of microprocessor clock frequency, f_C, from approximately one megahertz in the early 1970's to greater than one gigahertz in the past several years has been driven primarily by feature size and consequent latency reductions due to transistor scaling. In addition, circuit innovations have promoted increasing clock frequencies. Again, the key observation is that scaling has been a most effective means of increasing both IPC and f_C and consequently the performance, IPS, of a microprocessor.

The salient conclusion of the preceding review of microchip productivity, N, and performance, IPS, is that scaling has been *the* most effective means for their enormous advancements. Scaling has been the most potent "fuel" energizing the microchip engine, which has been the most powerful driver of the information revolution.

Throughout the nearly five-decade history of the silicon microchip, its "pacing" technology has been microlithography, which enables scaling. For example, in 1960 the minimum feature size, F, of a microchip transistor was approximately 25 μm; by 2000, F had scaled down over two decades to a value of 0.25 μm; and in 2005 transistor printed gate length is 45 nanometers, nm, and copper interconnect half pitch is 80 nm [11]. In addition, current field effect transistor gate oxynitride insulator thickness is in the 1.5 nm range. These 2005 transistor and interconnect dimensions clearly indicate that silicon microchips have entered the 1.0–100 nanometer domain of nanotechnology [15].

The entry of the microchip into the realm of nanotechnology has been accomplished by exploiting a "top-down" approach. Transistor and interconnect dimensions have been sequentially scaled down for more than four decades through a continuing learning process. However, viewing the development of silicon technology from this perspective alone could be misleading. It is revealing to recognize that modern silicon microchip manufacturing begins with a 300-millimeter (mm) diameter wafer that is sliced from a single crystal ingot of silicon, which is 1–2 meters in length. The density of atoms in this ingot is $5 \times 10^{22}/\text{cm}^3$ and the atomic spacing is 0.236 nm. Perhaps the most interesting feature of this ingot is that it is entirely *"self-assembled"* atom-by-atom during its growth by the Czochralski process [16]. This process has been used for volume production of silicon crystals since the mid-1950s. It is patently *"bottom-up"* nanotechnology. Consequently, in 2005, silicon microchips exploit a quintessential *fusion* of top-down and bottom-up nanotechnology. This fusion has been and remains paramount to the success of microchip technology.

1.3 In Prospect

In projections regarding the prospects of nanotechnology as applied to gigascale and terascale levels of integration for future generations of microchips, it is interesting to consider a scenario that postulates a continuing fusion of top-down and bottom-up approaches. Without a virtually perfect single-crystal starting material it is difficult to project batch fabrication of billions and trillions of sub-10 nm minimum feature size binary switching elements (i.e. future transistors) in a low cost microchip. It is equally difficult to imagine the purposeful design, verification and testing of a multi-trillion transistor computing chip without a disciplined top-down approach. Consequently, this particular prospective is based on the premise of a fusion of top-down and bottom-up nanotechnology with the target of advancing the information revolution for another half-century or more. Discussion of the prospects of nanotechnology begins with an assessment of the most serious obstacles now confronting silicon microchip technology as it continues to progress more deeply into the nanotechnology space. Subsequently, a tentative projection of the salient challenges and opportunities for overcoming these obstacles through nanotechnology and more specifically through carbon nanotube technology is outlined.

A selected group of grand challenges that must be met in order to sustain the historic rate of progress of silicon microchip technology includes the following: 1) field effect transistor (FET) gate tunneling currents a) that are increasing rapidly due to the compelling need for scaling gate insulator thickness and b) that serve only to heat the microchip and drain battery energy; 2) FET threshold voltage that rolls-off exponentially below a critical value of channel length and consequently strongly increases FET subthreshold leakage current without benefit; 3) FET subthreshold swing that rolls-up exponentially below a critical channel length and consequently strongly reduces transistor drive current and therefore switching speed; 4) critical dimension tolerances that are increasing with scaling and therefore endangering large manufacturing yields and low cost chips; 5) interconnect latency and switching energy dissipation that now supercede transistor latency and switching energy dissipation and this supercession will only be exacerbated as scaling continues; 6) chip power dissipation and heat removal limitations that now impose *the* major barrier to enhancement of chip performance; and 7) rapidly escalating design, verification and testing complexity that threatens the economics of silicon microchip technology.

Although the preceding grand challenges appear daunting, prospects for meeting them are encouraging due to the exciting opportunities of nanotechnology as eloquently summarized in the words of Professor Richard Feynman [17]: "There is plenty of room at the bottom." In 1959 he articulated an inspiring vision of nanotechnology [17]: "The principles of physics, as far as I can see, do not speak against the possibility of maneuvering things atom by atom. It is not an attempt to violate any laws; it is something, in principle, that can be done; but in practice, it has not been done because we are too big."

Several relatively recent advances in nanotechnology reveal encouraging progress toward fulfillment of Feynman's vision. First among these advances was the invention of the scanning tunneling microscope in 1981 by Binnig and Rohrer [18]. This novel measurement tool is capable of imaging individual atoms on the surface of a crystal and thus providing a new level of capability to understand what is being built "atom by atom." A second major advance was the discovery of self-assembled geodesic nanospheres of 60 carbon atoms in 1985 by Smalley [19]. A third was the discovery of self-assembled carbon nanotubes in 1990 by Iijima [20]. A fourth was the demonstration, by two separate teams, of carbon nanotube transistors in 1998 [21, 22]. The latter three of these advances deal with carbon nanostructures, which currently represent the particular area of nanotechnology that has been most widely investigated as a potential successor (or extender) of mainstream silicon microchip technology. Consequently, this discussion now focuses on carbon nanotube (CNT) technology as a prime example of the prospects of nanotechnology.

Key challenges that carbon nanotube technology must meet if it is to prove useful for gigascale and terascale levels of integration can be summarized succinctly in two words: *precise control.* Precise control must be achieved of: 1) CNT transistor placement; 2) CNT transistor semiconductor properties or chirality; 3) precise control of CNT interconnect placement; 4) precise control of CNT interconnect metallic properties or chirality; and 5) precise control of semiconductor and metallic junctions. A historical analogy serves to elucidate the comparative state-of-the-art of CNT technology. This analogy suggests that the current status of CNT technology is comparable to that of early semiconductor technology between the 1947 invention of the point contact transistor [1] and the 1958 invention of the silicon monolithic integrated circuit or microchip [3]. A lack of the necessary degree of control to fabricate a monolithic integrated circuit is reflected in the two striking scanning electron micrographs illustrated in Fig. 1.1 [23]. The conclusion of this analogical comparison is that the first critical step in the advancement of CNT technology has been demonstrated but not (yet) the second.

Based on progress to date several rather promising characteristics of CNT transistors and interconnects can be identified. The first of these is the potential for CNT transistors with a subthreshold swing, S, less than the fundamental limit of $S = (kT/q)\ln 2 = 60\,\text{mV/decade}$ on FET transistor subthreshold swing at room temperature, where k is Boltzman's constant, T is temperature in degrees Kelvin and q is the electronic charge. CNT transistors with room temperature $S \approx 40\,\text{mV/decade}$ have recently been reported [24]. The benefits of smaller S are manifold. A performance improvement is in prospect due to the opportunity to reduce binary signal swing and thus reduce transistor latency. A reduction in switching energy dissipation is quite feasible due to a reduced binary signal swing and supply voltage. A reduction in static energy dissipation is expected due to a smaller subthreshold

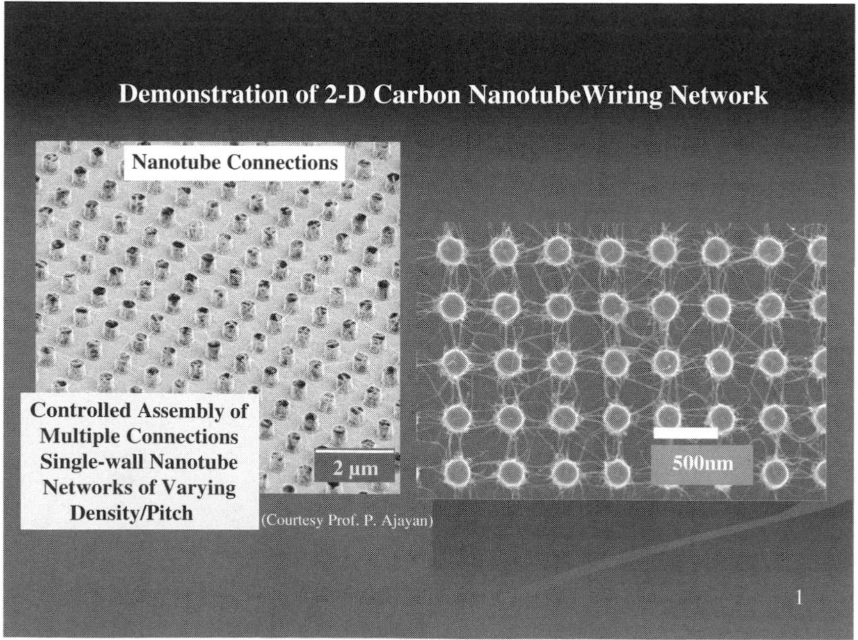

Fig. 1.1 Demonstration of 2D carbon nanotube wiring network

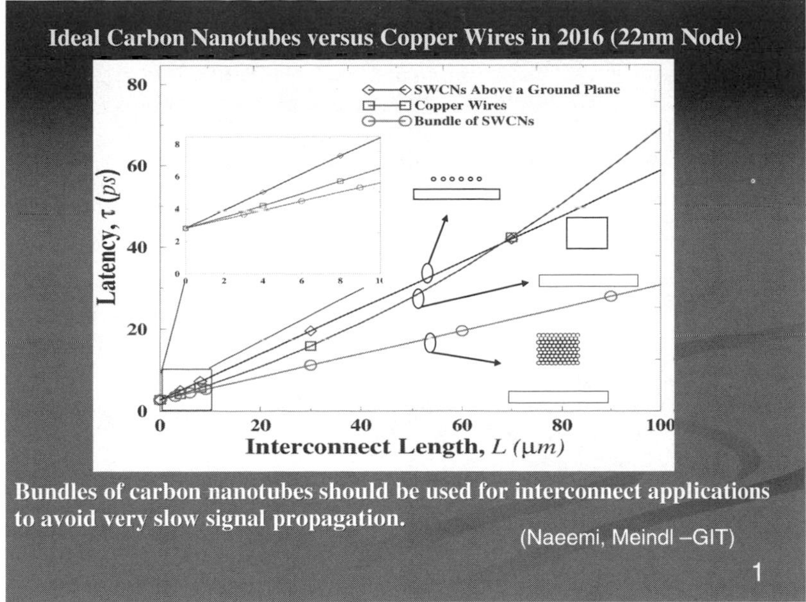

Fig. 1.2 Ideal carbon nanotubes compared with copper wires in 2016 (22 nm node)

leakage current resulting from a reduced S. A second promising characteristic of CNT transistors would be smaller transistor gate and channel lengths. Shorter channels should reduce carrier transit time and thus device switching latency. A third major advantage would be CNT interconnect with smaller latency than copper wires due to ballistic carrier transport in nanotubes in contrast to the multiple scattering of carriers in polycrystalline copper interconnects. A comparison of interconnect latency versus length for both CNTs and copper wires is illustrated in Fig. 1.2 [25]. A rather demanding requirement that Fig. 1.2 reveals is that for CNT interconnects to achieve smaller latency than copper wires at the 22 nm node of silicon microchip technology, projected for 2016 by the ITRS [11], precise control of placement and chirality of a bundle of 100 CNTs each 2 nm in diameter appears to be necessary.

In summary, the potential advantages of CNT technology discussed above could result in substantial improvements in microchips including greater speed, reduced dynamic and static energy dissipation as well as smaller size and therefore lower cost.

1.4 Conclusion

The key conclusion that emerges from the foregoing retrospective and prospective reviews of nanotechnology is that apparently it represents our best prospect for continuing the exponential rate of advance of the information revolution. Recent participation of representatives of corporations, universities and governments in the US, Europe and Japan in the First International Conference on Nanotechnology confirms this conclusion [26]. The implications of continuing this exponential rate of advance to the mid-21st century and beyond are utterly profound. Perhaps the most magnificent prospect is that through continued rapid development of a global information infrastructure, a global educational system and a thriving global economy, the quality of life of all people of the world may be enhanced to unprecedented levels!

References

1. Ross, I. M. (1998). The Invention of the Transistor. *Proceedings of the IEEE: Special Issue: 50th Anniversary of the Transistor, 28*, 7–28.
2. Seitz, F., & Einspruch, N. (1998). *Electronic Genie: The Tangled History of Silicon.* Urbana and Chicago, IL: University of Illinois Press.
3. Kilby, J. S. (1959). *U.S. Patent No. 3,138,743.* Washington, DC: U.S. Patent and Trademark Office.
4. Moore, G. E. (1965, April 19). Cramming More Components onto Integrated Circuits. *Electronics, 38, No. 8.*
5. Takai, Y., et al. (1999). *A 250Mb/s/pin 1GB Double Data Rate SDRAM with a Bi-directional Delay and an Inter-bank Shared Redundancy Scheme. IEEE International Solid-state Circuits Conference, February 15–17,* (pp. 418–419). Augusta, ME: The J. S. McCarthy Co.
6. Thompson, S., et al. (2002, May 16). 130nm Logic Technology Featuring 60nm Transistors, Low-K Dielectrics and Cu Interconnects. *Intel Technology Journal: Semiconductor Technology and Manufacturing, 6*(2), 5–13.
7. Meindl, J. (1993). Evolution of Solid-State Circuits: 1958-1992-20??, Digest of Papers, *IEEE International Solid-State Circuits Conference, February 24–26,* (pp. 23–26).
8. Moore, G. E. (2003). *No Exponential is Forever: But "Forever" can be Delayed! IEEE International Solid-State Circuits Conference, February 9–13,* (20–23). Augusta, ME: J. S. McCarthy Printers.
9. Moore, G. E. (2003). *Progress in Digital Integrated Electronics, Technical Digest of IEEE International Electron Devices Meeting,* p. 11, Dec. 1975.
10. Meindl, J. D. "Theoretical, Practical and Analogical Limits in ULSI", Technical Digest, IEEE International Electron Devices Meeting, pp. 8–13, Dec. 1983.
11. Tokyo, J. (2004). *International Technology Roadmap for Semiconductors Update.* December 1.
12. Hennessy, J. L., & Patterson, D. A. (1990). *Computer Architecture: A Quantitative Approach,* San Mateo, CA: Morgan Kaufmann Publishers, Inc.
13. Naffziger, S., et. al. (2005). *The Implementation of a 2-core Multi-Threaded Itanium®-Family Processor, IEEE Solid-State Circuits Conference, February 6–10,* (pp. 182–183). Lisbon Falls, Maine: S^3 Digital Publishing, Inc.
14. Pham, D., et al. (2005). *The Design and Implementation of a First-Generation CELL Processor, IEEE Solid State Circuits Conference, February 6–10,* (pp. 184–185). Lisbon Falls, Maine: S^3 Digital Publishing, Inc.
15. Nanoscale Science, Engineering and Technology Subcommittee. (2004). *The National Nanotechnology Initiative Strategic Plan.* Washington, DC: National Science and Technology Council.
16. Huff, H. R. (1997). Twentieth Centery Silicon Microelectronics, *ULSI Science and Technology/1997,* ECS PV 97-3, 53–117.
17. Feynman, R. P. (1992). There's Plenty of Room at the Bottom. *Journal of Microelectromechanical Systems.* 1(1), 60–66.

18. Binnig, H. G., Rohrer, C. G., & Weibel, E. (1981). Tunneling through a controllable vacuum gap. *Applied Physics Letters, 40*(2), 178–180.
19. Kroto, H. W., Heath, J. R., O'Brien, S. C., Curl, R. F., & Smalley, R. E. (1985). C_{60}: Buckminsterfullerene. *Nature, 318*, 162–163.
20. Ijiima, S. (1991). Helical microtubules of graphitic carbon. *Nature, 354*, 56–58.
21. Tans, Sander J, Verschueren, Alwyin R. M., Dekker, C. (1998). Room-temperature transistor based on a single carbon nanotube. *Nature, 393*, 49–52.
22. Martel, R., Schmidt, T., Shea, H. R., Hertel, T., & Avouris, Ph. (1998). Single- and multi-wall carbon nanotube filed-effect transistors. *Applied Physics Letters, 73*(17), 2447–2449.
23. Jung, Y., et al. (2003). High-Density, large-Area Single-Walled Carbon Nanotube Networks on nanoscale Patterned substrates. *Journal of Physical Chemistry*, 107, 6859–6864.
24. Appenzeller, J., Lin, Y. M., Knoch, J., & Avouris, Ph. (2004). Band-to-Band Tunneling in Carbon Nanotube Field-Effect Transistors. *Physical Review Letters, 93*(19), 196805-1-196805-4.
25. Naeemi, A., Sarvari, R., & Meindl, J. D. (2004). *Performance Comparison between Carbon Nanotube and Copper Interconnects for GSI. IEEE International Electron Devices Meeting. December 13–15,* 29.5.1–29.5.4.
26. San Francisco, C. A. (2005). *First International Nanotechnology Conference on Communication and Cooperation.* June 1–3.
27. Kirihata, T., et al. (1999). *A 390mm^2 16 Bank 1 Gb DDR SDRAM with Hybrid Bitline Architecture. IEEE International Solid-State Circuits Conference, February 15–17,* (pp. 422–423). Augusta, ME: The J. S. McCarthy Co.
28. Noyce, R. N. (1959). *U.S. Patent No. 2,981,877.* Washington, DC: U.S. Patent and Trademark Office.

Chapter 2
Synthesis of Oxide Nanostructures

Chenguo Hu, Hong Liu and Zhong Lin Wang*

Abstract Growth of oxide nanostructures is an important part of nanomatirials research, and it is the fundamental for fabricating various nanodevices. This chapter introduces the four main growth processes for synthesizing oxide nanostructures: hydrothermal synthesis, vapor-liquid-solid (VLS), vapor-solid (VS) and composite-hydroxide mediated synthesis. Detailed examples will be provided to illustrate the uniqueness and applications of these techniques for growing oxide nanowires, nanobelts and nanorods.

Keywords: Hydrothermal synthesis · Vapor-liquid-solid · Vapor-solid · Composite-hydroxide mediated · ZnO · $BaTiO_3$ · Nanobelts, Nanowires, Nanorods

Abbreviation

CHM-Composite hydroxide mediated, **MMH**-Microemulsion-mediated hydrothermal, **VS**-Vapor solid, **VLS**-Vapor liquid solid, **HRTEM**-High resolution transmission electron microscope, **XRD**-X-ray Diffraction

2.1 Introduction

Functional oxides are probably the most diverse and rich materials that have important applications in science and technology for ferromagnetism, ferroelectricity, piezoelectricity, superconductivity, magnetoresistivity, photonics, separation, catalysis, environmental engineering, etc. [1] Functional oxides have two unique structural features: switchable and/or mixed cation valences, and adjustable oxygen deficiency, which are the bases for creating many novel materials with unique electronic, optical, and chemical properties. The oxides are usually made into

Z. L. Wang
School of Materials Science and Engineering Georgia Institute of Technology, Atlanta, GA 30332-0245, USA
e-mail: zhong.wang@mse.gatech.edu

P. J. Hesketh (ed.), *BioNanoFluidic MEMS.*

nanoparticles or thin films in an effort to enhance their surface sensitivity, and they have recently been successfully synthesized into nanowire-like structures. Utilizing the high surface area of nanowire-like structures, it may be possible to fabricate nano-scale devices with superior performance and sensitivity. This chapter reviews the general techniques used for growing one-dimensional oxide nanostructures.

2.2 Synthesis Methods

2.2.1 VS Growth

The vapor phase evaporation represents the simplest method for the synthesis of one-dimensional oxide nanostructures. The syntheses were usually conducted in a tube furnace as that schematically shown in Fig. 2.1 [2]. The desired source oxide materials (usually in the form of powders) were placed at the center of an alumina or quartz tube that was inserted in a horizontal tube furnace, where the temperatures, pressure, and evaporation time were controlled. Before evaporation, the reaction chamber was evacuated to $\sim$1–3$\times10^{-3}$ Torr by a mechanical rotary pump. At the reaction temperature, the source materials were heated and evaporated, and the vapor was transported by the carrier gas (such as Ar) to the downstream end of the tube, and finally deposited onto either a growth substrate or the inner wall of the alumina or quartz tube.

For the vapor phase evaporation method, the experiments were usually carried out at a high temperature (>800°C) due to the high melting point and low vapor pressure of the oxide materials. In order to reduce the reaction temperature, a mixed source material, in which a reduction reaction was involved, was employed. For example, Huang et al. [3] obtained ZnO nanowires by heating a 1:1 mixture of ZnO and graphite powders at 900–925°C under a constant flow of Ar for 5–30 minutes. In addition, the reaction temperature can be further reduced when the low melting point metal that is the cation of the final oxide compound was heated in an oxidized atmosphere.

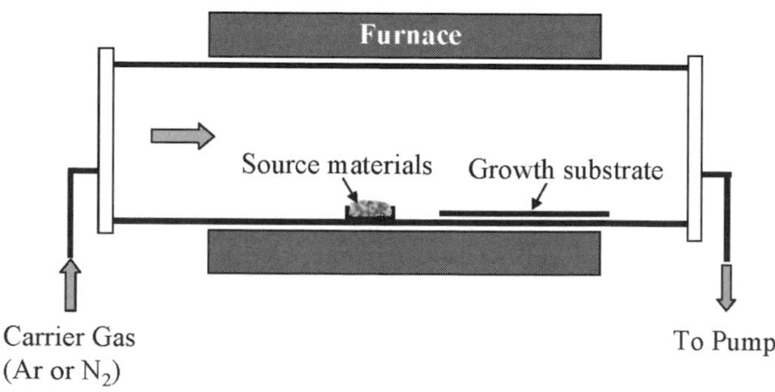

Fig. 2.1 Schematic experimental setup for the growth of one-dimensional oxide nanostructures via an evaporation-based synthetic method

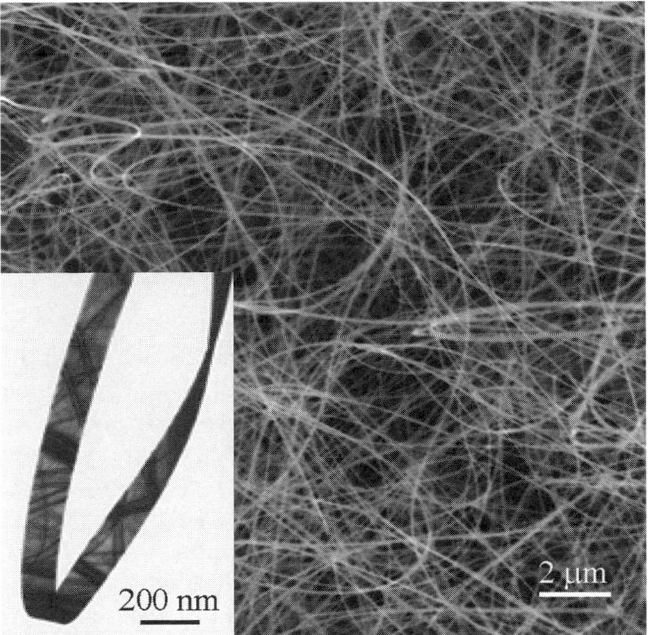

Fig. 2.2 SEM image of ZnO nanobelts. The inset is a TEM image showing the morphological feature of the nanobelts

Figure 2.2 shows the vapor-solid process synthesized ZnO nanobelts. The as-synthesized nanobelts have extremely long length and they are dispersed on the substrate surface. The nanobelt has a rectangular cross-section and uniform shape. The quasi- one dimension structure and uniform shape are a fundamental ingredient for fabrication of advanced devices.

2.2.2 VLS Growth

The growth of one-dimensional oxide nanostructures via vapor phase evaporation may occur with or without catalyst. The feature of the catalyzed-grown nanowires is that a catalyst nanoparticle is always present at one end of the nanowires. The function of the catalyst during nanowire growth is to form a low melting point eutectic alloy with the nanowire materials, which acts as a preferential site for absorption of gas-phase reactant and, when supersaturated, the nucleation site for crystallization. During growth, the catalyst particle directs the nanowire's growth direction and defines the diameter of the crystalline nanowires. The growth of the nanowires catalyzed by a catalyst particle follows a mechanism called vapor-liquid-solid (VLS), which was proposed by Wagner and Ellis in 1964 for silicon whisker growth [4].

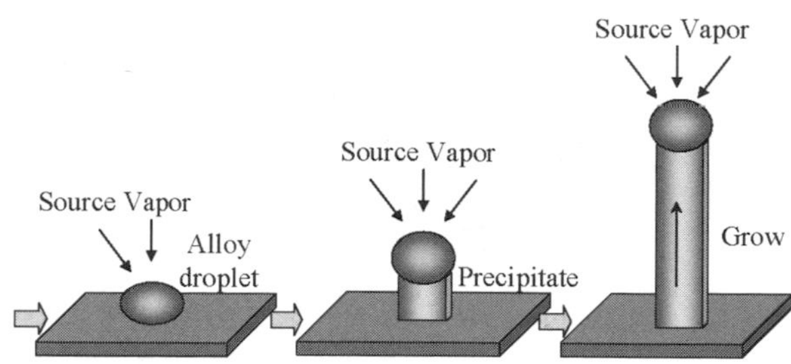

Fig. 2.3 Schematic diagram showing the growth process in VLS method

In the VLS process (Fig. 2.3), a liquid alloy droplet composed of metal catalyst component (such as Au, Fe, etc.) and nanowire component (such as Si, III–V compound, II–V compound, oxide, etc.) is first formed under the reaction conditions. The metal catalyst can be rationally chosen from the phase diagram by identifying metals in which the nanowire component elements are soluble in the liquid phase but do not form solid solution. For the 1D oxide nanowires grown via a VLS process, the commonly used catalysts are Au [3], Sn [5], Ga [6], Fe [7], Co [8], and Ni [9]. The liquid droplet serves as a preferential site for absorption of gas phase reactant and, when supersaturated, the nucleation site for crystallization. Nanowire growth begins after the liquid becomes supersaturated in reactant materials and continues as long as the catalyst alloy remains in a liquid state and the reactant is available. During growth, the catalyst droplet directs the nanowire's growth direction and defines the diameter of the nanowire. Ultimately, the growth terminates when the temperature is below the eutectic temperature of the catalyst alloy or the reactant is no longer available. As a result, the nanowires obtained from the VLS process typically have a solid catalyst nanoparticle at its one end with diameter comparable to that of the connected nanowires. Thus, one can usually determine whether the nanowire growth was governed by a VLS process form the fact that if there present a catalyst particle at one end of the nanowire.

Figure 2.4 shows an array of ZnO nanowire arrays grown by VLS approach on sapphire substrate. The distribution of the Au catalyst determines the locations of the grown nanowires, and their vertical alignment is determined by the epitaxial growth on the substrate surface.

2.2.3 *Hydrothermal Synthesis*

Hydrothermal synthesis appeared in 19th century and became an industrial technique for large size quartz crystal growth in 20th century [10]. Recent years, hydrothermal synthesis method has been widely used for preparation of numerous kinds of inorganic and organic nanostructures.

Hydrothermal synthesis offers the possibility of one-step synthesis under mild conditions (typically <300°C) in scientific research and industrial production [11]. It involves a chemical reaction in water above ambient temperature and pressure in a sealed system. In this system, the state of water is between liquid and steam,

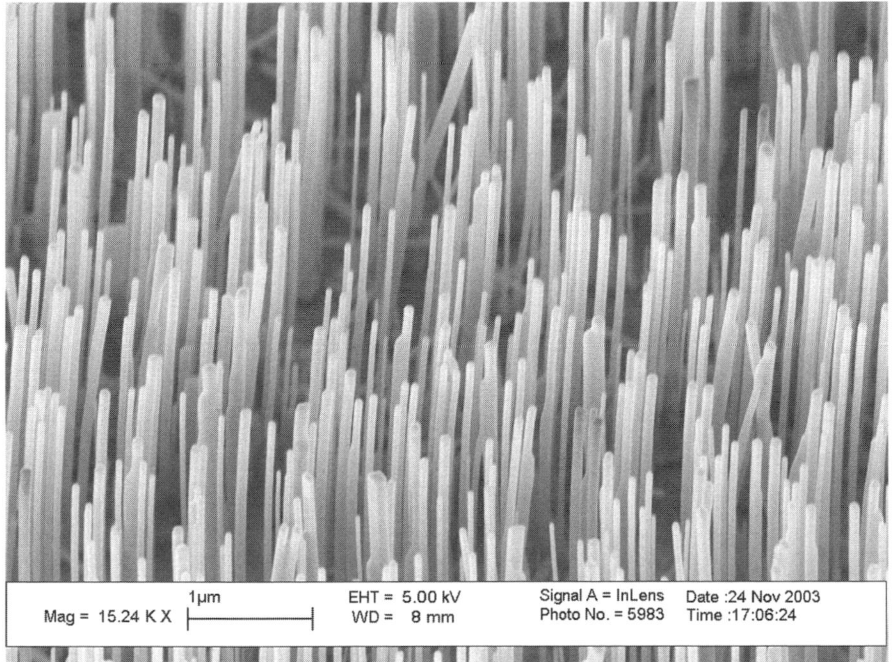

Fig. 2.4 Aligned ZnO nanowires grown by a VLS process

and called as supercritical fluid (Fig. 2.5a). The solubility to the reactants and transportation ability to the ions in the liquid of such a fluid is much better than that in water. Therefore, some reactions that are impossible to carry on in water in ambient atmosphere can happen at a hydrothermal condition. Normally, hydrothermal synthesis process is a one-step reaction. All the reactants with water are added into the autoclave. The reaction occurs in the sealed autoclave when the system is heated, and the nanostructures can be obtained after the autoclave cooled down.

During the reaction, temperature of the reaction system and the pressure in the autoclave are very important for the reaction results, such as the phase and morphology of the product. The amount of water percentage in the vessel determines the prevailing experimental pressure at a certain temperature [12]. In hydrothermal systems, the dielectric constant and viscosity of water decrease with rising temperature and increase with rising pressure, the temperature effect predominating [13, 14]. Owing to the changes in the dielectric constant and viscosity of water, the increased temperature within a hydrothermal medium has a significant effect on the speciation, solubility, and transport of solids. Formation of metal oxides through a hydrothermal method should follow such a principal mechanism: the metal ions in the solution react with precipitant ions in the solution and form precipitate, and the precipitate dehydrate or decompound in the solution at a high temperature and form crystalline metal oxide nanostrucutres [15].

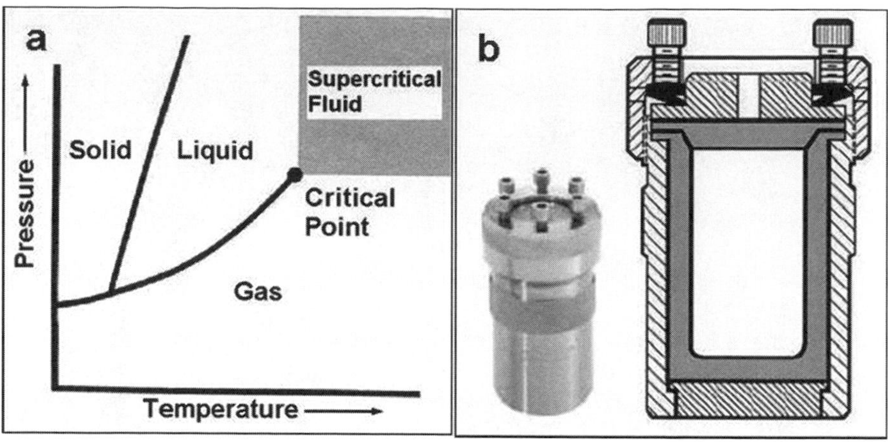

Fig. 2.5 Phase Diagram of water; b. An autoclave for synthesis of oxide nanostructure

Although the chemical reaction mechanism of growth of oxide nanostructures is very similar to the growth of large size quartz by a hydrothermal method, the autoclave used for synthesis of oxide nanostructure is much simple compared with that for growth of quartz. Figure 2.5b shows a typical autoclave for synthesis of oxide nanostructure. Because the synthesis apparatus and controlling process is very simple, the hydrothermal route has been used for preparation of oxide nanostructures both in research and in industry.

Simple oxides and complex oxides can be synthesized through hydrothermal method by designing special chemical reactions and at proper conditions.

2.2.3.1 Simple Oxide Nanostructures Synthesized by a Hydrothermal Method

For some oxide nanostructures, the synthesis approach is very simple. Here we take synthesis of MnO_2 nanowires as an example of the synthesis method for simple oxides [16]. In this synthesis, the reactants were $MnSO_4{\cdot}H_2O$ and $(NH_4)_2S_2O_8$, without any catalyst or template. Both the reactant are put into a Teflonlined stainless steel autoclave, sealed, and maintained at 120°C for 12 h. After the reaction was completed, α-MnO_2 nanowires diameters 5–20 nm and lengths ranging between 5 and 10 μm can be obtained (Fig. 2.6a and b). By adding little amount of $(NH_4)_2SO_4$ into the reactants, β-MnO_2 nanowires with diameters 40–100 nm and lengths ranging between 2.5 and 4.0 μm can be obtained (Fig. 2.6c and d).

Hydrothermal method is a marvelous method for synthesis of oxide nanostructures. For some oxides, put oxide powder with some hydroxide, such as NaOH or KOH, with some water into the autoclave. After heating, some nanostructures with interesting morphology can be obtained. Mn_2O_3 powder is treated in NaOH solution by a hydrothermal process at 170°C for over 72 hours, MnO_2 nanobelts can be obtained [17].

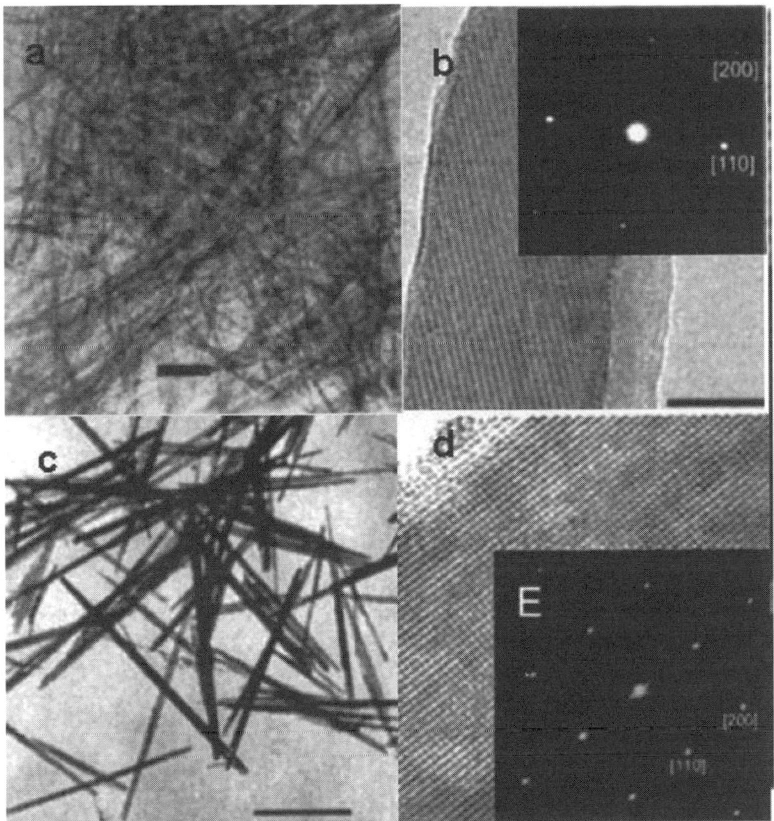

Fig. 2.6 Morphology of MnO_2nanostructures synthesized through a hydrothermal method [16]. a., b. α-MnO_2 nanowires; c., d. β- MnO_2 nano rods

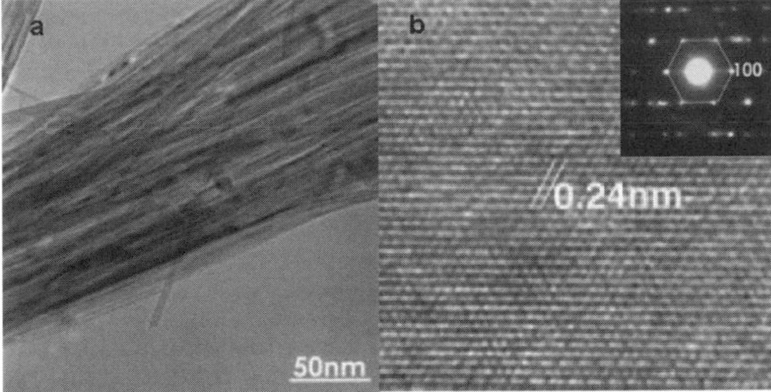

Fig. 2.7 MnO_2 nanobelts synthesized through a hydrothermal method by using Mn_2O_3 as starting material [17]

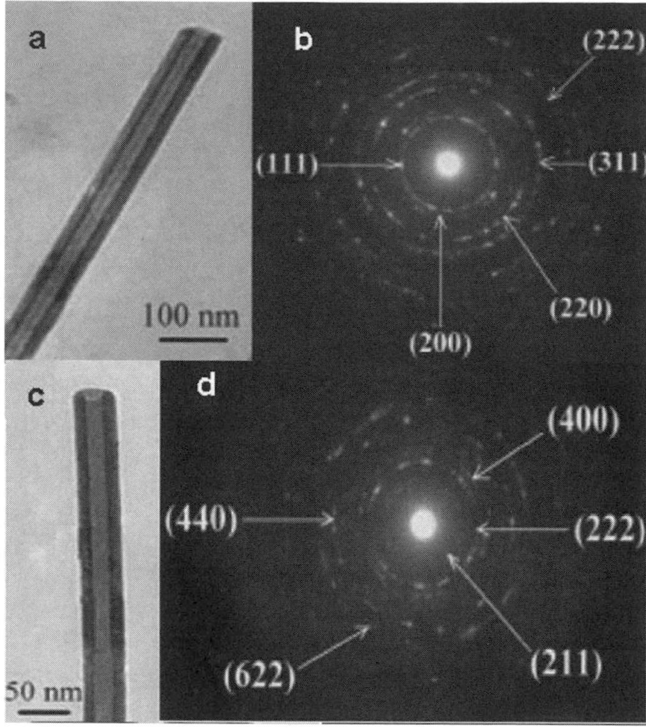

Fig. 2.8 Morphology of Tb_4O_7 (a, b) and Y_2O_3nanotubes synthesized through a hydrothermal method and calcining [18]

By using this method, Tb_4O_7 and Y_2O_3 powders can be transformed into $Tb(OH)_3$ and Y(OH)3 nanotubes through a hydrothermal method at 170°C for 48 h, and then convert into $Tb_4O_7$and Y_2O_3nanotubes by calcining the hydroxide nanotubes at 450°C for 6 h [18].

For some oxides, it is difficult to obtain nanoparticles without surfactant or template. Normally, to get better morphology of oxide nanostructures, some organic regents are often added into the reactant system as surfactants or chemical templates. Otherwise, it is very difficult to get nanoparticles with special morphology. Therefore, some modified hydrothermal methods have appeared for synthesis of some special nanostructures of oxides.

Combining microemulsion technique and hydrothermal method, a modified hydrothermal method, so called a microemulsion-mediated hydrothermal (MMH) method has been suggested [19]. TiO_2 nanorods and nanospheres can be obtained by this method. For synthesis of TiO_2 nanostructures, a kind of solution was formed by dissolving tetrabutyl titanate into hydrochloric acid or nitric acid, and the solution was dispersed in an organic phase for the preparation of the microemulsion medium. Thc aqucous cores of water/Triton X-100/hexanol/cyclohexane microemulsions were used as constrained microreactors for a controlled growth of titania particles

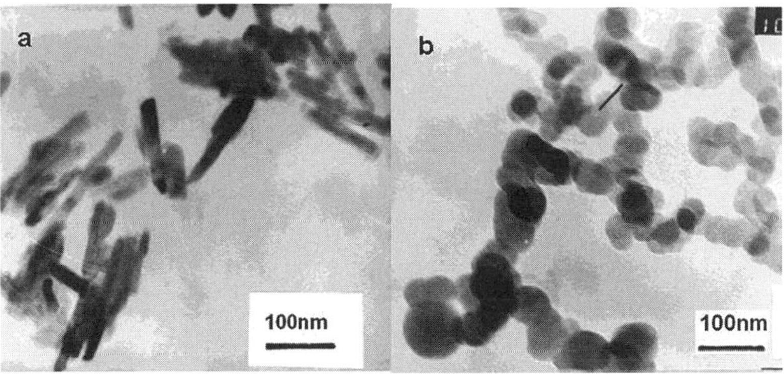

Fig. 2.9 TiO_2 nanorods (a) and nano-spheres synthesized through a MMH method [19]

under hydrothermal conditions. Figure 2.9 is the morphology of TiO_2 nanostructures synthesized by MMH method.

In recent years, some progress has been made in modified hydrothermal synthesis approach. The most significant progress for modified hydrothermal synthesis route should be the synthesis of hollow oxide nanospheres published recently.

Metal oxide Fe_2O_3, NiO, Co_3O_4, CeO_2, MgO, and CuO hollow spheres that are composed of nanoparticles have been explored using hydrothermal synthesis [20]. As shown in Fig. 2.5a, after the hydrothermal treatment of mixtures of carbohydrates with different metal salts in water in sealed steel autoclaves at 180°C, carbon spheres with the metal precursors tightly embedded in the microsphereswere obtained. The removal of carbon directly results in hollow spheres of the corresponding metal oxide that are composed of nanoparticles with high surface areas, as shown in Fig. 2.10, the SEM micrographs of the hollow spheres.

2.2.3.2 Complex Oxide Nanostructures Synthesized by a Hydrothermal Method

Except for the simple oxide system, hydrothermal method has applied to synthesize some complex oxide nanostructures. Two successful examples for synthesis of complex oxides through hydrothermal method are synthesis of $ZrGeO_4$ nanoparticles and $ZnAl_2O_4$ nanorods.

$Al(NO_3)_3{\cdot}9H_2O$, $Zn(NO_3)_3$ and aqueous ammonia was used as raw materials for synthesis of $ZnAl_2O_4$ nanostructures. The product obtained from the hydrothermal reaction at 200°C for 20 hours and following calcination at 750°C for 5 hours is consist of nanorods 20 nm in diameter and several hundreds nanometers in length (Fig. 2.11) [21].

Another important example for synthesis of complex oxide nanostructures is synthesis of $ZrGeO_4$ [22]. Single-phase zircon- and sheelite-type $ZrGeO_4$ were selectively synthesized from the reaction of a $ZrOCl_2$ solution and GeO_2 under mild hydrothermal conditions at 120–240°C by pH control of the solution via

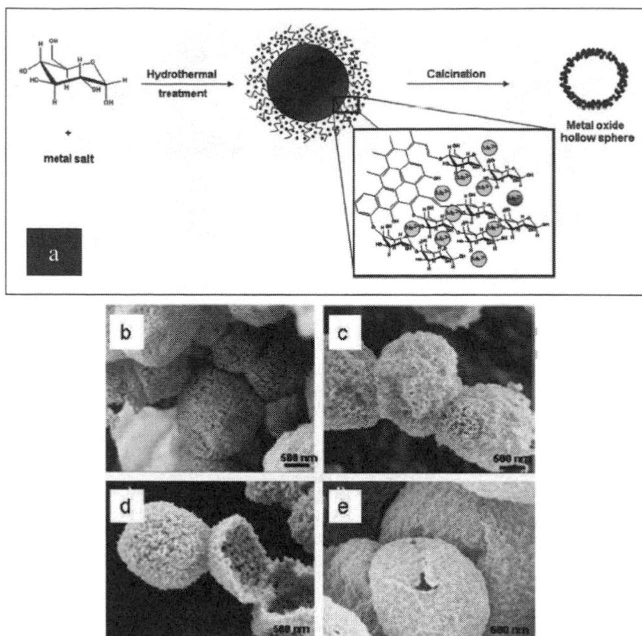

Fig. 2.10 Schematic illustration of the synthesis of metal oxide hollow spheres from hydrothermally treated carbohydrate and metal salt mixtures (a) and SEM images of NiO (b), Co_3O_4(c), CeO_2, and (d) MgO hollow spheres [20]

homogeneous generation of a hydroxide ion through the decomposition of urea. The morphology of obtained $ZrGeO_4$ is varied with the amount of urea added in the reactant solution. Figure 2.12 shows the morphology of $ZrGeO_4$ nanoparticles synthesized by hydrothermal method. Cube-like and rhombohedron-like particles can be obtained in the solution with different urea content.

Hydrothermal synthesis method has been used for quartz crystal growth for almost one century, and becomes a very important synthesis process for synthesis of nanostructured materials, such as, zeolites, mixed oxides, and layered oxides

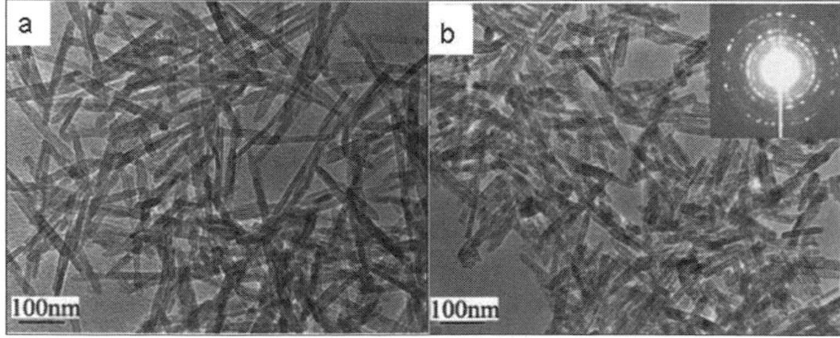

Fig. 2.11 $ZnAl_2O_4$ nanorods synthesized by a hydrothermal method, a. before calcinations; b after calcinations at 750°C for 5 hours [21]

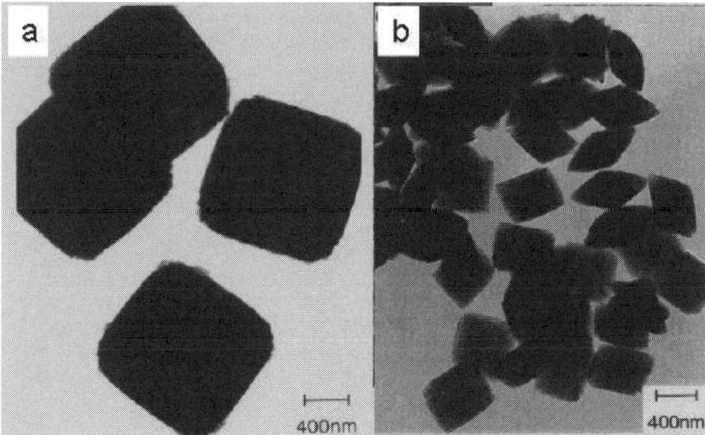

Fig. 2.12 TEM micrographs of zircon-type ZrGeO precipitates; (a) in the absence of urea, (b) in the presence of 0.1 mol/dm^3 urea [22]

in recent decades. This synthesis method will attract more attention because of its novel reaction mechanism and wide application in synthesis of some oxide be extensively studied, because of its low-cost and novel mechanism for synthesis of some nanomaterials with special nanostrucures.

2.2.4 Composite-Hydroxide-Mediated Technique

Composite-hydroxide-mediated (CHM) technique is an effectively universal new approach to synthesize nanostructures of scientific and technological importance, which is first invented by Liu, Hu and Wang [23, 24]. The method is based on a reaction of source materials in a solution of composite-hydroxide eutectic under temperature of higher than 165°C and normal atmosphere without using organic dispersant or capping agent. Although the molting points of both pure sodium hydroxide and potassium hydroxide are over 300°C, T_m = 323°C for NaOH, and T_m = 360°C for KOH, the eutectic point for mixed NaOH/KOH=51.5:48.5 is only about 165°C, as is shown in the schematic phase diagram (Fig. 2.13). So, nanocrystals can be grown at ~200°C or lower and avoid high pressure ambient, which is essential condition in hydrothermal synthesis. The as-produced nanomaterials are single crystalline with clean surface, which is most favorable for further modification in bio-uses. This methodology provides a one-step, convenient, low cost, nontoxic and mass-production route for synthesis of nanostructures of functional oxide materials of various structure types.

The CHM method offers one-step synthesis under mild conditions (typically >165°C, ambient pressure) for scientific research and industrial production. It involves a chemical reaction in melted hydroxides in a vessel. In this system, solution state of composite-hydroxide serves as a reaction medium, something like water

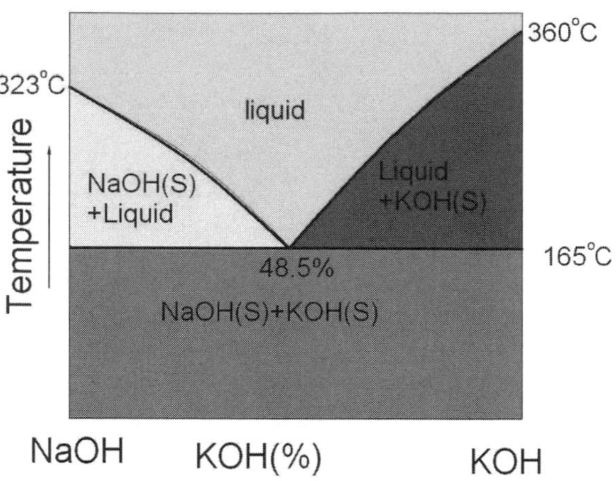

Fig. 2.13 Phase diagram of NaOH–KOH. Molting points of potassium hydroxide is 323°C. Molting point of sodium hydroxide is 360°C. The eutectic point at NaOH/KOH=51.5:48.5 is only about 165°C. [23]

or organic solution in solution reaction method. As it is an ion solution, the solubility to reactants and transportation ability to ions in such liquid is much better than that in water. The melted NaOH and KOH behave not only as solvent also a reactant to participate in reactions, but they do not appear in final oxides, acting as catalysts. Therefore, some reactions that are impossible to carry out in aqueous solution in the atmosphere can happen at the solution state of the hydroxides. Normally, the synthesis process of the CHM method is a one-step reaction. All the reactants with mixed hydroxides are added into the Teflon vessel. The reaction occurs in the vessel when it is heated, and the nanostructures can be obtained after the vessel cooled down.

The preparation steps of nanomaterials by the CHM method is illustrated in Fig. 2.14.

It is very easy way to synthesis some simple oxides by the CHM method. Take the synthesis of CeO_2 nanoparticles as an example to show how to prepare simple oxide nanomaterials by the CHM method [25]. In a typical experiment, 20 g of mixed hydroxides (NaOH:KOH=51.5:48.5) is placed in a 25 ml covered Teflon vessel. Then, 0.1g $Ce(NO_3)_3$ is added into the Teflon vessel. The vessel is put into a furnace, which is preheated to 190°C. After the hydroxides being totally molten, the molten hydroxide solution is stirred by a platinum bar or by shaking the covered vessel to ensure the uniformly of the mixed reactants. After reacting for 48 hours, the vessel is taken out and cooled down to room temperature. Then, deionized water is added to the solid product. The product is filtered and washed by deionized water to remove hydroxide on the surface of the particles. The produced CeO_2 particles under the condition of 190°C for 48 hours are shown in Fig. 2.15(a). The CaF_2structured small particles assemble into big particle. The synthesized Cu_2O nanowires by the CHM method are also shown in Fig. 2.15(b). (Hu, et. al., to be published)

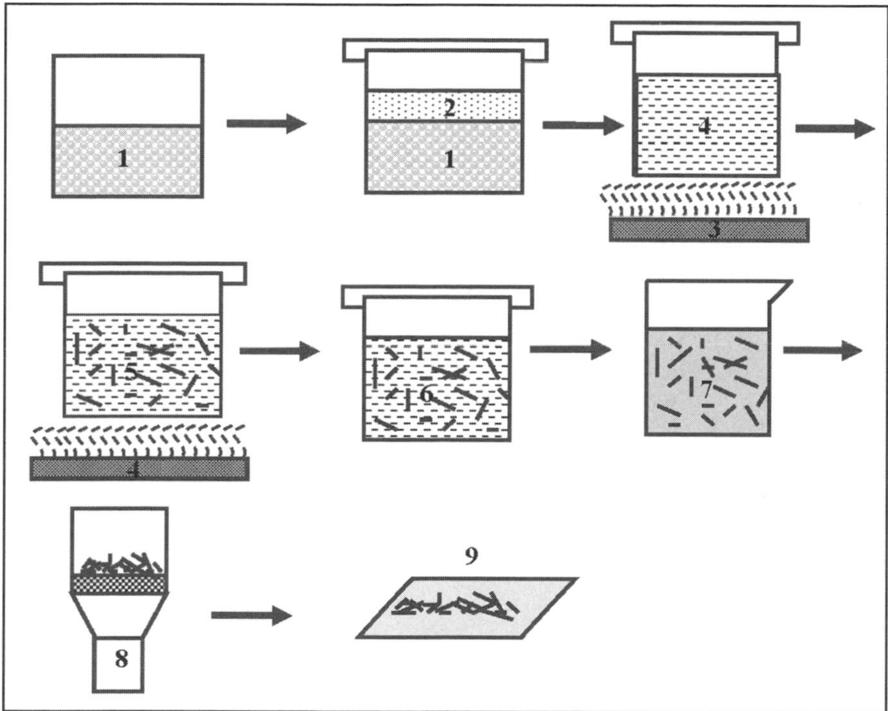

Fig. 2.14 Preparation steps of synthesis of complex oxides. 1 complex hydroxides, 2 source materials, 3 heating, 4 mixed solution of melten complex hydroxides and source materials, 5 form of nanostructures and growth, 6 stop heating, 7 cooling, 8 washing and filtrating, 9 as-produced nanomaterials

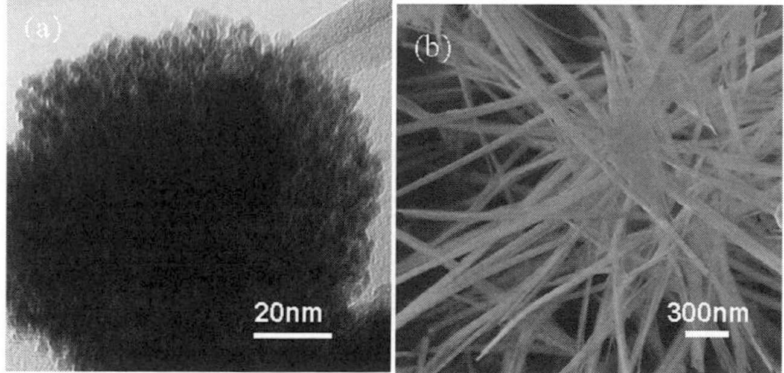

Fig. 2.15 (a) TEM image of ultrafine CeO_2 nanoparticles with size of 3–6 nm. (b) SEM image of Cu_2O nanowires. (Hu et al., to be published)

2.3 Hydroxides Mediated Synthesis of Complex Oxides

Complex oxides with structures such as perovskite, spinel, and garnet have many important properties and applications in science and engineering, such as ferroelectricity, ferromagnetism, colossal magnetoresistance, semiconductor, luminance, and optoelectronics [26,27]. Nanostructures of complex oxides have attracted much attention recently because of their size induced novel properties. Although some synthesis methods are successful for fabricating single-cation oxide nanocrystals [2, 28, 29], only a limited amount of work is available for synthesizing nanostructures of complex oxides (with two or more types of cations) because of difficulties in controlling the composition, stoichiometry and/or crystal structure. The existing techniques rely on high pressure, salt-solvent mediated high temperature, surface capping agent, or organometallic precursor mediated growth process [30, 31], and the types of oxides that can be synthesized are rather limited. Therefore, seeking a simple approach for low-cost, lower-temperature, large-scale, controlled growth of oxide nanostructures at atmospheric pressure is critical particularly for exploring zero- and one-dimensional complex oxide based nanostructures in nanodevices and nanosystems. However, the CHM method provides a new route to synthesize complex oxides.

Take the synthesis of two families of complex oxides, perovskite (ABO_3; $A_xA'_{1-x}BO_3$; $AB_x B'_{1-x} O_3$) and spinel (AB_2O_4), to illustrate the principle of source materials for the CHM method. The sources for A and A' cations are from metallic salts, such as nitrates, chlorates, creosote, or acetates, and etc., and the sources for B and B' cations are from oxides with valence states that match to those present in the desired product to be synthesized.

2.3.1 *Perovskites*

The perovskite structure ABO_3, constitutes one of the most basic and important structures in solid-state science. This is not only because of its relative simplicity, but also due to the fact that the structure leads itself to a wide variety of chemical substitutions at the A, B and O sites, provided the ionic radius and the charge neutrality criteria are satisfied. In addition, many members of this family are found to be useful in various technological applications. This is a direct consequence of their wide spectrum of interesting physical properties such as electrical, magnetic, dielectric, optical and catalytic behaviors.

Our first example of perovskite (ABO_3) is $BaTiO_3$, an important ferroelectric material [32]. The synthesis follows the following steps. (1) An amount of 20 g of mixed hydroxides (NaOH:KOH=51.5:48.5) is placed in a 25 ml covered Teflon vessel. (2) A mixture of anhydrous $BaCl_2$ and TiO_2 at 0.5 mmol each is used as the raw material for reaction. (3) The raw material is placed on the top of the hydroxide in the vessel. The vessel is put in a furnace, which is preheated to 200°C. (4) After the hydroxides being totally molten, the molten hydroxide solution is stirred by a platinum bar or by shaking the covered vessel to ensure the uniformly of the mixed reactants. (5) After reacting for 48 hours, the vessel is taken out and cooled down to room temperature. Then, deionized water is added to the solid product. The product is filtered and washed by first deionized water and then hot water to remove hydroxide on the surface of the particles.

X-ray diffraction (XRD) measurement proved that the as-synthesized product is tetragonal $BaTiO_3$ (P4 mm, JCPD 81-2203) (Fig. 2.16a). Scanning electron microscopy (SEM) image of the powder shows that the particles are nanocubes or nanocuboids with 30–50 nm in sizes (Fig. 2.16b), and energy dispersive X-ray analysis (EDS) shows that the presence of oxygen, barium, and titanium. Electron diffraction (ED) and high-resolution transmission electron microscope (HRTEM)

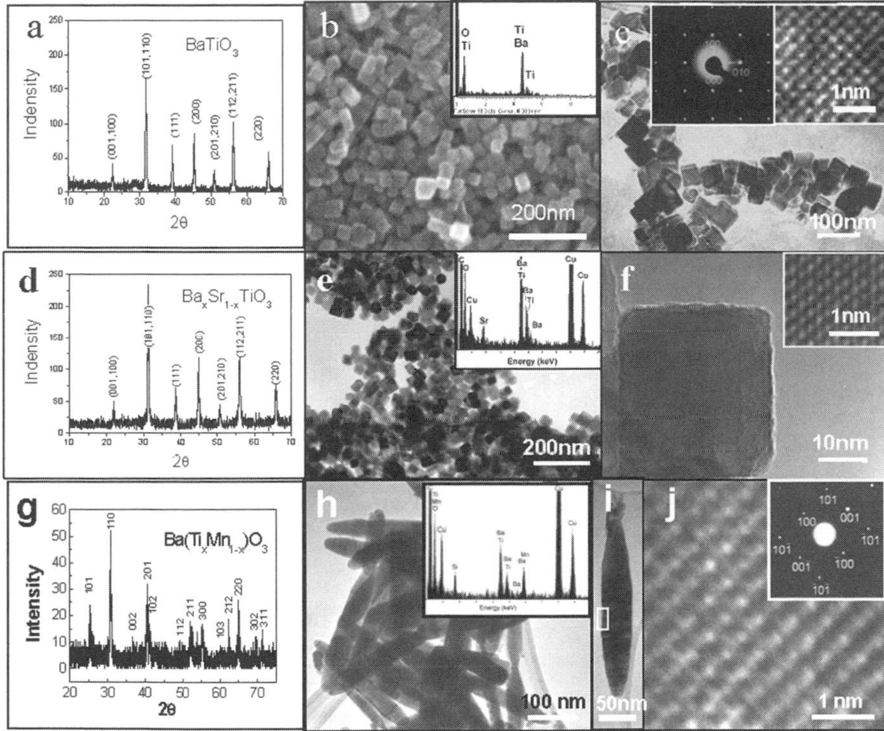

Fig. 2.16 Perovskite (a–c) $BaTiO_3$ and (d–f) $Ba_xSr_{1-x}TiO_3$ nanocubes and (g–j) $Ba(Ti_xMn_{1-x})O_3$ synthesized by the CHM approach. (a) XRD pattern of $BaTiO_3$ nanopowder. (b) SEM image of $BaTiO_3$ nanocubes; inset is EDS of the nanocubes showing the presence of Ba, Ti and O. (c) TEM image of $BaTiO_3$ nanocubes, insets are electron diffraction pattern and HRTEM image of a nanocube, showing its single-crystal structure. (d). XRD pattern of $Ba_xSr_{1-x}TiO_3$ nanopowder. (e). TEM image of $Ba_xSr_{1-x}TiO_3$ nanopowder; inset is EDS of the nanocubes showing the presence of Ba, Sr, Ti and O. The Cu signal came from the TEM grid. (f). A single-crystal $Ba_xSr_{1-x}TiO_3$ nanocube and its corresponding HRTEM image (inset). (g) XRD pattern of $BaTi_xMn_{1-x}O_3$ nanopowder. (h) TEM image of the nanostructure. (i) A single-crystal nanostructure and (j) its HRTEM image as well as its electron diffraction pattern (inset). [23]

images show that the nanocubes are single crystal and the three crystal faces are {100} planes (Fig. 2.16c and inset).

A possible reaction mechanism for the synthesis of $BaTiO_3$ in hydroxide solution is suggested as follows. During the reaction, hydroxides play a role not only as a solvent, but also as a reactant to participate in reaction. In the molten hydroxide, TiO_2 reacts with NaOH/KOH and forms a hydroxide-soluble Na_2TiO_3/K_2TiO_3. The simple chemical reaction (where M denotes Na or K) is as follows:

$$2MOH + TiO_2 \rightarrow M_2TiO_3 + H_2O \tag{2.1}$$

At the same time, $BaCl_2$ reacts with hydroxide to form $Ba(OH)_2$, which is dissolved in the hydroxide solution:

$$BaCl_2 + 2MOH \rightarrow Ba(OH)_2 + 2NaCl \tag{2.2}$$

The M_2TiO_3 from process (1) reacts with $Ba(OH)_2$ produced in process (2) and forms an indissoluble solid $BaTiO_3$:

$$M_2TiO_3 + Ba(OH)_2 \rightarrow BaTiO_3 + 2MOH \tag{2.3}$$

The Gibbs free energy of the above three steps for the formation of $BaTiO_3$ at 200°C is calculated to be –24.16 Kcal/mol. Because the viscosity of hydroxide is large, the formation of $BaTiO_3$ nanostructure is slow and it is not easy for the nanostructures to agglomerate. This is likely the key for receiving dispersive single crystalline nanostructures during the reaction without using surface capping material. The hydroxides mediate the reaction, but they are not part of the final nanostructures.

The second example perovskite of $(A_xA'_{1-x}\,BO_3)$ is $Ba_{0.5}Sr_{0.5}TiO_3$ to explore the applicability of this method for synthesis of complex perovskites with partially chemical substitution at the *A* site. Follow the same procedures as used for receiving $BaTiO_3$ except replacing the source cation supplying materials by a mixture of $BaCl_2$, $SrCl_2$ and TiO_2 at 0.5, 0.5 and 1.0 mmol, respectively. XRD pattern shows that the received product is a pure perovskite $Ba_{0.5}Sr_{0.5}TiO_3$ phase (Fig. 2.16d). TEM measurement demonstrated that the powder product is nanocubes with about 30–40 nm in sizes (Fig. 2.16e). EDS measurement shows that the ratio of Ba to Sr is ~1:1, demonstrating the controllability in chemical composition. HRTEM observation proved that $Ba_{0.5}Sr_{0.5}TiO_3$ nanocubes are single crystals (Fig. 2.16f and inset). However, there are some defects such as atomic disorders in the crystal because strontium and barium share the same sites in the crystal, which possibly results in substitution point defects. For both of $BaTiO_3$ and $Ba_{0.5}Sr_{0.5}TiO_3$, the crystal face is clean and sharp, and no amorphous layer is present, because no organic reagent or capping material was introduced during the synthesis. The perovskite nanocubes with clean surfaces are desirable for investigating ferroelectricity at nano-scale and for building functional components. The mechanism about the formation of $Ba_{1-x}Sr_xTiO_3$ is described (where M denotes Na or K) as follows:

$$2MOH + TiO_2 \rightarrow M_2TiO_3 + H_2O \tag{2.4}$$

$$(1\text{–}x)BaCl_2 + xSrCl_2 + 2MOH \rightarrow Ba_{1-x}Sr_x(OH)_2 + 2MCl \tag{2.5}$$

$$M_2TiO_3 + Ba_{1-x}Sr_x(OH)_2 \rightarrow Ba_{1-x}Sr_xTiO_3 + 2MOH \tag{2.6}$$

The third example of perovskite ($AB_{1-x}B'_xO_3$) is $BaTi_{0.5}Mn_{0.5}O_3$ to explore the applicability of this method for synthesis of complex perovskites with partially chemical substitution at the *B* site. When 50% of atoms at the Ti sites in barium titanate is substituted by Mn, $BaTi_{0.5}Mn_{0.5}O_3$ is received, which is a high dielectric constant material. A mixture of $BaCl_2$, MnO_2 and TiO_2 at 0.422, 0.211 and 0.211 mmol, respectively, is used as the source material for the synthesis. XRD measurement shows that the crystalline structure of the material is the same as $BaMnO_3$ (Fig. 2.16g), and EDS shows the atomic ratio of Mn to Ti is close to 1.0 (inset of Fig. 2.16h). The morphology of $BaTi_{0.5}Mn_{0.5}O_3$ is different from that of $BaMnO_3$ (Fig. 2.20) [24] and $BaTiO_3$(Fig. 2.16h). The products are elliptical nanorods about 40 nm in width, 20 nm in thickness, and 500 nm in length. ED and HRTEM show that each nanobelt is a single crystal (Fig. 2.16 3j) with a flat plane of (010). The growth direction is [101]. The mechanism about the formation of $Ba(Ti_xMn_{1-x})O_3$is described (where M denotes Na or K) as follows:

$$BaCl_2 + 2MOH \rightarrow Ba(OH)_2 + 2MCl \tag{2.7}$$

$$xTiO_2 + (1\text{–}x)MnO_2 + 2MOH \rightarrow M_2(Ti_xMn_{1-x})O_3 + H_2O; \tag{2.8}$$

$$Ba(OH)_2 + M_2(Ti_xMn_{1-x})O_3 \rightarrow Ba(Ti_xMn_{1-x})O_3 + 2MOH. \tag{2.9}$$

2.3.2 Spinel

Ferromagnetic spinel structured complex oxide is chosen as an example to demonstrate the extensive applicability of the CHM method. To synthesize spinel Fe_3O_4 ($Fe^{2+}Fe_2^{3+}O_4^{2-}$) nanostructure, a mixture of anhydrous $FeCl_2$ and Fe_2O_3 at 0.5 mmol each was used as the source material for providing Fe^{2+} and Fe^{3+} cations at the desired atomic ratio. Synthesis temperature and time were 200°C and 72 hours, respectively. XRD and EDS show that the product is cubic Fe_3O_4 (JCPDS 89-3854) (Fig. 2.17a and inset in Fig. 2.17b). In the product, most particles are nanocubes about 250 nm in sizes, and nanocuboids about 250 nm in short sides and 300–400 nm in long sides. From ED patterns of single particles, we can see that the nanocubes and nanocuboids are single crystals. The faces of the nanocubes are the {100} crystallographic planes (Fig. 2.17c and d). The growth direction of the nanocuboids is [121] (Fig. 2.17e and f).

$CoFe_2O_4$ nanocrystals are synthesized as an example to show the substitution at A site of spinel structured complex oxide (AB_2O_4). A mixture of $Co(NO_3)_2{\cdot}6H_2O$ and Fe_2O_3 at 0.5 mmol each was used as the source material. XRD pattern demonstrated that the product is cubic $CoFe_2O_4$ (JCPDS 22-1086) (Fig. 2.17g), as

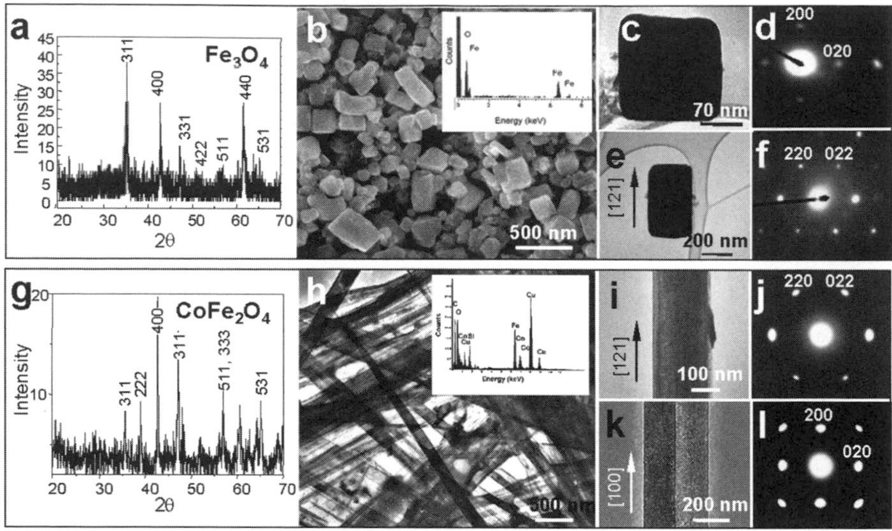

Fig. 2.17 Spinel (a–f) Fe_3O_4 nanoparticles and (g–l) $CoFe_2O_4$ nanobelts synthesized by the CHM approach. (a) XRD pattern of Fe_3O_4; (b) SEM image of Fe_3O_4 nanoparticles, and EDS pattern (inset). (c) A cube-like nanoparticle and (d) its electron diffraction pattern. (e). A Fe_3O_4 cuboids and (f) its diffraction pattern. (g). XRD pattern of $CoFe_2O_4$ nanobelts. (h) Morphology of the nanobelts and the corresponding EDS spectrum (inset) showing the presence of Co, Fe and O. The Si signal came from the TEM grid and holder. (i) A single-crystal nanobelt growing along [121] and (j) its electron diffraction pattern. (k) A nanobelt growing along [100] and (l) its electron diffraction pattern. [23]

supported by EDS microanalysis (inset in Fig. 2.17h). The morphology of $CoFe_2O_4$ is nanobelts with about 20–40 nm in thickness, 150–250 in width, and more than 20 μm in length (Fig. 2.17h). ED shows that there are two kinds of belts growing along different directions, [121] and [100] (Fig. 2.17i, j, k and l). The suggested formation mechanism of ferromagnetic MFe_2O_4(M=Co, Fe, Ni, Co) spinel nanostructures is as follows:

$$MCl_2 + 2NaOH \rightarrow M(OH)_2 + 2NaCl;$$

or

$$M(NO_3)_2 + 2NaOH \rightarrow M(OH)_2 + 2NaNO_3 \tag{2.10}$$

$$Fe_2O_3 + NaOH \rightarrow Na_2Fe_2O_4 + H_2O \tag{2.11}$$

$$M(OH)_2 + Na_2Fe_2O_4 \rightarrow MFe_2O_4 + 2NaOH. \tag{2.12}$$

Furthermore, we have also successfully synthesized $FeAl_2O_4$ by the CHM method to display the substitution at B site of spinel structured complex oxide (AB_2O_4). (Hu et al. to be published)

2.3.3 Hydroxide

Hydroxide nanostructure has many potential applications [33]. The surface hydroxyl groups may act as active sites for possible surface modification treatment through condensation reactions with amino acids or biologically active molecules, and thus, hydroxide nanostructure may have potential in the field of biological labeling. In addition, the similarity of the crystal structure and lattice constants suggests that doped hydroxide nanostructure could be prepared by a similar growth process, as lattice mismatching would not be a serious concern. Meanwhile, since hydroxides can be easily converted into oxides or sulfides through sulfuration, the hydroxide or co-doped hydroxide nanostructrues can act as important precursor to oxide or sulfide nanostructures.

Our investigations demonstrate that the CHM approach not only can synthesize simple and complex oxides nanostructures, but also can produce hydroxide nanostructures under normal atmosphere pressure. Taken the synthesis of lanthanum hydroxide ($La(OH)_3$) as an example [34]. To prepare $La(OH)_3$ nanostructrue, 0.1g $La(CH_3COO)_3$ with adding 1 ml deionized water is put into 18 g mixed hydroxides (NaOH:KOH=51.5:48.5) in a covered Teflon vessel and heating them at 200°C for 48 h in a furnace. When the vessel was cooled down to room temperature, the solid product was washed and filtered by deionized water. And then the product is washed by diluted HCl solution of pH 1.2 to remove other by hydroxides. The cleaned $La(OH)_3$ nanobelts are obtained after twice deionized water washing. To obtain the La_2O_3, we have tried calcinations of the $La(OH)_3$ nanobelts from 300 to 700°C. The pure La_2O_3 nanobelts could be successfully obtained by calcining the $La(OH)_3$ nanobelts at 690°C for 6 h in air [34].

X-ray diffraction (XRD) of the obtained $La(OH)_3$ product is shown in Fig. 2.18a. All of the peaks can be perfectly indexed as a pure hexagonal phase ($P6_3/m$ (176), JCPDS-361481) of $La(OH)_3$ with lattice constants a = 6.528 Å and c = 3.858 Å. The morphology of the obtained $La(OH)_3$ product was characterized by scanning electron microscopy (SEM). Figure 2.18b–c gives the SEM and TEM images of $La(OH)_3$, displaying the belt-like structure with typical widths of 30 to 200 nm, thickness of 5–30 nm, and length up to a few millimeters. The diffraction pattern and HRTEM image demonstrate the nanobelts are single crystalline and growth direction is [110].

2.3.4 Sulphides

Many metal elements can combine with sulphur to form stable crystalline semiconductor phases that exhibit a variety of unique optical and electrical properties [35]. Such metal sulphide semiconductors spend a large range of electronic energy band gap and often possess a substantial exciton binding energy. Therefore, they have attracted considerable technological and scientific interest [36, 37]. The metal sulphide semiconductors possess a variety crystalline phases depending largely on the

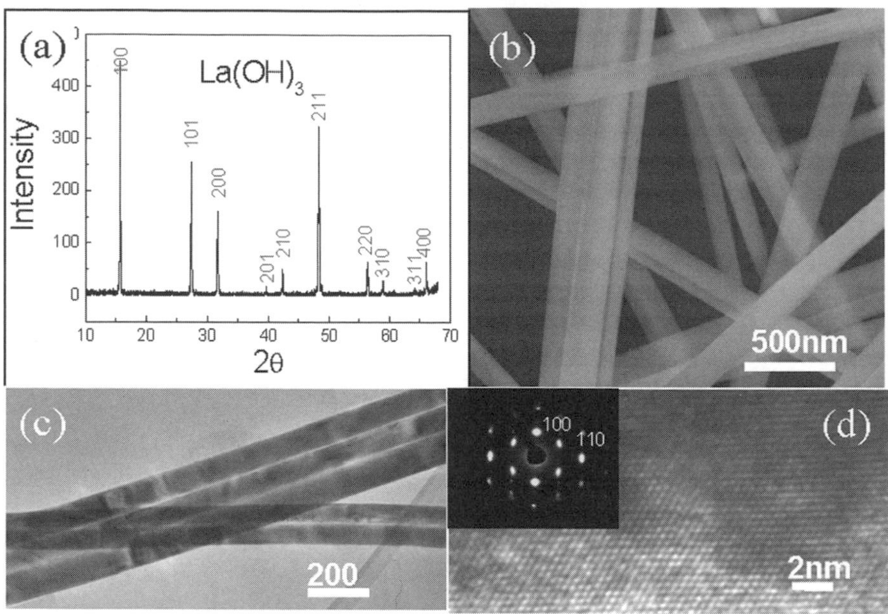

Fig. 2.18 (a) A typical XRD pattern of the as-synthesized $La(OH)_3$ product. (b) SEM images of the $La(OH)_3$ nanobelts, (c) TEM image of the $La(OH)_3$ nanobelts. (d) HRTEM image and electron diffraction (inset d), indicating the nanobelt is single-crystalline with growth direction of [110]. [34]

atomic radius ratios and electronegativity differences of the constituent atoms of the semiconductors [38].

Metal sulphide quantum dots have been the subject of extensive research [39]. Their applications in biomolecular imaging, profiling, and drug targeting have been developed quickly [40]. It has been well established that confinements of electrons and holes in the quantum dots change their physical and chemical properties in a profound way. Salient size-dependent properties have been observed, and hence the size constitutes and new parameter one can use to design, tune, and control the attributes of the so-called quantum dots using chemical colloidal techniques. In contrast to the conventional vacuum deposition techniques based on sophisticated instrumentation [41], the simplicity of the synthetic methodology and the possibility of large-scale chemical synthesis greatly facilitated the sulphide quantum dot research.

In spite of synthesis for compound involved oxygen, the CHM method can also give an easy way to synthesize metal sulphides (MS). The sources for M cation is from metallic salts, such as nitrates, chlorates, creosote, or acetates, and etc., and the sources for S cation is from sulfur powder or sulf-composite with valence states that match to those present in the desired product to be synthesized. Take CdS as an example of synthesis of sulphides. 0.5 mmol $CdCl_2{\cdot}2.5H_2O$ and 10 mmol of sulfur fine powder were put into 18 g homogeneously mixed hydroxides (7.8 g NaOH and

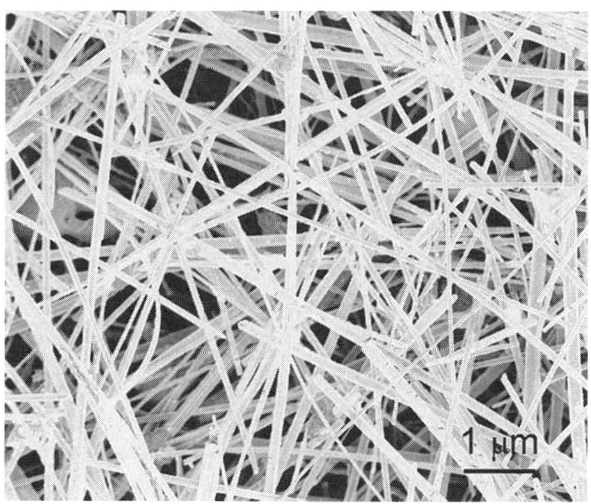

Fig. 2.19 SEM image of CdS nanowires. (Liu et al., to be published)

10.2 g KOH) in a covered Teflon vessel and heating them at 200°C for24 h in a furnace. When the vessel was cooled down to room temperature, the solid product was washed and filtered by deionized water. The cleaned product is ready for characterizing. Figure 2.19 shows the SEM image of the as-produced CdS. (Liu et al. to be published)

2.3.5 Other Kinds of Nanomaterials

From the above, it demonstrates the CHM approach is such a powerful method to synthesis extensive nanomaterials. We have synthesized variety nanomaterials by now, such as complex oxides of $BaTiO_3$, $SrTiO_3$, $Ba_{1-x}Sr_xTiO_3$, $BaTi_{0.5}Mn_{0.5}O_3$, $BaMnO_3$, $Ba_{1-x}Sr_xMnO_3$, $BaCeO_3$, Fe_3O_4, $CoFe_2O_4$, $CuAlO_2$, $Pb_2V_2O_7$, hydroxides of $La(OH)_3$, $Cd(OH)_2$, $Zn(OH)_2$, $Dy(OH)_3$, $Mg(OH)_2$, sulphides of ZnS, PbS, CdS, Cu_2S, Bi_2S_3, simple oxides of Cu_2O, CeO_2, Bi_2O_3, BaO, NiO, selenide of CdSe, ZnSe, PbSe, fluorides of SrF_2, CaF_2, mono- metals of Ni, Ag, Pd, Sb, Mg, Cu, Bi, etc. (Hu et al., to be published). We can conclude the CHM method is one of general methods to synthesize nanomaterials.

The morphology of nanomaterials can be easily controlled under varying reaction conditions by the CHM approach, such as heating temperature or/and heating time during the synthesis process. We have synthesized systematically $BaMnO_3$ under different conditions. The results are shown in Fig. 2.20. The size of the $BaMnO_3$ nanorods is width of 30 nm and length of less than 100 nm when we synthesize them at 200°C for 24 hours (Fig. 2.20a). But, the length can be greatly increased when we prolong the growth time under same temperature, as is shown in Fig. 2.20b, exhibiting the length of 200–400 nm for 120 hours growth. However, both the length and width can be enlarged at lower the growth temperature, and the width and length

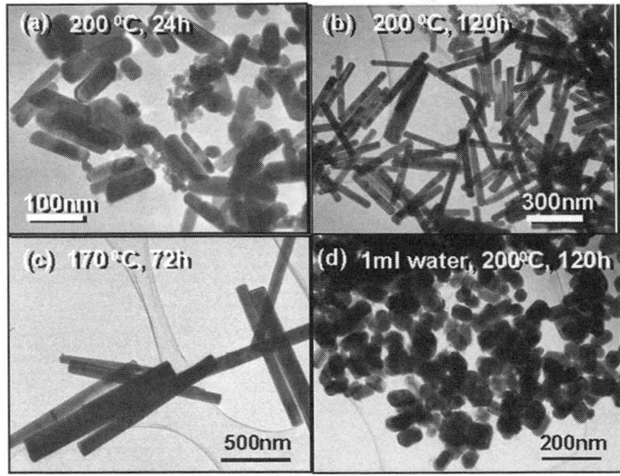

Fig. 2.20 TEM images of $BaMnO_3$ grown at different time and temperature. (a) at 200°C for 24 hours, (b) at 200°C for 120 hours, (c) at 170°C for 72 hours and (d) at 200°C for 120 hours, adding 1 ml water. [24]

can be reached 50–100 nm and 600–1000 nm respectively at 170°C for 72 hours (Fig. 2.20c). More interesting, the shape can be changed when we add 1 ml water under the temperature of 200°C for 120 hours growth, displaying the particle shape with diameter of 50 nm. It demonstrates that the composite-hydroxide-mediated rout is more suitable to grow one dimensional nanostructure than hydrothermal synthesis.

2.4 Discussion

It is obvious that the composite-hydroxide-mediated approach has many advantages including simplicity, ease of scale-up, and low costs, as it avoids high temperature and high pressure. The as-obtained nanomaterials possesse regular shape and clean surface attribute to no surfactant involved in the process, which benefit to further modification in chemical application or bio-uses. The CHM method is superior in synthesis of composite-oxides in comparison with other methods.

In order to compare the morphology control of the composite-hydroxide-mediated approach, we have carried series experiments to investigate influence by adding some amount of water in the CHM approach (Hu et al., to be published). Figure 2.21 shows SEM images of $La(OH)_3$ synthesized in 25 ml Teflon vessel at 200°C with 0.05 g $La(CH_3COO)_3$ under different conditions, (a–b) 4.5 g mixed alkali, 18 ml water for 48 hours in an autoclave , (c–d) 9 g mixed alkali, 5 ml water for 48 hours in an autoclave, (e–f)18 g mixed alkali, 1 ml water for 4 hours, (g–h) and (the inset) 18 g alkali 1 ml water for 12 hours. We can clearly see the growth time can adjust the size of nanobelts (Fig. 2.21e–h) and amount of water can adjust the shape and size of the nanobelts (Fig. 2.21a–f). We intend to believe that a small amount of water added in mixed alkali acts as an adjuvant solvent, not a media, such as 1 ml water, and there is no high pressure in Teflon vessel during the crystal growth. However, when 5 ml and 18 ml water is added in Teflon

Fig. 2.21 SEM images of $La(OH)_3$ synthesized in 25 ml Teflon vessel at 200°C with 0.05 g $La(CH_3COO)_3$ under different conditions. (a–b) 4.5 g mixed alkali, 18 ml water for 48 hours, (c–d) 9 g mixed alkali, 5 ml water for 48 hours, (e–f)18 g mixed alkali, 1 ml water for 4 hours, (g–h and inset) 18 g alkali 1 ml water for 12 hours. (Hu et al., to be published)

vessel and then put the Teflon vessel into autoclave, the crystals may grow under high pressure owing to the evaporation of the water in the sealed Teflon vessel. The condition of 4.5 g mixed alkali with 18 ml water can be regarded as a typical hydrothermal method. The resultant products are not well-shaped floc with smaller size, indicating the hydrothermal method cannot produce nanobelts of $La(OH)_3$ here. If the condition change into 9 g mixed alkali with 5 ml water, which is between hydrothermal method and the CHM approach, still there may has higher pressure in sealed Teflon vessel. But as the amount of water is small, the pressure is not as high as that with 18 ml water. The resultant products are short nanobelts (Fig. 2.21c–d). It demonstrates once again that the CHM rout is more suitable to grow one dimensional nanostructure than hydrothermal synthesis.

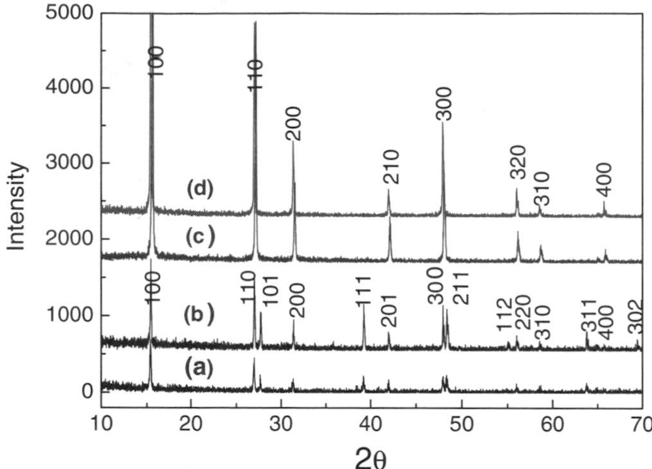

Fig. 2.22 XRD of $La(OH)_3$ synthesized in 25 ml Teflon vessel at 200°C with 0.05 g $La(CH_3COO)_3$ under different conditions. (a) 4.5 g mixed alkali, 18 ml water for 48 hours, (b) 9 g mixed alkali, 5 ml water for 48 hours, (c)18 g mixed alkali, 1 ml water for 4 hours, (d) 18 g alkali 1 ml water for 12 hours. (Hu et al., to be published)

Figure 2.22 shows XRD of $La(OH)_3$ synthesized in 25 ml Teflon at 200°C with 0.05 g $La(CH_3COO)_3$ under different conditions, (a) 4.5 g mixed alkali, 18 ml water for 48 hours, (b) 9 g mixed alkali, 5 ml water for 48 hours, (c)18 g mixed alkali, 1 ml water for 4 hours, (d) 18 g alkali 1 ml water for 12 hours. From the XRD pattern, we can see all of the peaks in Fig. 2.22a–d can be perfectly matched the XRD card of JCPDS-361481 and indexed as a pure hexagonal phase $P6_3/m$ (176) of $La(OH)_3$ with lattice constants a = 6.528 Å and c = 3.858 Å, except the existence of more crystalline facets when adding 5 ml and 18 ml water. The intensity of the XRD is much stronger for the sample with adding 1 ml water than those of samples with adding 5 ml and 18 ml water. These results indicate that the nanocrystal synthesized by the CHM approach have strongly selected growth direction and better crystallization in comparison with hydrothermal method.

2.5 Summary

The main synthesis strategies for functional nanowires and nanoparticles introduced here are VL and VLS growth, hydrothermal synthesis and composite-hydroxide mediated technique. VL and VLS method need high vacuum and high temperature with/without catalysts in the process, which can provide cleanly one dimensional nanomaterials. But the cost of the materials synthesized by these methods is high and the species is limited. The hydrothermal method is extensively used in synthesis of nanomaterials. Hardly can it be scale-up without considering the high cost of sealed vessel attributed to the high pressure in process. The CHM technique is new and promising in comparison with the traditional synthesis methods of VL/VLS and hydrothermal methods. It can synthesize many kinds of nanomaterials, especially composite oxides with clean surface and regular shape. The CHM method has advantages of simple, low cost, ease of scale-up, as it avoids effectively high temperature and high pressure. It can be developed and obtained more achievement in further investigation.

References

1. Wang ZL, Functional and Smart Materials – Structural Evolution and Structure Analysis, Plenum Press (1998).
2. Pan ZW, Dai ZR, Wang ZL, Nanobelts of semiconducting oxides. Science, 2001: 291: 1947–1949.
3. Huang MH, Wu YY, Feick H, Tran N, Weber E, Yang PD, Catalytic growth of zinc oxide nanowires by vapor transport. Adv. Mater., 2001: 13: 113–116.
4. Wagner RS, Ellis WC, Vapor-liquid-solid mechanism of single crystal growth. Appl. Phys. Lett., 1964: 4: 89–90.
5. Gao PX, Wang ZL, Substrate atomic-termination induced anisotropic growth of ZnO nanowires/nanorods by VLS process. J. Phys. Chem. B, 2004: 108: 7534–7537.
6. Pan ZW, Dai ZR, Ma C, Wang ZL, Molten Gallium as A Catalyst for the Large-Scale Growth of Highly Aligned Silica Nanowires. J. Am. Chem. Soc., 2002: 124: 1817–1822.
7. Liang CH, Zhang LD, Meng GW, Wang YW, Chu ZQ, Preparation and characterization of amorphous SiOx nanowires. J. Non-Crystalline Solids, 2000: 277: 63–67.
8. Zhu YQ, Hsu WK, Terrones M, Grobert N, Terrones H, Hare JP, Kroto HW, Walton DRM, 3D silicon oxide nanostructures: from nanoflowers to radiolarian. J. Mater. Chem., 1998: 8: 1859–1864.
9. Han WQ, Kohler-Redlich P, Ernst F, Ruhle M, Growth and microstructure of Ga_2O_3 nanorods. Solid State Communications, 2000: 115: 527–529.
10. Spezia G, Pressure is chemically inactive regarding the solubility and regrowth of quartz, Atti. Accad. Sci. Torino, 1909: 44: 95–107.
11. Walton RI, Subcritical solvothermal synthesis of condensed inorganic materials. Chem. Soc. Rev., 2002: 31: 230–238.
12. Rabenau A, Rau H, Crystal growth and chemical synthesis under hydrothermal. conditions. Philips Tech. ReV., 1969: 30: 89–96.
13. Toedheide K, In Water: a Comprehensive Treatise. Vol. 1, Franks F (ed.), Plenum: New York, pp. 463–514 (1972).
14. Seward TM, Metal complex formation in aqueous solutions at elevated temperatures and pressures. Phys. Chem. Earth, 1981: 13–14: 113–132.
15. Zhang KF, Bao SJ, Liu X, Shi J, Hydrothermal synthesis of single-crystal VO_2(B) nanobelts. Materials Research Bulletin, 2006: 41: 1985–1989.
16. Wang X, Li YD, Selected-Control Hydrothermal Synthesis of a- and b-MnO_2 Single Crystal Nanowires. J. Am. Chem. Soc., 2002: 12: 2880–2881.
17. Ma R, Bando Y, Zhang L, Sasaki T, Layered MnO_2 Nanobelts: Hydrothermal Synthesis and the Electrochemical Measurements. Adv. Mater., 2004: 16: 918–922.
18. Fang YP, Xu AW, You LP, Song RQ, Yu JC, Zhang HX, Li Q, Liu HQ, Hydrothermal synthesis of. rare earth (Tb, Y) hydroxide and oxide nanotubes. Adv. Funct. Mater., 2003: 13: 955–960.
19. Wu MM, Long JB, Huang AH, Luo YJ, Feng SH, Xu RR, Microemulsion-mediated hydrothermal synthesis and characterization of nanosize. rutile and anatase particles. Langmuir, 1999: 15: 8822–8825.
20. Titirici MM, Antonietti M, Thomas A, A Generalized Synthesis of Metal Oxide Hollow Spheres Using a Hydrothermal Approach. Chem. Mater., 2006: 18: 3808–3812.

21. Shen SC, Hidajiat K, Yu LE, Kawi S, Simple Hydrothermal Synthesis of Nanostructured and Nanorod Zn–Al Complex Oxides as Novel Nanocatalysts. Adv. Mater., 2004: 16: 541–545.
22. Hirano M, Morikawa H, Hydrothermal Synthesis and Phase Stability of New Zircon- and Scheelite-Type $ZrGeO_4$. Chem. Mater., 2003: 15: 2561–2566.
23. Liu H, Hu CG, Wang ZL, Composite-Hydroxide-Mediated Approach for the Synthesis of Nanostructures of Complex Functional-Oxides. Nano Letters, 2006: 6: 1535–1540.
24. Hu CG, Liu H, Lao CS, Zhang LY, Davidovic D, Wang ZL, Size-Manipulable Synthesis of Single-Crystalline $BaMnO_3$ and $BaTi_{1/2}Mn_{1/2}O_3$Nanorods/Nanowires. J. Phys. Chem. B, 2006: 110: 14050–14054.
25. Hu CG, Zhang ZW, Liu H, Gao PX, Wang ZL, Direct synthesis and structure characterization of ultrafine CeO_2 nanoparticles. Nanotechnology, 2006: 17: 5983–5987.
26. Cohen RE, Origin of ferroelectricity in perovskite oxides. Nature, 1992: 358: 136–138.
27. Anderson PW, Abrahams E, Superconductivity theories narrow down. Nature, 1987: 327: 363–363.
28. Yin Y et al., Formation of Hollow Nanocrystals through the nanoscale kirkendall effect. Science, 2004: 304: 711–1714.
29. Yang L-X, Zhu Y-J, Wang W-W, Tong H, Ruan M-L, Synthesis and Formation Mechanism of Nanoneedles and Nanorods of Manganese Oxide Octahedral Molecular Sieve Using an Ionic Liquid. J. Phys. Chem. B, 2006: 110: 6609–6614.
30. Wang X, Zhuang J, Peng Q, Li YD A general strategy for nanocrystal synthesis. Nature, 2005: 437: 121–124.
31. Hyeon T, et al. Synthesis of Highly Crystalline and Monodisperse Cobalt Ferrite Nanocrystals. J. Phys. Chem. B, 2002: 106: 6831–6833.
32. Wang ZL, Kang ZC Functional and Smart Materials. Chapter 3, Plenum Press (1998).
33. Wang X, Li YD Synthesis and characterization of Lanthanide hydroxide single crystal nanowires. Angew. Chem. Int. Ed., 2002: 41: 4790–4793.
34. Hu CG, Liu H, Dong WT, Zhang YY, Bao G, Lao CS, Wang ZL (2007) $La(OH)_3$and La_2O_3nanobelts: synthesis and physical properties. Adv. Mater. 2007: 19: 470–474.
35. Mukesh J (ed.) II–VI semiconductor Compounds. World Scientific, Singapore (1993).
36. Larach S (ed.) Photoelectronic Materials and Devices. Van Nostrand, Princetion, NJ; (1995) Handbook of Optics. Vol. 1, McGrqw-Hill (1965).
37. Ueta M, Kanzaki H, Kobayashi K, Toyozawa Y, Hanamura E, Excitonic Processes in Solids. vol. 60, Springer Series in Solid State Sciences, Springer, Berlin (1986).
38. Madelung O (ed.) Numerical Data and Funtional Relationships in Science and Technology-Physics of Non-tetrahedrally Bonded Elements and Binary Compounds. III, vol. 17e, Springer, Berlin. (1983).
39. Kamat PV, Meisel D (ed.) Semiconductor Nanoclusters-Physical, Chemical, and Catalytic Aspects. Elsevier Science B. V., Amsterdam (1997).
40. Chan WCW, Nie SM, Quantum dot bioconjugates for ultrasensitive nonisotopic detection. Science 1998: 298: 2016–2018.
41. Petroff PM, Lorke A, Imamoglu A, Epitaxially self-assembled quantum dots. Phys. Today 2001: 54: 46–51.

Chapter 3
Nanolithography

Raghunath Murali

Abstract Direct-write electron beam lithography (EBL) has emerged as a key lithographic technique to fabricate nanometer structures. EBL has a resolution down to a few nanometer and does need a mask. A wide variety of EBL machines are available depending on the application: mask making machines, direct-write tools, SEMs fitted with a pattern generator, and R&D machines. This chapter presents topics of interest to a reader involved in fabricating Bio-Nano-Fluidic MEMS devices and systems. CAD file preparation and machine design basics are briefly reviewed. Resist technology and proximity effect is discussed in detail since they have a major impact on the e-beam lithography process. Other lithographic methods including ion-beam, X-ray, electron projection and AFM-based methods are also discussed.

Abbreviations

EBL-electron beam lithography, RET-reticle enhancement technique, CD-critical dimension, PMMA-poly (methyl methacrylate), MIBK-methyl iso butyl ketone, IPA-iso propanol, CAR-chemically amplified resist, DUV-deep ultra violet, HSQ-hydrogen silsesquioxane

3.1 Introduction

Optical lithography has made possible the tremendous scaling of semiconductor devices, all the way from tens of microns in the 1970s to tens of nanometers in current manufacturing. This 1000X scaling has been made possible by improvements in lenses, machine design, mask-making, and light-sources. But the current 193-nm optical lithography, which makes possible sub-100 nm features, entails the use of expensive masks thus making it more suitable for production-oriented processes than research-oriented processes. Optical exposure tools used in research labs have

R. Murali
School of Electrical and Computer Engineering, Georgia Institute of Technology, Atlanta, GA 30332, USA

P. J. Hesketh (ed.), *BioNanoFluidic MEMS.*

more limited capabilities but use cheaper masks and allow feature definition down to a half micron or so. To expose features below a half micron without the use of expensive optical lithography, a variety of options exist: electron-beam lithography, imprint lithography, X-ray lithography, ion-beam lithography, and AFM-based lithography. Among these alternative lithography techniques, electron beam lithography will be the focus of this chapter.

Optical lithography has seen continuously shrinking source wavelengths – 465 nm, 365 nm, 248 nm and now 193 nm. This is because of the Raleigh scattering limit that imposes a limit on the patterned line-width. Concurrently, enormous improvements have been made in the construction of lenses resulting in better numerical aperture. Reticle enhancement techniques (RET) have further improved the imaging resolution of current generation optical steppers. Some of the RETs include optical proximity correction, phase shift masking, immersion lithography, and double exposure. The semiconductor industry has circumvented the use of 157-nm lithography to avoid the steep expense of process development using a completely new exposure process. So 193-nm lithography using RETs have been pushed as far as possible. Sub-30 nm features has been demonstrated by IBM in 2006 using 193-nm lithography; but 193 nm lithography may not be sufficient for the 32 nm generation. The enabling lithography technique for this generation is expected to be EUV lithography, which has a 13.5 nm source. At this wavelength, most materials absorb light and so reflective optics is needed. Also, the entire system needs to be housed in vacuum since dirt particles can absorb EUV radiation. Masks used in EUV lithography will be reflective, and defect density and mask uniformity requirements will be much tighter than before. Thus the cost of EUV lithography is expected to be steep.

While optical lithography for sub-100 nm features is a recent development, electron-beam lithography (EBL) has existed since the 1970s and sub-20 nm features were demonstrated as far back as 1981 using this technique [1]. Most existing EBL machines scan a beam of electrons to get the desired patterns although projection EBL too exists. Early EBL machines were just an SEM fitted with a pattern generator. More recent machines, on the other hand, are much more sophisticated and offer a range of features that makes precision lithography of large samples and patterns possible. This will be discussed in a later section. Nano-imprinting is a technique that is gathering momentum as an attractive complement to the serial writing process of electron-beam lithography and is discussed in chapter 12. In the following sections, electron-beam will be discussed in more detail. Ion beam, X-ray and laser scanning lithography will be briefly touched upon; due to the limited use/capabilities of these lithographic techniques, the reader is referred to existing literature for more information on these topics. Figure 3.1 shows a comparison of the resolution limits for each lithography type.

Electron beam lithography uses a finely focused beam of electrons that is deflected over a substrate coated with an e-beam sensitive resist. Computer-controlled deflection and blanking of the beam is done in accordance with the pattern being written. The e-beam resist is exposed in the area where the beam of electrons hits the resist; this exposure may result in either molecular scission

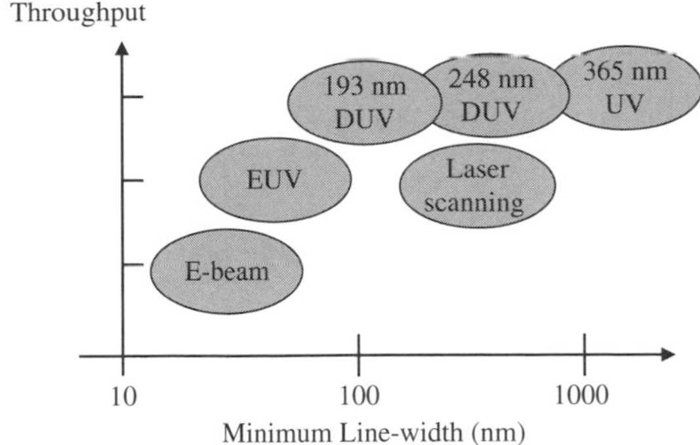

Fig. 3.1 Comparison of resolution limits of different types of lithography

(positive resist) or linking (negative resist). In positive resists, the area exposed dissolves faster in the developer than the unexposed area whereas in negative resists the reverse is true. The desired pattern is obtained in the resist after the develop process.

After the develop step, the resist pattern can be transferred to the substrate in a variety of ways. Three such methods are shown in Fig. 3.2 – lift-off, etch and ion-implantation. Directional metal deposition (e.g. in a filament evaporator, or an e-beam evaporator) results in metal being deposited into the trenches as well as on top of the resist (but very little on the sidewalls). A solvent wash will dissolve the resist and thus lift-off any metal on the resist. For this process to work successfully, the ratio of metal thickness to resist thickness needs to be at least 1:2. An undercut profile is helpful for the liftoff process and can be obtained by appropriate dose selection or by use of multi-layer resists. Figure 3.3 shows the effect of increasing dose on the sidewall profile. It can be seen that prudent dose selection is required to achieve the desired sidewall profile. A plasma etch can be used to transfer the pattern into the substrate. The different plasma etch systems that can be used include reactive ion etch (RIE), inductively coupled plasma (ICP) etch, and electron cyclotron resonance (ECR) plasma etch. Ion-implantation can also be a follow-on step to lithography, especially in processes that require transistor devices.

E-beam lithography offers many advantages like (1) needing no mask, and the pattern is input as a CAD file; this leads to a fast turn-around time, (2) precise

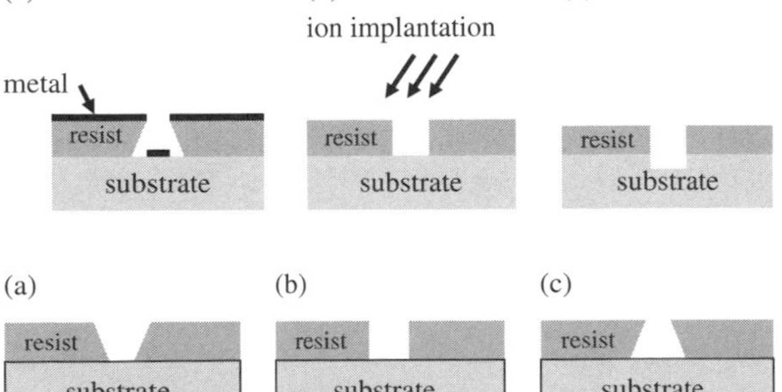

Fig. 3.2 Possible post-lithography steps – (a) lift-off, (b) ion-implantation, and (c) etch

Fig. 3.3 Effect of dose on the sidewall profile; the dose in increasing from (a) through (c)

overlay, resulting in relaxed design rules and higher yield; sub-pixel overlay with e-beam lithography has been demonstrated, (3) high resolution, (4) accommodates a variety of sample sizes, and this is beneficial to a multi-user facility where the users might have widely different requirements, and (5) geometric pattern correction, leading to better patterning accuracy. But disadvantages include (1) writing time – since direct-write e-beam lithography is a serial process, the writing time increases with an increase in the pattern density, (2) charging effect – if writing on insulators, this results in distorted patterns unless taken care of, and (3) the proximity effect which makes necessary dose/shape correction.

3.2 Pattern Preparation

In e-beam lithography, the pattern is input as a data file to the control computer. This file is usually in the GDSII format which is the format of choice for a large number of chip foundries as well as mask manufacturers. Some systems allow the pattern to be input as an ASCII file though such a format is more suitable for patterns that have a high degree of symmetry. Many software packages are available for preparing computer-sided design (CAD) files: Cadence and Mentor tools on the high-end, Design Workshop, Tanner Research L-Edit and AutoCAD on the intermediate-level, and free tools like the Magic layout editor. Some of these programs (e.g. AutoCAD) cannot output in GDSII format and a format-converter (such as LinkCAD) will have to be used to convert the pattern file to GDSII format. Some of these programs (e.g. Cadence tools) can output directly in the machine specific (e.g. JEOL, Leica) format which is advantageous since the fractured pattern can be viewed at the same time as pattern creation. Some higher end tools require add-on programs to perform the fracturing. When there is field-stitching involved (i.e. when the patterned area exceeds one field), care needs to be taken to place all sensitive areas of the pattern (e.g. a sub-100 nm cantilever beam, or a sub-50 nm gate) away from the field-boundaries.

One advantage that electron beam lithography offers is the precise dose control possible within a given pattern. To take advantage of this feature, different portions of the pattern are assigned to different layers/datatypes while creating a CAD file; note that this is different from just assigning different device layers – for e.g. p-doping, metallization, gate, etc., in a CMOS process. During pattern exposure in the e-beam lithography machine, these different datatypes can be given different doses. This intra-layer dose differentiation is useful for proximity effect correction as well as grayscale lithography. The JBX-9300FS system from JEOL offers up to 1024 levels of dose modulation.

3.3 Electron Beam Lithography System Design

Depending on the application need, an electron beam lithography system can have various requirements for resolution, accuracy and throughput. A university R&D system will have stringent requirements on resolution but accuracy and throughput will not be as important. A mask making machine will need good accuracy, moderate throughput but resolution is not as important since projection lithography masks are usually 4X the printed size. A direct write system that is used to pattern wafers will need to have a good throughput and accuracy, and a moderate resolution. A comparison of different systems is shown in Table 3.1.

Figure 3.4 shows a block diagram of an e-beam lithography system. The principal components of such a system are as follows.

1. Electron Optics: this part of the system is responsible for beam formation and consists of many components including the electron source and deflector.
2. Stage: a high precision stage is necessary for high-resolution patterning. High-accuracy stages usually contain a laser interferometer that enables relative accuracy down to sub-nm levels. The stage sub-system also consists of a leveling system that levels the substrate to be written. Advanced systems may contain a height-detection system to correct for uneven substrate height.
3. Load/Unload: this sub-system is responsible for substrate load/unload. The load (and unload) process is usually a 2-step process: the cassette is first transported to a load-lock chamber, and after this chamber reaches a preset vacuum level, the cassette is loaded into the stage chamber. To handle multiple cassettes, robotic handlers are available and can automate exposure jobs for multiple wafers.

Table 3.1 Comparison of electron beam lithography system types

	High-end system	Mid-end system	Converted-SEM
Applications	Direct-write, mask-making	Direct-write	Direct-write
Examples	JEOL JBX-9300 FS, Vistec VBR-300	Raith 150	JC Nabity NPGS
Acc. voltage	50–100 kV	up to 30 kV	Depends on SEM (usually up to 30 kV)
Beam diameter	6–8 nm	2 nm	2 nm
Field size*	1 mm	0.1 mm	0.1 mm+
Field stitching accuracy	20 nm	40 nm	20–100 nm (with registration marks)
Resolution**	20 nm	30 nm	20 nm
Overlay accuracy	20 nm	40 nm	20–100 nm
Sample size	up to 300 mm wafers	up to 200 mm wafer	Depends on SEM chamber size
Automatic height detection	Yes	Yes	No
Deflection speed	50 MHz	10 MHz	5 MHz
Stage accuracy	<1 nm	<2 nm	depends on SEM
Cost	$5 million	$ 1 million	$40 K, excluding SEM

*Depends on accelerating voltage.
**Field size also depends on the desired resolution.

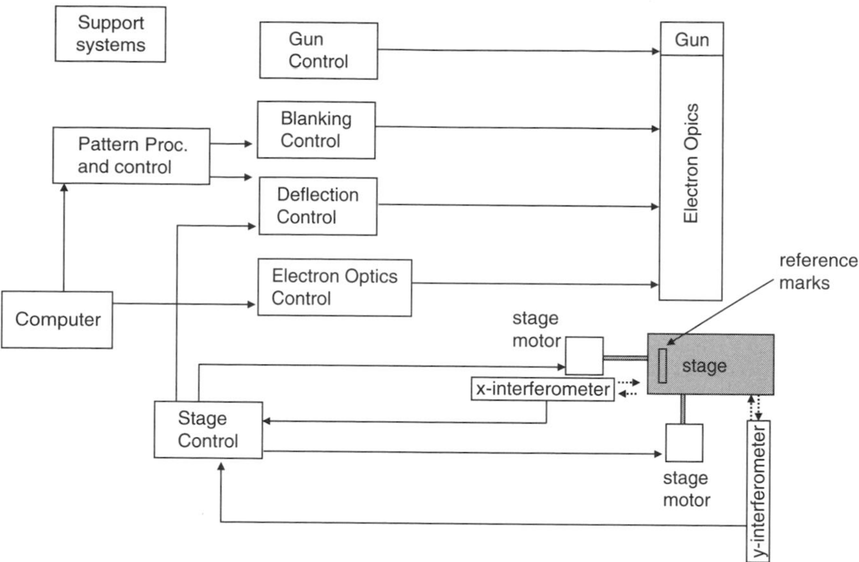

Fig. 3.4 Block diagram of an e-beam lithography system

4. Control Computer: the control computer is responsible for a variety of functions including pattern compilation, job scheduling, pattern transfer to deflection electronics, real-time system monitoring/logging, and remote control of all the above sub-systems.
5. Support System: this includes vacuum pumps, temperature control of stage and electronics, power supplies, etc.

3.3.1 Electron Optics

The electron optics system consists of an electron source, lenses for beam formation, deflectors, limiting apertures, and blanker. The electron source (also known as an electron gun) can be of the following types: thermionic emission, thermal field emission, or cold field emission. The thermal field emission type is widely used due to its good emission characteristics such as brightness, small energy broadening of electrons, and small beam diameter. An example electron optical column is shown in Fig. 3.5. This configuration is similar to the setup in scanning electron microscopes. A spray aperture collects stray electrons from the source. Three lenses are used before the beam is focused onto the substrate. A limiting aperture is used to cut off fringe electrons. The spot produced on the substrate will have a Gaussian intensity distribution and thus this type of system is known as a Gaussian beam system. To turn the beam off in areas that are not patterned, a beam blanker is used. This is usually electrostatic in nature and can be placed before a limiting/spray aperture. To correct for an elliptical beam (an effect called astigmatism), a stigmator is used (which is a n-pole element of opposite electric/magnetic fields).

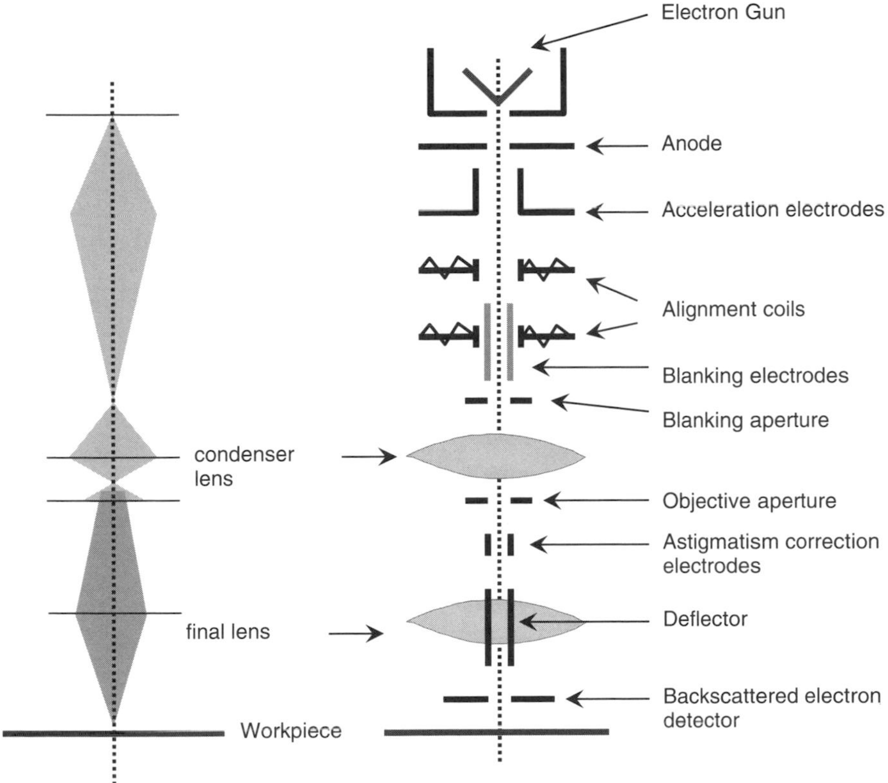

Fig. 3.5 An example electron optical column in an e-beam lithography system (*courtesy of JEOL Ltd.*)

To deflect the beam thus produced by the electron optical system, electromagnetic deflection is preferred over electrostatic deflection; electromagnetic coils provide a lower deflection distortion as well as better stability. Most systems utilize a multistage deflection system since such a setup offers both low distortion as well as large field size. In a multistage deflection system, a field is divided into subfields. A main deflector deflects the beam to the middle of a sub-field, whereas a sub-deflector scans the beam within the sub-field. For example, in the JEOL JBX-9300FS direct write system operating at 100 kV, the field size is 500 um, whereas the sub-field size is 4 um; the corresponding deflection frequencies are 2 MHz and 50 MHz, respectively [2]. Before the start of pattern writing, the deflectors are taken through a series of calibrations that ensure patterning accuracy. Adjustments to the focusing lens and stigmator are also needed to ensure a small beam diameter and a round beam.

3.3.2 Stage Control

The deflector optics has a limited deflection capability, and thus field size is usually on the order of a millimeter. To write patterns larger than a millimeter on a side, stage movement will be needed. The stage movement could be of two types: step and repeat or continuous movement. In the step and repeat method, after a field is written, the stage steps to the next field to be written. This results in loss of time if there are many fields to be written. In the continuous movement method, the stage is continuously moving and the beam is blanked in areas in which there is no pattern. Such a method has a high throughput but the writing time is independent of pattern density; also the system design is more complicated.

3.3.3 Beam Scanning

The electron beam can be scanned in two different ways – raster scan or vector scan. In the raster scan method, the beam is scanned continuously across the substrate and is blanked in areas where there is no pattern. This method has a write time independent of pattern density; also the dose can be varied in a limited manner in different regions. This type of scanning is usually combined with a continuously moving stage. Stitching between scan lines needs to be accurately controlled and thus the system design becomes complicated; however the electron optics is simpler since the deflection field size can be smaller. In the vector scanning method, the beam is scanned only in the area that needs to be patterned. The write time is directly proportional to pattern density and the dose can be varied over a wider range. However, field size needs to be large and thus the electron optics design becomes more complex.

3.3.4 Beam Shapes

Systems with various beam shapes exist – Gaussian beam, fixed shaped beam, variable shaped beam, and cell projection. The Gaussian beam system has a sharply focused electron beam and the intensity distribution of this beam is Gaussian. In the shaped beam system, a rectangular beam with fixed/varying size is created by passing through two square apertures. In the cell projection system, various unit cells are placed around the central aperture. A pre-shaped square beam is then deflected onto one of the unit cells resulting in the desired pattern. For a more detailed discussion on electron optics and system design refer to [3].

3.4 Electron Beam Resists

Electron-beam resist selection is an important part of the e-beam lithography process since they impose limits on line-width resolution, speed of writing, pattern uniformity, and pattern transfer. The ideal e-beam resist should have (1) a high sensitivity (on the order of a few tens of uC/cm^2 at 100 kV), (2) long shelf life (a few months), (3) long post-exposure stability (a few weeks), (4) low pattern distortion in vacuum, post-exposure (less than a few nm), (5) negligible resist swelling and distortion during develop, (6) high contrast, (7) high etch selectivity to the underlying substrate, and (8) high resolution. Critical dimension (CD) control is of importance to many applications including mask making, and optical elements. Sensitivity to variation of linewidth, dose, beam size, and develop time is important for any process where tight CD control is needed. Dose latitude depends on resist contrast and absorbed energy density; the absorbed energy density in turn depends on the accelerating voltage. The performance of a develop process is characterized by sensitivity, contrast, dose latitude, surface roughness, and resolution.

Resists can be classified broadly into positive and negative resists. In positive resists, e-beam irradiation causes chain scission; this results in lowering of the molecular weight of the exposed part. The lowered molecular weight leads to a differential solubility in a developer (usually an organic solvent). One of the earliest positive resists was poly(methyl methacrylate) or PMMA, Fig. 3.6. PMMA resist has one of the best resolutions and line-widths smaller than 10 nm have been reported [4]. PMMA has been extensively studied in the literature and the effect of PMMA molecular weight, developer, developer concentration, and develop temperature have been well documented.

A typical PMMA resist process is as follows. 950 K PMMA [from MicroChem Corp., Newton MA] is spin-coated on a substrate to obtain a resist thickness of a few hundred nm; the thickness is usually decided by the minimum linewidth required – aspect ratios of 1:3 (line-width to resist thickness) are common but high aspect ratio structures (up to 1:10) have been achieved. The spin-coat is followed by a hot-plate or oven bake to evaporate the solvent. This is followed by e-beam lithography, with doses from 50 to 300 uC/cm^2 for acceleration voltages from 50 to 100 kV. Resist develop is done by a puddle or spray develop in a 1:1 MIBK/IPA developer. For a high resolution process, a low sensitivity is needed because of the shot noise limit whereas for large patterns, a high sensitivity process is needed. A big shortcoming of PMMA is its poor etch resistance to various gases in a plasma etch system.

Different developers have been reported to give good results (for a PMMA resist process) including MIBK/IPA, IPA/Water, Acetone, and MIBK. Different developers and develop times result in different contrasts, and sensitivities. Mixtures of 1:3 MIBK/IPA and 1:1 MIBK/IPA have been widely used as developers for PMMA. After exposure, the chain-scission caused by electrons leaves behind a distribution

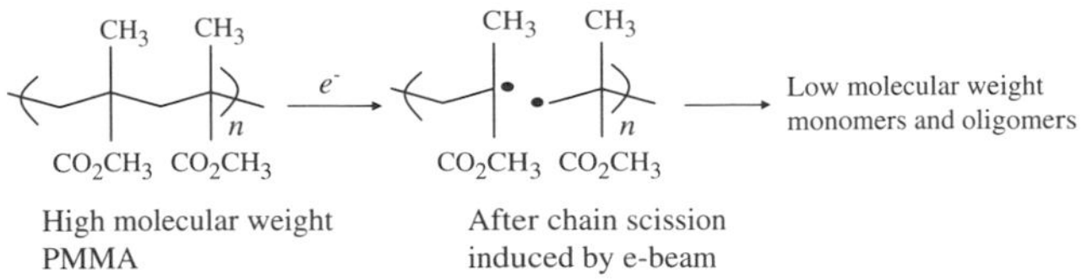

Fig. 3.6 Chemical structure of PMMA resist

of molecular weights in the resist. The selected developer solution will need to dissolve a range of low molecular weights leaving behind the higher molecular weight (i.e. unexposed) component. IPA is a weak solvent for low molecular weight PMMA, and it has been used as a developer for very thin (< 50 nm) PMMA layers; thicker PMMA layers need addition of MIBK to open up the range of molecular weights that are dissolved. Mixtures of IPA and water have been found to have a higher contrast, resolution, and process latitude than MIBK/IPA [5,6]. Development at low temperatures has been proposed as a means to achieve higher resolution. In [7], a develop at 6–10°C is found to give sub-10 nm resolution at 30 kV. In [8], cold develop is seen to result in sub-10 nm resolution.

Many improvements on PMMA have been attempted to increase it's etch resistance and improve its sensitivity. Such developments led to PBS, PMPS and SNS. But none of these achieved both the goals for sub-micron patterns. In the mid-1990s, a positive chain-scission resist based on poly(methyl-a-chloroacrylate-co-a-methylstyrene) was developed by Zeon Chemicals. This family of resists (ZEP 520 and ZEP 7000) has shown good sensitivity (200 uC/cm^2 for 100 kV) as well as good resolution (sub-20 nm at 100 kV); its etch selectivity is better than PMMA and it has found wide use in the mask-making industry. Amyl-acetate and xylene can be used as developers for ZEP. Hexyl acetate has been reported to be a better developer to obtain improved resolution in ZEP 520 [9]. A typical ZEP 520 process is as follows: spin-coat 300 nm of resist, followed by a post-apply bake for 2 min. at 180°C; expose with a base dose of 200 umC/cm^2 at 100 kV, develop in amyl-acetate for 2 min,. followed by a rinse in IPA for 2 min.

Bi-layer resist systems have are useful for applications such as T-gate fabrication and liftoff. A bi-layer resist process usually uses a stack of a low-sensitivity positive resist on top of a high-sensitivity positive resist. E-beam exposure of a pattern results in stronger development of the underlying layer thereby causing an undercut. This undercut aids in easy lift-off. A common bilayer stack is that using the copolymer methyl methacrylate (MMA) methyacrylic acid (MAA) and PMMA where the copolymer is the high-sensitivity layer. A good developer for this stack is 1:1 MIBK/IPA. Another possible stack is high molecular weight PMMA on top of a lower molecular weight PMMA; the low molecular weight PMMA has a higher sensitivity than the top layer and thus an undercut is formed after development. A bilayer stack can also be formed by ZEP/PMMA [10]. A trilayer stack can be formed by low sensitivity resist/metal/high sensitivity resist. After developing the top layer, the metal can be etched in a plasma, followed by development of the bottom layer. Alternatively, a trilayer stack can also be composed of three different resists [11].

Apart from the positive resists discussed above, there exist a variety of other positive resists that are suitable for e-beam lithography. These include EBR-9, PBS, and UV-5. EBR-9 is a high-sensitivity resist that has good process latitude but poor resolution. Poly butane 1-sulfine (PBS) has high sensitivity but needs tight process control during development. Resists made for the DUV spectrum are generally also sensitive to e-beam. Examples of e-beam sensitive resists that are also DUV sensitive include PMMA, ZEP 520 and UV5. Another class of resists that is of importance, especially in the mask making industry, is chemical amplified resists

Table 3.2 Comparison of resists used for e-beam lithography

Resist	Type	Typical sensitivity (at 100 kV, in uC/cm^2)	Developer	Resolution
PMMA	Positive	300	MIBK/IPA, IPA/Water, Acetone	sub-5 nm
P(MMA MAA)	Positive	150	MIBK/IPA	sub-100 nm
ZEP 520	Positive	150	Amyl-acetate, Xylene	20 nm
UV5	Positive	50	Shipley CD 26	50 nm
NEB 31	Negative	50	TMAH	40 nm
HSQ	Negative	800	TMAH	7 nm
ma-N 2403	Negative	400	MF 319	sub-80 nm

(CAR). These include the NEB 31 and NEB 22 from Sumitomo, FEN 270 from Fuji, and EN-024M from TOK. CARs have improved sensitivity thanks to a radiation sensitive photoacid generator (PAG); this PAG catalyzes molecular chain scission. But handling CARs is more complicated than for normal resists; CARs require strict processing control and also base-contamination filtration.

Most of the resists described above are positive resists. Negative resists are also an important part of e-beam lithography; some examples include Hydrogen silsesquioxane (XR-1541, from Dow Corning Co.) and ma-N 2400 (from Micro Resist Technologies). HSQ, originally meant for use as a flowable oxide for inter-layer dielectric applications, is e-beam sensitive and displays excellent resist characteristics such as high etch selectivity, good resolution (sub-7 nm), and small line-edge roughness. On the other hand, it has a short shelf life (a few months), is amenable to residue formation during the develop process, and might be sensitive to any delay between exposure and development. Various concentrations of TMAH have been found to be good developers; for e.g.2.5% TMAH for 60 s or 10% TMAH for 30 s or 25% TMAH for 5 s [12]. Hot develop and ultrasonic agitation has been found to result in reduced residue and increased resolution [13]. In [14] it has been found that a higher developer concentration and a lower bake temperature results in a better contrast. Table 3.2 shows a comparison of various e-beam sensitive resists.

3.5 Electron–Substrate Interaction and Proximity Effect

Energetic electrons entering a solid undergo scattering. Depending on the scattering angle, the scattering can be classified as forward and back scattering. Forward scattering involves small-angle scatterings and its effect is to broaden the beam. Back scattering involves large angle scatterings and may result in complete reversal of electron direction. Consider a beam of electrons incident at a point on a substrate that is coated with a thin film of a resist, Fig. 3.7. As the electrons enter the resist, they undergo forward scattering thereby broadening the beam. In the substrate, the back scattered electrons come back into the resist and deposit energy into the resist in unexposed areas. This beam broadening and exposure by backscattered electrons is collectively referred to as the proximity effect.

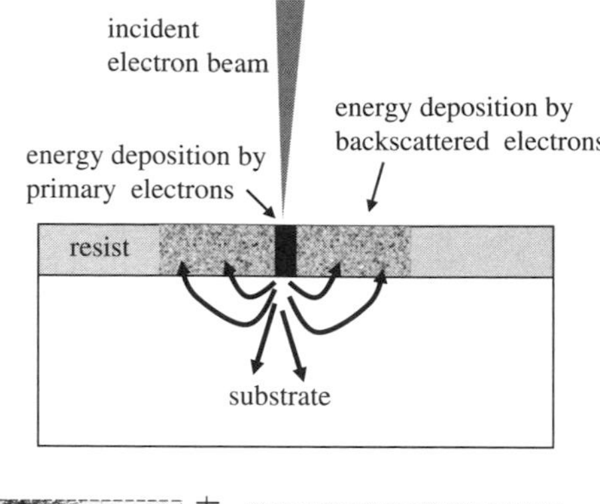

Fig. 3.7 Forward and back scattering of electrons resulting in the proximity effect

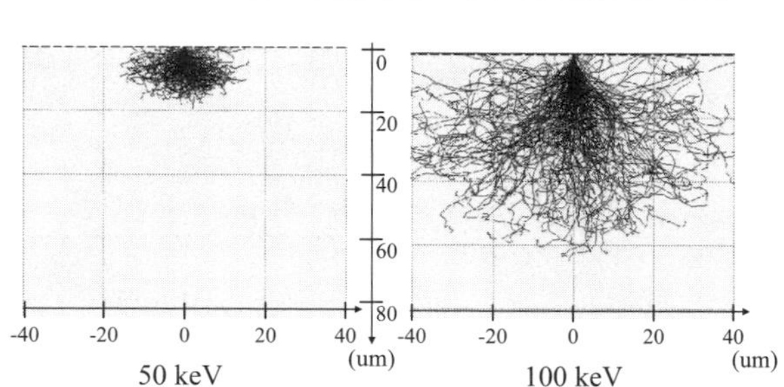

Fig. 3.8 Monte Carlo simulation of 50,000 electron trajectories at acceleration energies of 50 keV and 100 keV

Figure 3.8 shows a Monte Carlo simulation of 50,000 electron trajectories at different acceleration energies; it can be seen that as the acceleration energy increases, the beam broadening in the resist decreases but the backscattering effect increases. The extent of beam broadening in the resist is dependent on the acceleration voltage of electrons, the resist thickness, as well as the resist material itself. The average number of elastic events which an energetic electron suffers passing though a resist film is given as $P_e = 400.z_0(um)/E_0(keV)$ where z_0 is the resist thickness and E_0 is the accelerating voltage. The number of forward scattering events decreases with a decrease in resist thickness and an increase in acceleration voltage. Thus, the thinnest resist and the highest accelerating voltage usually result in the best resolution For 100 kV acceleration voltage and 50 nm resist thickness, the average number of scattering events per electrons comes out to be just 0.2. This means that 4 out of 5 electrons go through the resist without suffering any scattering. Thus, as acceleration energy is increased, the dose needed to expose the resist increases. Similarly, as the resist thickness is reduced, P_e decreases; thus thinner resists may need as much or a higher dose as thick resists for proper exposure. Another important factor in forward scattering is the incident beam diameter; this is dependent on the electron optics and beam current, and is also a limiting factor for the minimum line-width. Due to the many small angle scattering events, forward scattering increases the effective beam diameter. Empirically, it is given by the following formula:

$$d_f = 0.9(R_t/V_b)^{1.5} \tag{3.1}$$

where d_f is the effective beam diameter in nanometer, R_t is the resist thickness in nanometer and V_b is the beam voltage in kilovolt.

Backscattered electrons can deposit substantial energy in the resist, at distances far from the region of initial exposure. This effect is worsened as acceleration energy is increased since the increased energy results in increased electron range leading to a larger area of influence for backscattered electrons. The backscattering effect can be characterized by a backscattering coefficient (η); low atomic number substrates give the least backscattering whereas high atomic number substrates give the most. Another important parameter that influences backscattering is the range of electrons in the substrate. This range is dependent on the electron energy as well as the substrate material. The Bethe range is a useful indicator for the electron range in substrates. Figure 3.9 plots the Bethe range for various substrates and accelerating voltages. It can be seen that at 100 kV, electrons have a range of 40 um in Si and 10 um in Au. If patterns are placed closer than the Bethe range, then the backscattering contribution will influence the incident exposure. As Z increases, the number of backscattering events increase but the range decreases resulting in an increased backscattering effect; thus the backscattering effect is as much influenced by electron range in the substrate as the number of backscattering collisions.

The effect of forward and back-scattering can be modeled by Monte-Carlo simulations. In these simulations, it is assumed that the electrons continuously slow down as described by the Bethe equation, while undergoing elastic scattering, as described by the screened Rutherford formula. Such simulations average the effect of many thousand electrons to generate the deposited energy profile in the resist. An example of such a simulation was shown in Fig. 3.8. The accuracy of Monte Carlo simulations increases with the number of electrons simulated but this also increases the computation time. For fast analysis and proximity effect correction, it is more suitable to use analytical formulations; the parameters for analytical formulations are usually extracted from Monte Carlo simulations and experimental data. A double Gaussian [15–18] is often used to approximate the energy deposition profile which is also referred to as the point spread function.

$$f(r) = \frac{1}{1+\eta}\left(\frac{1}{\pi\alpha^2}\exp(-\frac{r^2}{\alpha^2}) + \frac{\eta}{\pi\beta^2}\exp(-\frac{r^2}{\beta^2})\right) \tag{3.2}$$

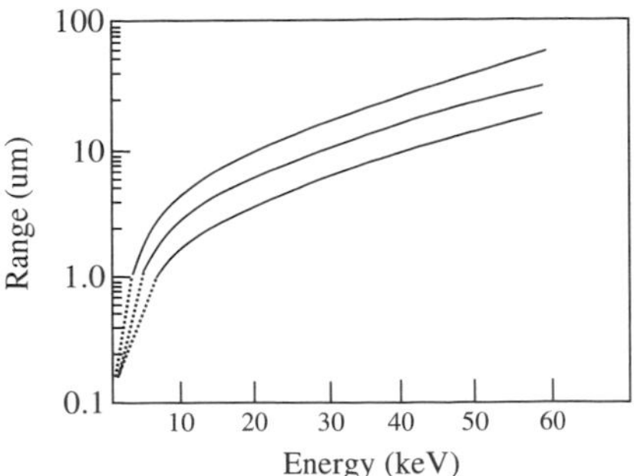

Fig. 3.9 Bethe range for various substrates and accelerating voltages

The two terms in the sum represent the forward and backscattered electrons, respectively. Here, η is the ratio of the backscattered energy to the forward-scattered energy, α is the forward scattering range parameter and β is the backscattering range parameter. The above equation is normalized so that

$$\int_0^\infty f(r)2\pi r\,\mathrm{d}r = 1 \tag{3.3}$$

Figure 3.10 shows the proximity function plotted as a function of the radial distance from the point of incident exposure; it can be seen that the deposited energy is mostly by forward scattered electrons and there is a small contribution from backscattered electrons. For some cases, a double Gaussian function may be insufficient for expressing the energy density profile. More complex functions are needed for certain types of substrates, and for multi-layer substrates. The models above are two dimensional versions of a three dimensional phenomenon. In general, the energy profile depends upon depth as well as radius. By averaging out the depth dependence, a two dimensional profile can be obtained out of a three dimensional profile. This results in a greatly reduced computation time for the exposure estimation and correction; true 3-D proximity effect modeling and correction has been proposed in [17].

A variety of proximity correction algorithms exist in the literature [18–21, 24]. They are based on various principles such as self-consistent calculations, fast-Fourier transform, background dose equalization, and hierarchical optimization. Any algorithm has to meet the stringent demands of being fast while at the same time resulting in accurate proximity correction; even with powerful computers, a complex pattern can take many hours to run through some of the above algorithms. Some e-beam machines can also perform proximity correction at the time of pattern writing by utilizing special hardware that can perform fast mathematical computations. The proximity correction software needs accurate information on the proximity parameters (α, β, η) to be able to correctly predict and correct for the proximity effect. Extraction of these parameters for a particular process usually involves exposing some test structures and evaluating the result [22, 23]; α is the toughest parameter to

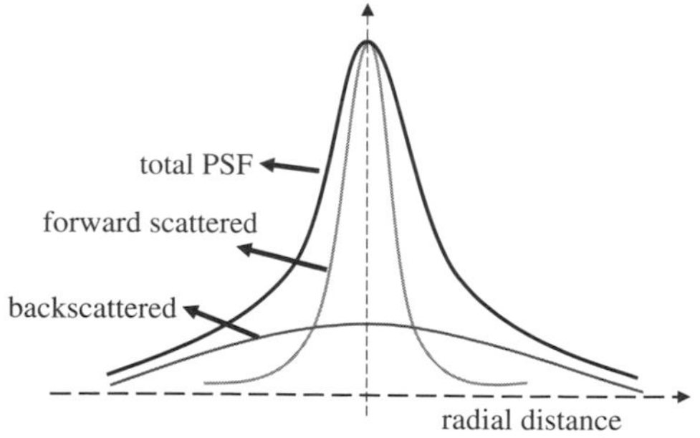

Fig. 3.10 Point spread function (PSF) to model the proximity effect; the PSF and its components are plotted as a function of the radial distance from the point of incident exposure

extract since its value is low in current generation machines. Direct beam-diameter measurements is one way to extract α. Monte-Carlo simulations can also be done to extract proximity parameters; usually information on the resist material components, substrate, and beam conditions are required. An example of a commercially available Monte Carlo simulation software is ProLith from KLA-Tencor. Table 3.3 shows the proximity effect parameters for a 0.5 um thick resist on a Silicon substrate. Commercial software packages that perform proximity correction include PROXECCO from Synopsys, and CAPROX from Sigma-C; a software package borne out of university research is PYRAMID [21]. Since the cost of many of these packages might be prohibitive for small-scale research labs, proximity correction is sometimes carried out manually. Manual correction involves splitting the pattern into different dose levels depending on the line-width and assigning line-width bias (i.e. shape modification) to critical areas of the pattern. With this approach, usually multiple lithography iterations are required before the desired CD control is achieved.

3.5.1 Critical Dimension Control

While minimum resolution of an isolated line is limited by forward scattering effects, an important aspect of patterning is to obtain the *desired* line-width. Line-width control, also known as critical dimension (CD) uniformity, is important for many applications, most notably, mask making. For a given set of beam conditions, CD uniformity is impacted by proximity effects as well as the resist process. The resist dose-depth profile gives a measure of the contrast which in turn determines resolution and side-wall angle. A high-contrast resist process with small sensitivity to dose variations is desired. Shot-pitch determines energy deposition profile in the resist and thereby impacts CD uniformity; shot-pitch is the distance between adjacent exposure spots/pixels. An improvement in the energy slope of the edge improves line definition and CD uniformity. This improvement can come from a variety of sources including better contrast of resist process, lower shot pitch, increased accelerating voltage, smaller beam diameter, or a combination of these sources. Figure 3.11 shows the impact of a reduced shot pitch on the energy deposition profile; as the number of pixels (in a cross-section of a line) is doubled from 2 to 4, the edge slope is seen to improve dramatically. Thus, for good line-width control, it is desirable to have at least 4 pixels to define the line-width. The beam

Table 3.3 Proximity parameters as a function of the beam energy for a 0.5 um resist on a silicon substrate. Values shown in brackets are extrapolations. The data is from [24]

Beam energy (keV)	α (um)	β (um)	η
5	1.33	[0.18]	[0.74]
10	0.39	[0.60]	[0.74]
20	0.12	2.0	0.74
50	0.024	9.5	0.74
100	0.007	31.2	0.74

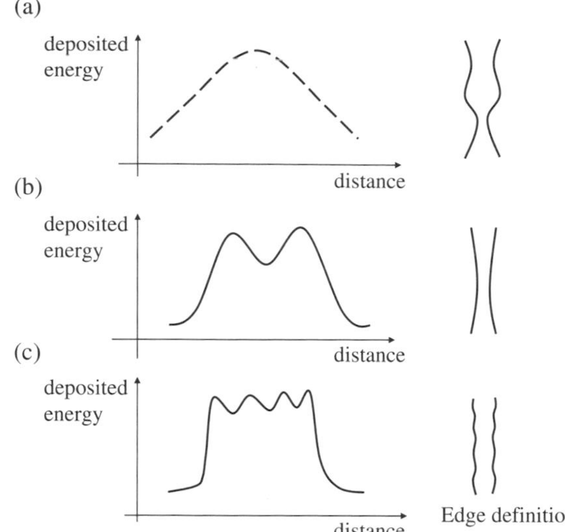

Fig. 3.11 Impact of shot pitch on the energy deposition profile and line edge roughness. In (a), the line is defined by 1 pixel in the lateral direction whereas in (b) and (c) it is defined by 2 and 4 pixels, respectively

diameter should be on the order of the shot pitch; if the beam is much smaller than shot-pitch, there would be non-uniformity in the deposited energy thus leading to poor line definition (Fig. 3.11).

Due to their nature, proximity effects will impact the CD of dense features than sparse features. For e.g., in the pattern shown in Fig. 3.12 if there is a uniform line/space (L/S) pattern close to a big pad, the L/S pattern closer to the pad will receive a larger dose than the one farther away from the pad. This results in an uneven CD in the L/S pattern. The way to resolve this would be to give a lower dose to lines closer to the pad than ones farther away from the pad. Alternatively, the lines closer to the big pad could get a width-bias (a reduction in width) to compensate for the backscattered dose contribution from the pad. A pattern with many shapes quickly becomes too complicated to analyze by simple calculations and thus proximity correction software becomes necessary. Once the pattern passes through proximity correction, the resulting pattern will usually be composed of multiple dose-levels.

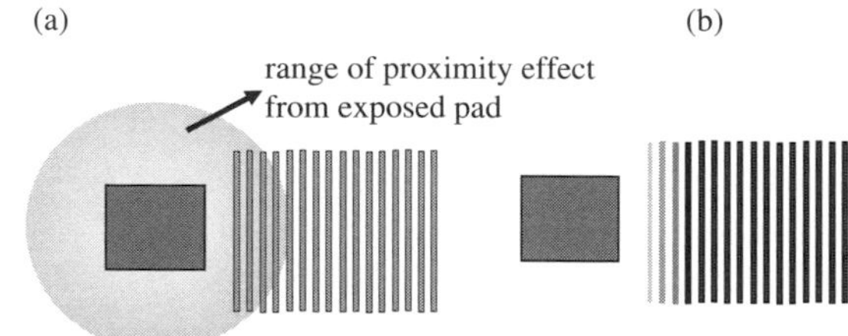

Fig. 3.12 Proximity effect: (a) line/space pattern close to a big pad; the lines closer to the pad experience a greater dose; (b) dose assignment after proximity correction – the lines closer to the pad are assigned a lower dose

3.6 Device Fabrication Examples

In this section, a few examples of device fabrication using e-beam lithography are given. Many fabrication flows require multiple lithography steps and thus layer-to-layer registration becomes necessary. Registration marks are commonly used for registration from one layer to the next. Most modern e-beam lithography systems have automatic mark detection and registration and can achieve sub-20 nm registration accuracy. The registration marks can be either a feature on the substrate or a trench in the substrate. A backscattered electron detector is usually used to detect the registration mark since backscattered electrons can provide topography information of the substrate even when the registration mark is coated with other materials (like oxides and resists). The process of signal detection is shown in Fig. 3.13. As the electron beam scans from left to right, the beam first hits the resist and substrate; this generates a signal in the backscattered electron detector. When the beam hits the Au portion on the substrate, the signal is increased because of the greater height of the Au mark. Higher atomic number materials (e.g. Au) on a lower atomic number substrate (e.g. Si, SiO_2) give the best contrast. The signal quality and contrast is also influenced by the height of the registration mark. Typically, about 50-100 nm of Au on Si provides a good signal for 100 kV accelerating voltage. Trenches and V-grooves in the substrate can also be used as registration marks. The trenches in the substrate need to have a depth on the order of a micron to obtain a good contrast, for 100 kV accelerating voltage. The width of the marks is on the order of a few microns.

Two types of registration marks are usually used for alignment – global marks and chip marks, Fig. 3.14. Global marks are for the whole wafer and their detection results in coarse alignment whereas chip marks are for each chip and is needed for high resolution alignment. Let us consider the example of fabricating nano-wires and contact pads (Fig. 3.15) on an insulating substrate. The process flow is shown in Fig. 3.15. Optical lithography is used to define 100 nm thick Au registration marks on the substrate. In a second optical lithography step, the pads are defined; if the pads are made of the same material as the registration marks, then only one optical

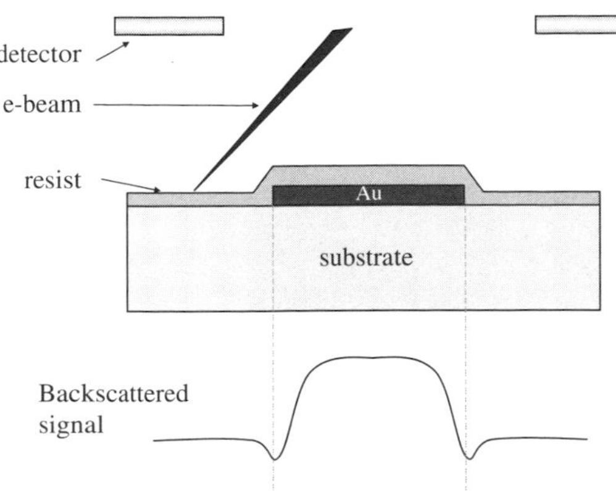

Fig. 3.13 The process of mark detection of a registration mark. The mark is made of Au on a Si substrate

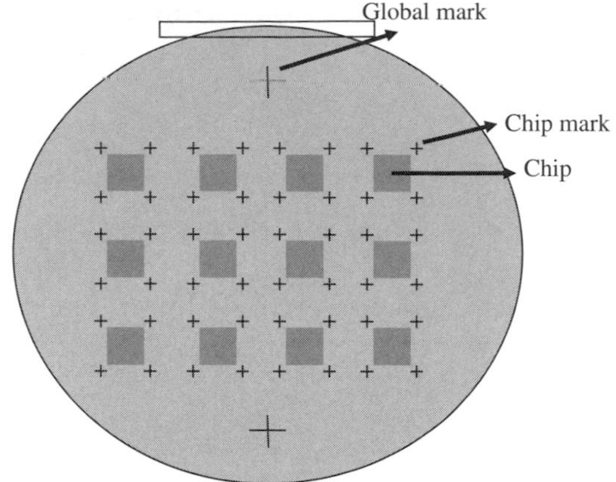

Fig. 3.14 Layout of global and chip marks on a wafer

lithography step is needed. This is followed by e-beam lithography of nanowires. Such a mix and match lithography process greatly reduces the write time of the e-beam lithography system. The resist thickness used in the e-beam lithography step is dependent on the resolution and height of the metallic nanowires. For e.g., for 50 nm wires, an allowable aspect ration of 4:1 translates to a resist height thickness of 200 nm. For a liftoff process, the metal thickness needs to be considerably lower

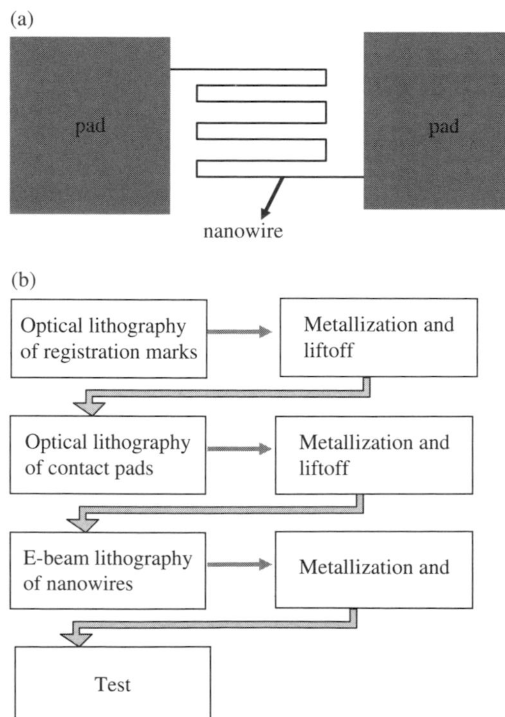

Fig. 3.15 Fabrication of nanowires using mix-and-match lithography – (a) pattern layout, and (b) process flow

than the resist thickness; the metal/resist thickness is usually in the range of 1:2 to 1:3. For a 50 nm wide line, the shot pitch needs to be 10 nm or less for good line definition, and the beam diameter should be considerably lower than the line width for good line-width control. Insulating substrates cannot conduct electrons and anti-charging layers are needed to manage charging effects. A thin layer of Au or Au/Pd (2 to 5 nm) on the e-beam resist is a good charge conducting layer that is also transparent to high energy electrons; after exposure, and before resist develop, the Au layer is dissolved by a KI/I etchant. Water soluble anti-charge layers are also available and these include Aquasave and Espacer; these are spun on the resist as a thin layer and washed away before resist develop.

The electron beam has a depth of focus of less than a few hundred microns; this value decreases for increasing beam current. Substrate height variations can severely affect the CD control and resolution of e-beam lithography. Many systems have an automatic height measurement/correction mechanism; the height is measured using an optical beam, and the electron beam focus is changed based on this height measurement. Chip-to-chip height correction is more effective than an average height measurement of the substrate since the substrate may have a bow/tilt. In the absence of an automatic height measurement system, a manual focus on the substrate is needed to obtain high resolution patterning.

Fine dose control in e-beam lithography makes possible the fabrication of 3D structures. The fabrication of 3D microstructures is important to many applications including diffractive optics, photonic elements, and MEMS. Figure 3.16 shows blazed gratings fabricated in a 100 kV system using about 300 dose levels [25]. The grating pitch is 11 um and the depth is 120 nm; the features are fabricated on a 0.8 um thick, 950 K PMMA; for optical applications, the surface roughness needs to be low (< 5 nm) and thus tight process control is needed to achieve the desired characteristics.

3.7 E-beam Lithography Limits

Resolution and writing speed are limited by many factors: resist limitations, the electron beam, and beam control electronics. Resist sensitivity determines how many electrons are needed to expose a given pattern. Resolution is also impacted both by the proximity effect as well as shot noise. Primary and secondary electron scattering in the resist tends to broaden the exposure profile and thus limits resolution. Even though a resist has a specific sensitivity (for a given resist thickness, substrate, acceleration voltage, and develop process), in reality the electron beam is statistical in nature. This means that the number of electrons that the smallest lines will receive has a certain uncertainty associated with it. For e.g., for a resist requiring a dose of 400 uC/cm^2, on a 100 nm $\times$ 100 nm region, the number of electrons needed is (N=Dose/Area.q) which comes out to 2.5 $\times$ 10^5 electrons. Because of the large number of electrons involved, the uncertainty in the number of electrons can be approximated as $\sqrt{N}$. The signal to noise ratio is than $N/\sqrt{N}=\sqrt{N}$. To be

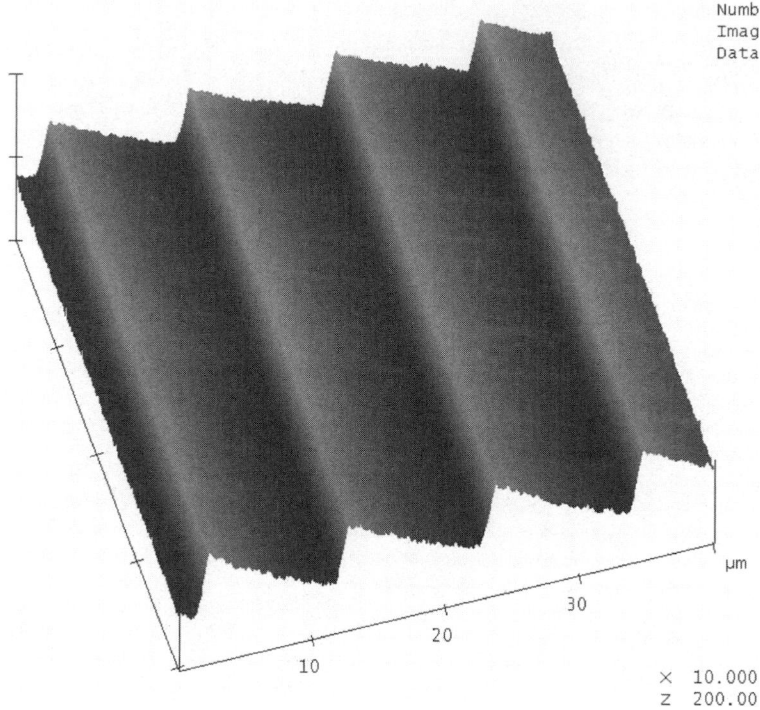

Fig. 3.16 Grayscale lithography – blazed gratings fabricated in a 100 kV system using about 300 dose levels

reasonably certain that the resist is exposed, this number should be at least 10, which imposes a minimum limit on the number of electrons needed for exposure, N=100. The minimum line-width can then be related to sensitivity as: $L_{min}=40/\sqrt{D}$, where L_{min} is in nm and D is in uC/cm^2. Thus, there is an inherent tradeoff between resist sensitivity and resolution.

The electron beam diameter is limited by spherical and chromatic aberrations of the lenses, the transverse velocity of the electrons as well as electron diffraction by a limiting aperture. Spherical aberrations result because the focusing fields of lenses are stronger closer to the lenses. Chromatic aberration results from the fact the electron lens affects electrons of different energies differently. Electron energies have a spread that is dependent on the type of source and thus chromatic aberrations depend on both the electron lenses as well as the incident beam of electrons. Due to electron emission properties, there is a transverse random component of velocity in the electron beam. While passing through a limiting aperture, the electron beam is diffracted but this is usually not a limiting factor for beam diameter. These four effects can be considered statistically independent and the total beam diameter is then the root mean square sum of the four components. Figure 3.17 shows the beam diameter and the various factors limiting it, as a function of the convergence angle, α.

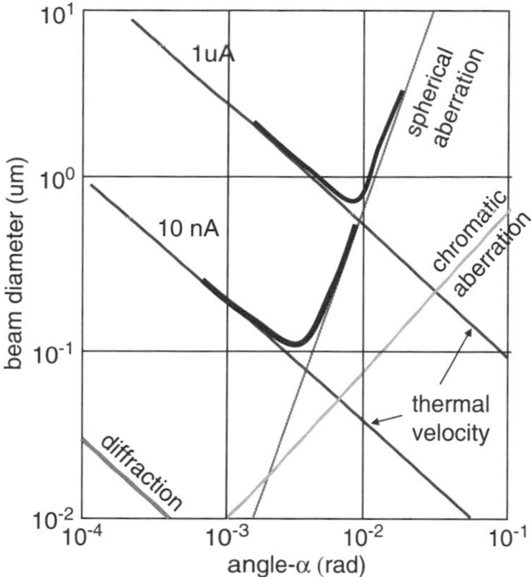

Fig. 3.17 Beam diameter and the various factors limiting it, as a function of the convergence angle, α

The electron-beam scanning system imposes its own limits. A limit on the patterned area arises because of the beam distortion as the beam is deflected over the area. Deflection of the beam over an area causes distortion of the spot shape as well as the beam diameter. Also, deflections that are supposed to be rectangular may result in non-rectangular deflections. As an example, consider the JEOL 100 kV system; this has a field size of 500 um × 500 um and a pixel-limit of 1 nm. As resolution requirements increase, the allowable field-size decreases since the tolerable error too decreases. The same JEOL system at 50 kV has a field-size of 1 mm × 1 mm since the resolution specification is lower at this accelerating voltage. For this system, there are 5×10^5 pixels in X and the same number in Y in each scan field. For a vector scan system that scans point-by-point, the DAC (that drives the deflection amplifier) resolution then needs to be at least 19 bits. Indeed, the system has a 20-bit DAC that can scan up to 2 MHz. DAC resolution and linearity might pose limits to achieving large field-size. There is also a tradeoff between the DAC resolution and deflection amplifier speed. Some systems have two DACs in series, one operating at a lower speed and that can address the whole field, the other operating at a much higher speed but able to address only a part of the field (sub-field). In the JEOL system described above, there is a second DAC, called the sub-deflector, that can scan at 50 MHz, and has a resolution of 16 bits thereby limiting its scan size to 4 um x 4 um. The two deflectors in combination offer a scan speed of 50 MHz and a scan field of 500 um x 500 um. The mechanical stage also imposes a limit since field-stitching accuracy needs to be on the order of pattern resolution. Laser interferometry can result in sub-nm stage accuracy. In addition, registration requirements also pose a limitation on the patterning capability of e-beam lithography. In production-grade machines, sub-20 nm registration accuracy has been demonstrated whereas in research machines, sub-pixel registration has been demonstrated.

Another limitation of the e-beam lithography process is the speed at which the patterns are written. This in turn depends on beam current and resist sensitivity. Beam current can be increased to speed-up the writing but this would result in a larger beam diameter. A highly sensitive resist could be used for faster writing but this would limit the minimum line width because of the shot-noise limit discussed previously. Thus speeding up the writing, either by increased current or increased resist sensitivity, might result in a loss of resolution.

3.8 Other Lithography Techniques

3.8.1 Ion Beam Lithography

The use of a finely focused beam of ions for lithography has been well studied. Focused ion beam (FIB) tools are commonly used in device analysis (e.g. SIMS), mask repair, interconnect rerouting, and advanced specimen preparation. FIB can be used for both etching (ion milling, chemical etching) as well as deposition (ion implantation, chemical deposition). Beam energy, current and spot size of these systems range from 10 to 100 kV, 10 pA-10 nA, and 8 to 200 nm, respectively [26–28]. Resolution is limited to about 30 nm. The invention of liquid metal ion sources has led to the use of a variety of ion species – Be+, B+, P+, Si+, Au+, AI+, Ga+, and Ar+. Three methods of using FIB for lithography exist: (i) direct ion milling – the FIB, using heavy ion sources, can be used to mill either the substrate itself or a hard mask such as Au, (ii) resist patterning with wet development – similar to e-beam lithography, energy deposition by ions results in selective dissolution of the resist; light ions such as H+, He+, and Be+ need to be used to obtain good penetration of the ion beam into the resist, (iii) resist exposure using dry development – here the resist is exposed to a FIB and then placed in an reactive ion etch chamber; implanted regions etch at a slower rate than other regions; Ga+ implanted spin-on-glass in oxygen plasma is an example.

3.8.2 X-ray Lithography

X-rays are very short wavelength radiation that have been successfully used for lithography; the wavelength of interest to lithography is 1–1.5 nm and this overcomes the diffraction limit of optical lithography [29–31]. Proximity X-ray lithography can achieve sub-0.1 um resolution. But because of extensions to 193-nm lithography, and the push toward EUV lithography, X-ray lithography (XRL) has not gained popularity. One of the most powerful sources for XRL is the compact synchrotron or electron storage ring. A synchrotron is expensive, requires a large area for installation, and safety is an issue. Other sources of X-rays include the

X-ray tube and laser-heated plasma source; but these sources are not good enough for large scale wafer production and the emitted radiation is not collimated thereby resulting in a lateral magnification error. XRL overcomes many problems of optical lithography; XRL has no field size limitation, can provide high aspect ratios in resist, overcomes the depth of focus problems, and are immune to defects caused by particles present on the mask. An X-ray mask blank consists of a thin membrane of an X-ray transparent material (1–2 um thick) supported on a frame of the same material. An X-ray absorber layer (about 0.5 um thick) is deposited onto the membrane. This is followed by resist coat, e-beam lithography of a 1x mask pattern, dry etch of the absorber layer, and resist strip. Since the membrane is thin, proximity effects are minimal. But patterning and mask inspection is more challenging since proximity XRL uses a 1X mask.

3.8.3 Electron Projection Lithography

In this type of lithography, the mask is a membrane made of areas that are opaque or transparent to an incident electron beam. But high energy electrons deposit energy in the mask causing mask heating and distortion; this in turn results in patterning distortion and overlay errors. The scattering with angular limitation projection electron beam lithography (SCALPEL) technique [32] uses masks with scattering contrast to overcome these issues; but this system imposes a tradeoff between resolution and throughput (similar to that in a conventional electron beam lithography system). Masks for the SCALPEL system are 4X the minimum feature, and are made from a simple blank. Reticle enhancement techniques (RET) are not needed since EPL is a high resolution technique, and this reduces mask costs. Another type of EPL is the low-energy electron beam proximity lithography (LEEPL) [33]. In LEEPL, low-energy electrons with a 1x mask is used for lithography. Thus mask costs increase but column design is simpler and pattern distortion correction is easier for low-energy electrons.

3.8.4 Dip Pen Nanolithography

Dip Pen Nanolithography (DPN) was first demonstrated in 1999 [34]. DPN uses an atomic force microscope (AFM) to pattern molecules on specific areas of a surface. DPN is most commonly used to pattern self-assembled monolayers (SAMs) [35]. Recently, a parallel DPN lithography was demonstrated and this process can make DPN competitive with other optical and stamping lithographic methods used for patterning large areas on metal and semiconductor substrates. DPN can be used to deliver many different types of inks simultaneously to a surface in any desired configuration. Mask-based lithography and imprinting are limited in this regard. Another DPN technique, called thermal DPN (tDPN) uses an AFM cantilever with a built-in tip heater, along with an ink that is solid at room temperature [36]. Heating the tip melts the ink and allows its deposition onto a surface with great accuracy; turning off the heater in the tip stops the ink deposition. For the case of polymers, the method can control both the physical dimensions and orientation of the material.

3.8.5 Laser Direct Write

In the semiconductor industry, laser beams are used for trimming, via etching, ablation and deposition. In the mask industry, lasers have been used for mask making since the late '80s [37]. Recently, lasers have also found use in the direct writing of wafers [38–40]. Two types of architectures are popular in laser write systems. The first is raster-scan type where the beam scans over the substrate and is modulated according to pattern data by an acousto-optical modulator. Beam scanning is done by either a high-speed rotating polygonal mirror or by an acousto-optic deflector. For e.g. the DWL 66-FS direct write system from Heidelberg instruments has a resolution of about 0.6 um [41], and an address grid down to 20 nm. It uses an acousto-optic modulator to modulate beam intensity; multiple beam intensities are allowed making it possible to fabricate gray-scale structures. A second type of laser writer architecture utilizes spatial light modulation (SLM). An SLM is usually a micro-mechanical device that can modulate the amplitude, phase or polarization of light reflecting off it. Example of SLMs include the digital micromirror device (DMD) that Texas Instruments uses in digital light projection (DLP) devices and the grating light valve (GLV) by Silicon Light Machines.

3.9 Further Reading

- Brewer, ed., *Electron-Beam Technology in Microelectronic Fabrication,* Academic Press (1980)
- M. A. McCord and M. J. Rooks in *Handbook of Microlithography, Micromachining, and Microfabrication: Vol. 1,* P. Rai-Choudhury, Ed., SPIE Press (1997).
- S. Wolf, and R. W. Tauber, *Silicon Processing for the VLSI Era, Vol 1 – Process Technology*, Lattice Press (1999).
- S. A. Campbell, *The Science and Engineering of Microelectronic Fabrication*, Oxford University Press (2001).

References

1. S. P. Beaumont, P. G. Bower, T. Tamamura, and C. D. W. Wilkinson, "Sub-20-nm-wide metal lines by electron-beam exposure of thin poly(methyl methacrylate) films and liftoff," *Applied Physics Letters,* 1981: 38, 436–9.
2. D. M. Tennant, R. Fullowan, H. Takemura, M. Isobe, and Y. Nakagawa, "Evaluation of a 100 kV thermal field emission electron-beam nanolithography system," *Journal of Vacuum Science and Technology B: Microelectronics and Nanometer Structures,* 2000: 18, 3089–3094.

3. G. Brewer, ed., *Electron-Beam Technology in Microelectronic Fabrication,* Academic Press (1980).
4. D. R. S. Cumming, S. Thoms, S. P. Beaumont, and J. M. R. Weaver, "Fabrication of 3 nm wires using 100 keV electron beam lithography and poly(methyl methacrylate) resist," *Applied Physics Letters,* 1996: 68, 322–4.
5. S. Yasin, D. G. Hasko, and H. Ahmed, "Comparison of MIBK/IPA and water/IPA as PMMA developers for electron beam nanolithography," *Microelectronic Engineering,* 2002: 61–62, 745–53.
6. M. J. Rooks, E. Kratschmer, R. Viswanathan, J. Katine, R. E. Fontana, Jr., and S. A. MacDonald, "Low stress development of poly(methylmethacrylate) for high aspect ratio structures," *Journal of Vacuum Science & Technology B: Microelectronics and Nanometer Structures,* 2002: 20, 2937–41.
7. H. Wenchuang, G. H. Bernstein, K. Sarveswaran, and M. Lieberman, "Low temperature development of PMMA for sub-10-nm electron beam lithography," in *2003 Third IEEE Conference on Nanotechnology. IEEE-NANO 2003.* San Francisco, CA, USA, 2003, pp. 602–5.
8. L. E. Ocola and A. Stein, "Effect of cold development on improvement in electron-beam nanopatterning resolution and line roughness," *Journal of Vacuum Science and Technology B: Microelectronics and Nanometer Structures,* 2006: 24, 3061–3065.
9. H. Namatsu, M. Nagase, K. Kurihara, K. Iwadate, and K. Murase, "10-nm silicon lines fabricated in (110) silicon," *Microelectronic Engineering,* 1995: 27, 71–74.
10. L. An, Y. Zheng, K. Li, P. Luo, and Y. Wu, "Nanometer metal line fabrication using a ZEP52050 K PMMA bilayer resist by e-beam lithography," *Journal of Vacuum Science and Technology B: Microelectronics and Nanometer Structures,* 2005: 23, 1603–1606.
11. S. C. Kim, B. O. Lim, H. S. Lee, D. H. Shin, S. K. Kim, H. C. Park, and J. K. Rhee, "Sub-100 nm T-gate fabrication using a positive resist ZEP520/P(MMA-MAA)/PMMA trilayer by double exposure at 50 kV e-beam lithography," *Materials Science in Semiconductor Processing,* 2004: 7, 7–11.
12. Nanolithography, Georgia Institute of Technology, http://nanolithography.gatech.edu.
13. Y. Chen, H. Yang., and Z. Cui, "Effects of developing conditions on the contrast and sensitivity of hydrogen silsesquioxane," *Microelectronic Engineering*, 2006: 83, 1119–23.
14. W. Henschel, Y. M. Georgiev, and H. Kurz, "Study of a high contrast process for hydrogen silsesquioxane as a negative tone electron beam resist," *Journal of Vacuum Science and Technology B: Microelectronics and Nanometer Structures*, 2003: 21, 2018–2025.
15. T. H. P. Chang, "Proximity effect in electron-beam lithography," *Journal of Vacuum Science and Technology,* 1976: 12, 1271–5.
16. M. Parikh and D. Kyser, "Energy deposition functions in electron resist films on substrates," *Journal of Applied Physics,* 1979: 50, 1104–11.
17. K. Anbumony and S. Y. Lee, "True three-dimensional proximity effect correction in electron-beam lithography," *Journal of Vacuum Science and Technology B: Microelectronics and Nanometer Structures,* 2006: 24, 3115–3120.
18. M. Parikh, "Corrections to proximity effects in electron beam lithography. I. Theory," *Journal of Applied Physics,* 1979: 50, 4371–7.
19. G. Owen and P. Rissman, "Proximity effect correction for electron beam lithography by equalization of background dose," *Journal of Applied Physics,* 1983: 54, 3573–81.
20. S. J. Wind, P. D. Gerber, and H. Rothuizen, "Accuracy and efficiency in electron beam proximity effect correction," *Journal of Vacuum Science and Technology B: Microelectronics and Nanometer Structures,* 1998: 16, 3262–8.
21. S. Y. Lee and B. D. Cook, "PYRAMID-a hierarchical, rule-based approach toward proximity effect correction. I. Exposure estimation," *IEEE Transactions on Semiconductor Manufacturing,* 1998: 11, 108–16.
22. A. Misaka, K. Harafuji, and N. Nomura, "Determination of proximity effect parameters in electron-beam lithography," *Journal of Applied Physics,* 1990: 68, 6472–9.

23. M. G. Rosenfield, S. J. Wind, W. W. Molzen, and P. D. Gerber, "Determination of proximity effect correction parameters for 0.1 um electron-beam lithography," *Microelectronic Engineering,* 1990: 11, 617–23.
24. G. Owen, "Methods for proximity effect correction in electron lithography," *Journal of Vacuum Science and Technology B: Microelectronics and Nanometer Structures*, 1990: 8, 1889–92.
25. R. Murali, D. K. Brown, K. P. Martin, and J. D. Meindl, "Process optimization and proximity effect correction for gray scale e-beam lithography," *Journal of Vacuum Science and Technology B: Microelectronics and Nanometer Structures,* 2006: 24, 2936–2939.
26. K. Arshak, M. Mihov, A. Arshak, D. McDonagh, and D. Sutton, "Focused ion beam lithography- overview and new approaches," in *Proceedings of the International Conference on Microelectronics*, Nis, Serbia and Montenegro, 2004, pp. 459–462.
27. F. Watt, A. A. Bettiol, J. A. Van Kan, E. J. Teo, and M. B. H. Breese, "Ion beam lithography and nanofabrication: A review," *International Journal of Nanoscience,* 2005: 4, 269–286.
28. J. Melngailis, A. A. Mondelli, I. L. Berry, III, and R. Mohondro, "A review of ion projection lithography," *Journal of Vacuum Science & Technology B: Microelectronics and Nanometer Structures,* 1998: 16, 927–57.
29. F. Cerrina, "Application of X-rays to nanolithography," *Proceedings of the IEEE,* 1997: 85, 644–51.
30. Q. Leonard, D. Malueg, J. Wallace, J. W. Taylor, S. Dhuey, F. Cerrina, B. Boerger, R. Selzer, M. Yu, Y. Ma, K. Myers, M. Trybendis, E. Moon, and H. I. Smith, "Development, installation, and performance of the x-ray stepper JSAL 5C," *Journal of Vacuum Science and Technology B: Microelectronics and Nanometer Structures,* 2005: 23, 2896–2902.
31. M. C. Peckerar and J. R. Maldonado, "X-ray lithography – an overview," *Proceedings of the IEEE,* 1993: 81, 1249–1273.
32. L. R. Harriott, "SCALPEL: Projection electron beam lithography," in *Proceedings of the IEEE Particle Accelerator Conference*, New York, NY, USA, 1999, pp. 595–599.
33. T. Utsumi, "The status of LEEPL: Can it be an alternative solution?," in *Proceedings of SPIE – The International Society for Optical Engineering*, Dresden, Germany, 2006, p. 628102.
34. R. D. Piner, J. Zhu, F. Xu, S. Hong, and C. A. Mirkin, "'Dip-Pen' nanolithography," *Science,* 1999: 283, 661–663.
35. L. M. Demers, D. S. Ginger, S. J. Park, Z. Li, S. W. Chung, and C. A. Mirkin, "Direct patterning of modified oligonucleotides on metals and insulators by dip-pen nanolithography," *Science,* 2002: 296, 1836–1838.
36. B. A. Nelson, W. P. King, A. R. Laracuente, P. E. Sheehan, and L. J. Whitman, "Direct deposition of continuous metal nanostructures by thermal dip-pen nanolithography," *Applied Physics Letters,* 2006: 88, 33104–1.
37. P. C. Allen, "Laser scanning for semiconductor mask pattern generation," *Proceedings of the IEEE,* 2002: 90, 1653–1669.
38. C. Schomburg, B. Hofflinger, R. Springer, and R. Wijnaendts-van-Resandt, "Economic production of submicron ASICs with laser beam direct write lithography," *Microelectronic Engineering,* 1997: 35, 509–12.
39. C. Yuen-Chuen, L. Yee-Loy, Z. Yan, X. Feng-Lan, L. Chin-Yi, J. Wei, and A. Jaeshin, "Development and applications of a laser writing lithography system for maskless patterning," *Optical Engineering,* 1998: 37, 2521–30.
40. M. Rieger, J. Schoeffel, and M. Bohan, "Laser direct-write-on-wafer lithography," *Microelectronic Manufacturing and Testing,* 1990: 13, 1–12.
41. R. W. Wijnaendts-van-Resandt and C. Buchner, "Super resolution lithography using a direct write laser pattern generator," in *Proc. SPIE – Int. Soc. Opt. Eng.*, Lindau, Germany, 1994, pp. 18–23.

Chapter 4
Nano/Microfabrication Methods for Sensors and NEMS/MEMS

Peter J. Hesketh

Abstract An introduction and overview of nanofabrication methods is provided, including physical vapor deposition, atomic layer deposition, focused ion beam processing and electroplating. Atomic layer deposition is an atomic layer-by-layer deposition process which produced conformal coatings with nanometer control of thickness. It has been demonstrated on a wide range of materials, including semiconductors, metals, ceramics and metal oxides. Focused ion beam fabrication uses a beam of metal ions to carry out precision etching, lithography, and CVD with nanometer scale feature sizes, providing versatile processing capabilities. Electroplating through polymer or porous alumina templates provides a route for the fabrication of nanowires by electroplating. These methods of nanofabrication are illustrated with selected examples including an impedance-based biosensor, an SECM probe, nanobowls, sensors, membranes, MEMS/NEMS, and nanowires with novel properties. These selected examples taken from recent research results demonstrate the exciting potential for development of new nanomanufacturing processes that present opportunities for nanodevice fabrication.

Keywords: Atomic layer deposition · Ion beam · Nanostructure · Electroplating · Nanowire · Biosensor · Chemical vapor deposition · Ultrathin film · Metal oxide · Ferromagnetic · Template · Membrane

4.1 Introduction

There are a number of well established fabrication processes that have been developed for integrated circuit manufacturing. The question is: which among them are the most suitable for nanodevice fabrication?

P. J. Hesketh
The George W. Woodruff School of Mechanical Engineering, Georgia Institute of Technology, Atlanta, GA 30332, USA, (404)385-1358
e-mail: peter.hesketh@me.gatech.edu

P. J. Hesketh (ed.), *BioNanoFluidic MEMS.*

Integration of nanodevice material synthesis—and fabrication with silicon device fabrication and CMOS integrated circuit substrates—presents significant challenges. One of them is that the processing of the nanomaterials or nanostructures may not be compatible with integrated circuit fabrication processes. The fabrication processes take place at either too high a temperature or in corrosive ambient, conditions incompatible with silicon substrates.

Selection or adaptation of the most appropriate of these integrated circuit fabrication processes depends upon the specific materials utilized in the final nano/microdevice [1, 2]. In particular, the application of electron beam lithography, described in Chapter 3, has been a cornerstone for enabling nanodevice and nanostructure fabrication. When such lithographic processes are used in conjunction with physical vapor deposition (PVD), chemical vapor deposition (CVD), or a combination of both, a variety of materials can be successfully processed on planar substrates. Ion beam lithography is an alternative method for nanolithography and fabrication. Ion beam lithography is capable of high resolution and has other significant advantages. Focused ion beam (FIB) methods are being used both for etching and deposition of a range of materials.

With this background in mind, it is time to examine what is in the current tool box for nanodevice fabrication.

Other fabrication processes for microdevices include sputtering, chemical vapor deposition, liquid phase epitaxy, molecular beam epitaxy, plasma based etching and deposition processes, rapid thermal processing and electron beam and ion beam milling. Excellent references on these topics include the books by Madou [3], Campbell [4] and Brodie [2]. In this chapter we will focus on just four processes we have selected for nanofabrication, that is specifically, atomic layer deposition, electroplating and focused ion beam milling.

4.2 Physical Vapor Deposition

Low pressure vapor deposition has a long history and is a primary method for the formation of both crystalline, polycrystalline, and amorphous thin films. The condensation of material from a vapor provides the opportunity to grow a pure material at a reduced pressure. Conditions under vacuum can reduce the presence of contaminants in the growing film, with provisions. It is a prerequisite that the arrival rate of the condensing vapor is more rapid than that of collisions with the surface from residual gas molecules in the chamber. (The boiling point is reduced at a low pressure.) Further details on evaporation processes, and on vacuum system design and operation, are presented in the excellent books by Mahan [5], Wolf and Tauber [6], and Campbell [4].

Figure 4.1 shows the essential components of a typical system for thermal evaporation. The substrate is held either at room temperature or at a fixed, elevated temperature by electrical heating. The higher temperature has a pronounced effect on the nucleation and growth of the film. Substrate heating can be used initially to elevate the substrate to a temperature high enough to desorb contaminants and

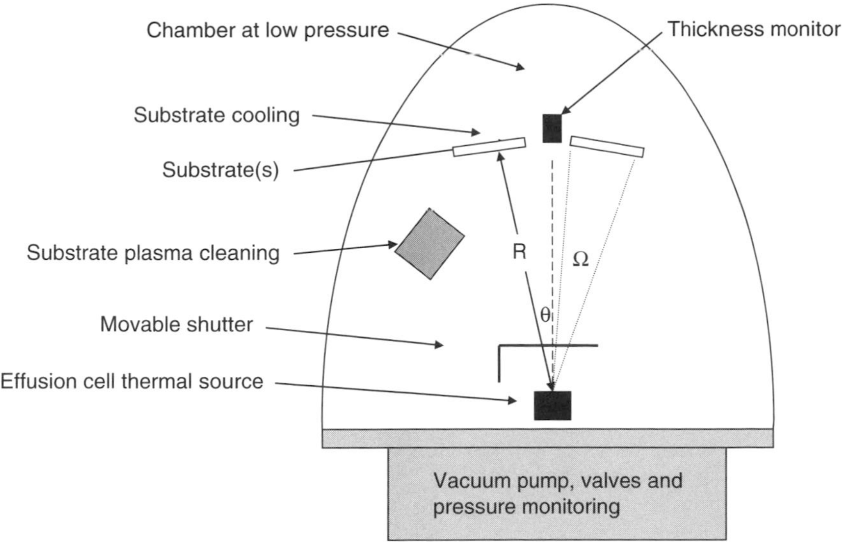

Fig. 4.1 Schematic diagram of evaporation system

water vapor while under vacuum. This important step increases the adhesion of the deposited film on the substrate.

By contrast, at room temperature, in ambient air, a layer of water is present on the surface of the substrate. This layer should be removed prior to the evaporation process. If not removed, it could react with the condensing vapor to form a thin film likely to negatively impact adhesion to the substrate. Take the case in which a noble metal is deposited on the substrate: one method of increasing adhesion is the evaporation of two layers of metal. The adhesion layer comprises a reactive metal, such as chromium, tantalum, titanium, or tungsten.

4.2.1 Vapor Pressure and Deposition Rate

The temperature required to produce a vapor pressure P_{eq} can be calculated based upon the enthalpy and entropy for vaporizations, that is:

$$P_{\text{eq}} = P^o \exp\left(\frac{\Delta_{\text{vap}} S_A^o}{R}\right) \exp\left(-\frac{\Delta_{\text{vap}} H_A^o}{RT}\right) \tag{4.1}$$

Where P^o is the standard pressure 10^5 Pa, $\Delta_{\text{vap}} S_A^o$ is the standard entropy of vaporization, $\Delta_{\text{vap}} H_A^o$ is the standard enthalpy of vaporization, R is the molar gas constant, and T is the absolute temperature [5]. Vapor pressure is estimated with the thermodynamic data presented in Table 4.1 [7]. For Aluminum at a temperature of 1000 K:

$$P_{\text{eq}} = 12.5\mu \text{ Pa} \tag{4.2}$$

Table 4.1 Standard enthalpies and entropies of vaporization for pure elements

Element	Melting point (K)	Standard enthalpy $\Delta_{vap}H^o_A$ (kJ)	Standard entropy $\Delta_{vap}S^o_A$ (J/K)	Reference
Aluminum	933	314	117.8	[7]
Chromium	2130	349	118.8	[7]
Gold	1338	324.4	–	Environmentalchemistry.com
Platinum	2045	510.5	–	Environmentalchemistry.com
Titanium	1939	438	124.6	[7]

The Hertz-Knudsen-Langmuir relation expresses the evaporation flux as a function of the mass, temperature, and vapor pressure [5]:

$$J = \frac{P_{eq} - P}{\sqrt{2\pi m \mathrm{KT}}} \tag{4.3}$$

Where m is the gram-molecular mass, T is the absolute temperature, K is Boltzmann's constant, P is pressure in chamber, and P_{eq} is the vapor pressure in Pascals.

Figure 4.2 shows the calculated vapor pressure of metals commonly encountered in evaporation processes. To achieve evaporation, the pressure generated by the source must be greater than the pressure in the chamber. The evaporation rate can be estimated based upon modeling the source as an effusion cell that has a vapor at equilibrium with the source. Thus Equation 4.3 can be used to estimate the

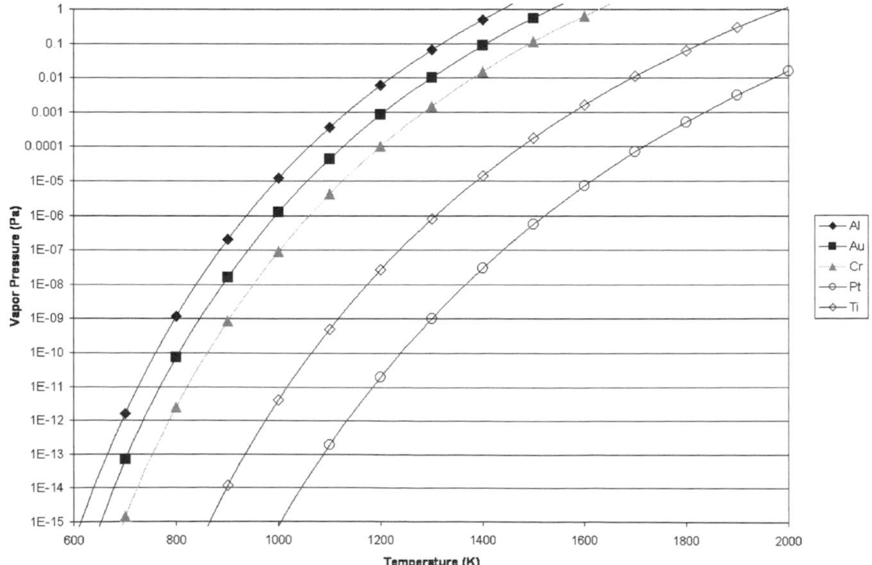

Fig. 4.2 Vapor pressure of metals commonly deposited by thermal evaporation

evaporation flux from the effusion cell, and hence the arrival at the surface of the substrate, given the solid angle Ω, based upon geometry (see Fig. 4.1). The exiting flux J_Ω, from an ideal effusion cell of area δ A, is calculated by:

$$J_\Omega = \frac{J\delta A \cos\theta}{\pi} \tag{4.4}$$

The arrival rate at the surface at a distance R from the source is expressed by:

$$j_s = \frac{J_\Omega \cos\phi}{R^2} \tag{4.5}$$

where j_S is the flux arriving at the substrate at an angle ϕ to the surface normal. For example, for aluminum the thermodynamic data indicate that the vapor pressure at 1500 K is 1.69 Pa. Hence the arrival rate at the source can be calculated as follows:

$$js = \frac{J\delta A}{\pi R^2} = 1.75 \times 10^{19}/\mathrm{m}^2/\mathrm{s} \tag{4.6}$$

for δ A $= 1$ cm^2, $R = 0.2$ m, and normal incidence. This arrival rate can be compared to that for residual gas (nitrogen) in the chamber at 0.13 mPa (1×10^{-6} Torr) by incorporating the Kinetic Theory of gases [8].

$$\text{Arrival rate for nitrogen} = 3.8 \times 10^{18}/m^2/s \tag{4.7}$$

To produce a pure film, it is important that this rate of evaporation be much greater than the arrival rate of impurities and water vapor molecules in the vacuum system.

The film thickness can be monitored by a quartz crystal microbalance whose frequency is decreased as the mass of evaporated material deposits on the surface. This system has to be corrected for the mechanical properties of the metal film. It is also quite usual to heat the chamber to reduce the amount of impurities present prior to evaporation. Initial evaporation against a shutter also removes impurities present on the source crucible.

The direction of the evaporation is basically a line-of-sight, as the mean free path between gas particle collisions is much greater than the distance between the source and substrate. The evaporation layer thickness can be determined using geometry. This is beneficial when defining structures with the lift-off process, or when defining comb electrodes. There is however a disadvantage. This disadvantage presents itself when uniform step coverage is required. This disadvantage is overcome using a planetary drive and rotation of the substrate with tilting. The constantly changing angle and position of the substrate with respect to the source provides a more uniform coating.

In many cases, sputtering is the preferred technique for improved step coverage. For refractory metals, an electron beam source is used to generate a higher local temperature than does a filament heating method. The former method is used for platinum deposition in the example presented below. The high voltage used in the e-beam source may generate X-ray exposure of the substrate, an additional consideration when high power is needed. Alloys can also be deposited if multiple electron beam sources are used and the deposition rate can be independently controlled for each source. Multiple layers can be evaporated without breaking the vacuum when additional sources are present in the system, such as an adhesion layer or diffusion barrier.

4.2.2 Ultrathin Film Growth

Condensation of atoms or molecules from the vapor phase takes place during film growth. An atomistic view of nucleation theory is obtained by considering the kinetics of the process, and by developing a rate equation for the formation of nuclei or clusters. This equation is based upon how adatoms diffuse to the nucleated islands on the surface [9].

The theory of nucleation by Gibbs, as applied to liquid droplets, can be applied—with caution [10]. The film's morphology is defined in the early stages of growth by the migration of the adatom along the surface, and the formation of nuclei or clusters. An adatom is able to diffuse across the surface, before it combines with others, to form an island or cluster.

The adatoms arrive at the surface and have a mobility dependant upon their kinetic energy. The interaction of the surface with the adatom, along with the temperature of the surface, have important influences on the nucleation and island growth of the film. Because the surface diffusion has activation energy, the temperature has an important influence on surface diffusion. As adatoms combine, they form nuclei or clusters. These are enlarged in size, forming islands. Then coalescence can take place, until a continuous film is formed. Excellent references on film growth [9] and models appropriate to epitaxial growth are presented by Hirth and Pound [11], and by Michely and Krug [12].

Figure 4.4 shows the initial stages of film growth. In these stages, nucleation is followed by an increase in the diameter of the islands. Secondary nucleation can occur in the gaps between the islands. When two islands meet, coalescence can occur, forming a new, larger island. Each island can have a different crystallographic orientation. As the gap between closely spaced islands decreases, the open space fills to form an interconnected array of islands. Such a film is porous and not yet a continuous film. The growing islands meet and form the basis for the initial grain size distribution in the resultant film.

Island films are often beneficial for catalysis in chemical sensors and fuel cells. In such applications, a smaller island diameter increases the surface area for triple contact between the gas phase, the catalytic metal (solid phase), and electrolyte (liquid phase). Figure 4.6a is an atomic force microscope (AFM) image of an ultrathin platinum film. The film was deposited by sputtering platinum onto a 0.5 nm titanium adhesion layer on silicon dioxide-coated, silicon substrates of a nominal thickness of 2.5 nm. This image illustrates the porous metal island structure.

While the number of density islands that form on a surface can be calculated by application of nucleation theory to film growth, that number is in practice strongly

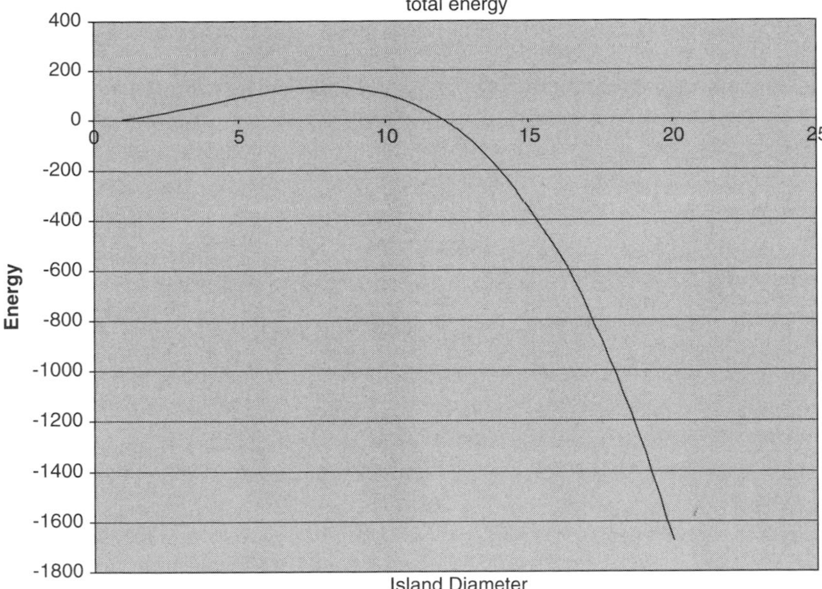

Fig. 4.3 Surface free energy as a function of island diameter, idealized plot neglecting any surface re-construction or interfacial strain

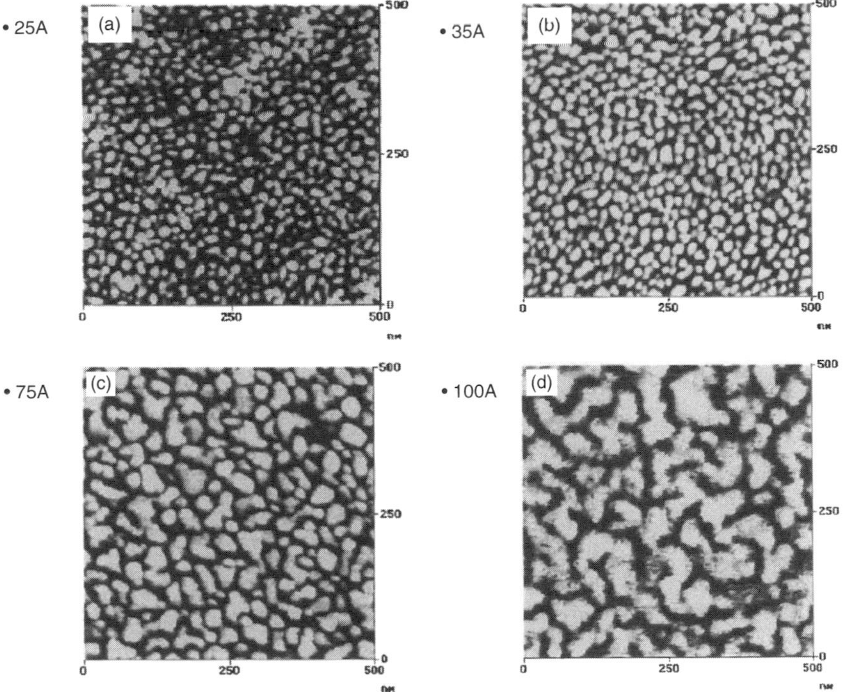

Fig. 4.4 Stages of nucleation and growth or evaporated gold film: (A) 2.5 nm, (B) 3.5 nm, (C) 7.5 nm, and (D) 10 nm [136]

influenced by other factors. One such factor is impurities. These can either enhance or depress adatom mobility, and thus slow or accelerate island growth. An adsorbate on the surface can influence the binding energy between the island and the substrate. This circumstance can modify the surface diffusion via a lowering of the energy barrier. The size of and space between the islands is not random: they depend upon the substrate surface and the adatom interactions. The binding energy associated with physisorption is typically much less than chemisorption, so that the adatoms can continue to move along the surface to find a suitable reaction site. Studies of nucleation and growth of platinum on single crystal surfaces indicate that each island has a capture radius dependant upon the kinetics of the process [12]. Figure 4.4 shows the process of nucleation and growth of gold on silicon dioxide. Coating the silica surface with mercapto trimethyl-ethyl-silazane (MTS) provides an additional surface reaction between the gold and sulfur terminated silane film. This coating process reduces the average island diameter.

Thermodynamically, the surface energy of the islands will determine their stability, as indicated in Fig. 4.3. If an island is smaller than the minimum critical radius, there is a possibility that the island could decrease in size. However, the kinetics

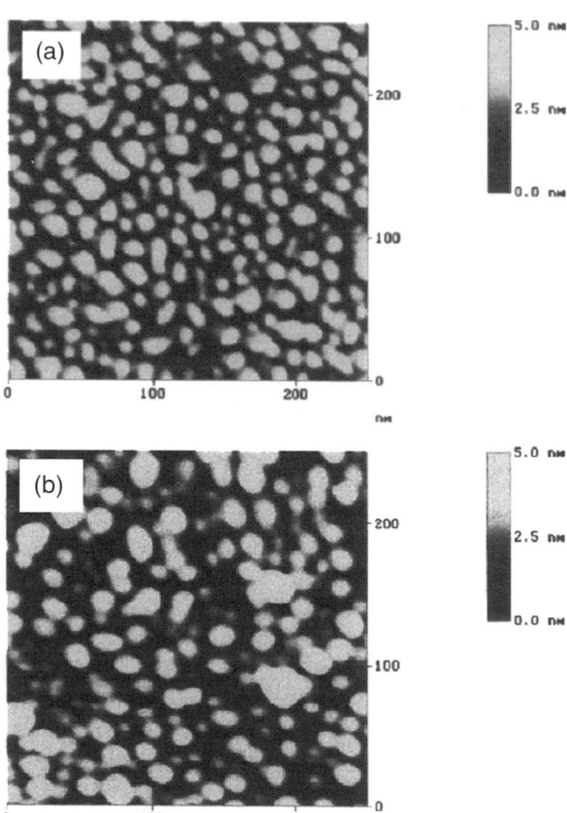

Fig. 4.5 FM images of the ultra thick gold film on silicon dioxide (A) before annealing, and (B) after annealing for 1 hr at 100 °C [136]

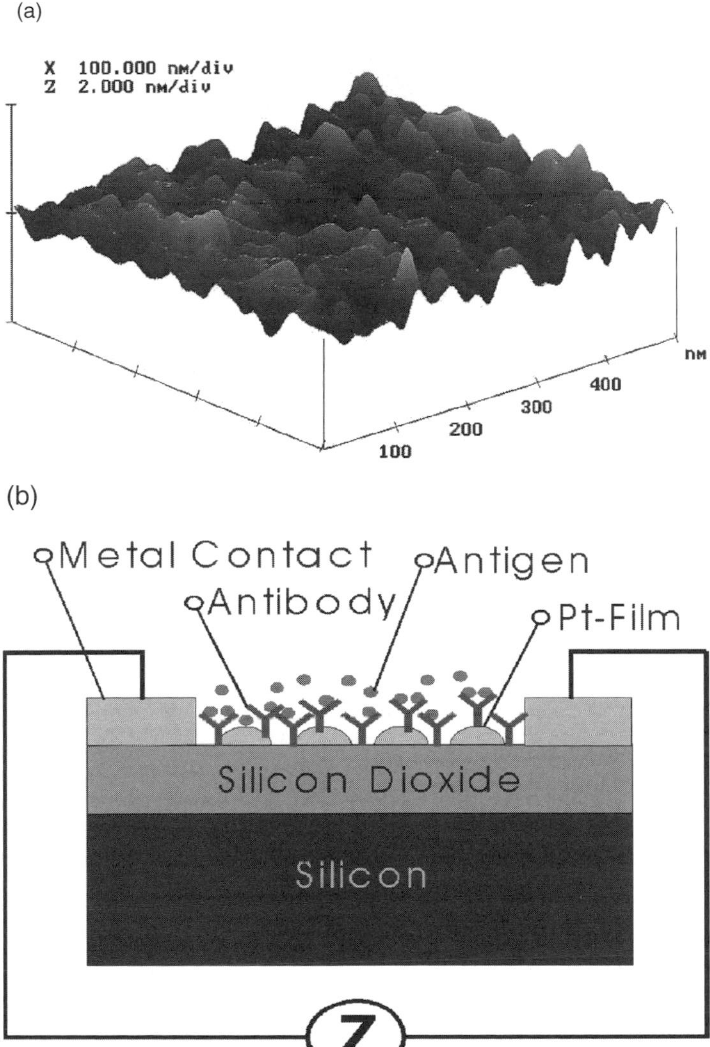

Fig. 4.6 Impedance based immunobiosensor. (A) AFM image of the 2.5 nm platinum island film measured with tapping mode Digital Instruments Nanoscope III, and (B) Schematic diagram of immunobiosensor. Electrical contact is made to the porous Pt island film by two gold electrodes approximately 8 mm apart [14]

of the process have a strong effect on growth and dissolution. Therefore a theory that includes the flux of arriving adatoms, presence of adsorbates, and substrate temperature (diffusion) provides more insight than does thermodynamic nucleation theory alone. The growth process might be quite different and can result in islands smaller or larger than the critical radius.

4.2.3 Example 1: Impedance-Based Immunobiosensor

Examples are provided to illustrate the benefits and limitations of particular fabrication approaches. One such case is that of a porous platinum film immunobiosensor. The detection of proteins with microsensors is a highly challenging endeavor. Impedance-based transduction of binding at a surface can be highly sensitive.

An immunoassay is based upon the selective binding of an antibody-antigen complex. Here, for the sensor transduction, selective antibody/antigen binding takes place on a solid phase, at the conductive, ultra-thin film electrode surface. This method alleviates the need for an enzymatic or fluorescence labeling of the antibody to effectuate the detection of the concentration of bound complexes.

Depositing an ultra-thin platinum film on the silicon dioxide surface is a method that has been used to construct a sensitive conductiometric immunosensor. This process is illustrated schematically in Fig. 4.6(b). The average film thickness is 2.5 nm and is characterized by an interconnected metal island structure.

Monoclonal antibodies were immobilized to the electrode surface through covalent linking with glycidoxypropyldimethylethoxy silane (GOP) [13]. The sensor impedance was monitored by an HP 4284A Precision LCR Meter with a 10 mV signal amplitude. The measurement revealed an increase of 55% at 20 Hz during the activation of the surface with anti-alkaline phosphatase. In addition, the binding of 40 attomole alkaline phosphates to the surface resulted in a further impedance change of 12% [14]. The response time is on the order of several seconds. However, the durability of the sensors is such that this measurement can only be used as a dosimeter. This restriction is due to the high binding constant of an antibody-antigen complex and the concomitant difficulty in regenerating the surface for subsequent measurements.

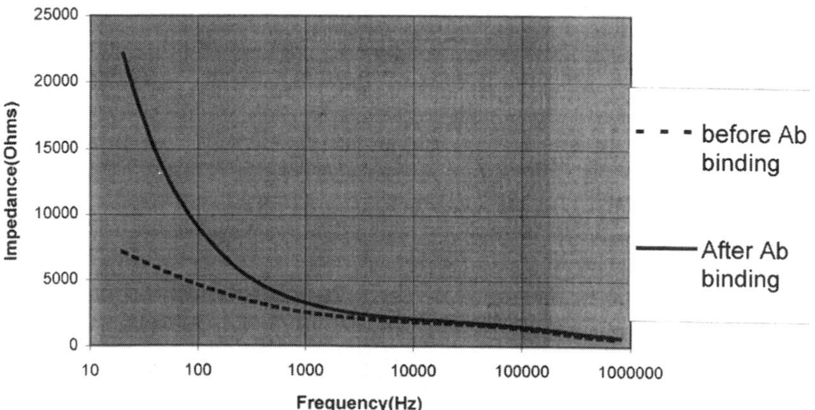

Fig. 4.7 The impedance of the ultrathin Pt-film is measured at a small 10 mV applied ac voltage. Measurements of the magnitude of impedance for the immobilization of primary antibody anti-alkaline phosphatase to the sensor surface through covalent immobilization [14]

4.3 Atomic Layer Deposition

4.3.1 Introduction

Atomic Layer Deposition (ALD), or Atomic Layer Epitaxy (ALE), is a growth process that allows films to be grown layer-by-layer. There has been a rapid increase in the number of published studies on this topic. Selected references on the subject are provided in this section. They provide a sampling of the range of materials that have been investigated to date.

ALD has several advantages over chemical vapor deposition and physical vapor deposition for thin film growth. These advantages include improved surface morphology, low pin-hole density, and control of stoichiometry and microstructure. A key benefit of ALD is the improved step coverage, particularly relevant for coating high aspect ratio structures. The process is based upon cycling the growth ambient between two or more precursors, such that the formation of the new phase is carried out layer-by-layer (see Fig. 4.8).

The first compound introduced absorbs on the surface, and then is removed from the ambient. Next comes the introduction of the second compound, facilitating a reaction on the solid phase. This step produces a film thickness measuring approximately a monolayer in thickness, achieved because it is produced by a self-limiting process.

The process is then repeated, so that, after many cycles, a controlled film thickness results with improved conformality and surface roughness, enabling control of nanometer-scale film thickness. The keys to these sequential processes are that:

- The reactions are self-limiting, and
- The thickness does not depend upon the exposure time nor on the partial pressure of the reactants.

Steps 1 to 4 make up a reaction cycle, as indicated in Fig. 4.8:

- Exposure to reactant A,
- Purge,
- Exposure to reactant B, and
- Purge.

The film growth per reaction cycle is known as the "growth per cycle," rather than the growth rate, the latter of which is time-derivative of the film thickness during growth.

A typical reactor configuration is shown in Fig. 4.9. The precursors in this case consist of a first step, which is $TiCL_4$or TMA, and that is followed by water, enabling films of alumina or titania to be produced. A wide range of materials, both amorphous and polycrystalline, can be grown by ALD. These materials include semiconductors, metal oxides, and metals.

A range of materials are reviewed by Ritala [15], Leskelä [16], and Ritala and Leskelä [17]. A review of early work is provided by Suntola [18].

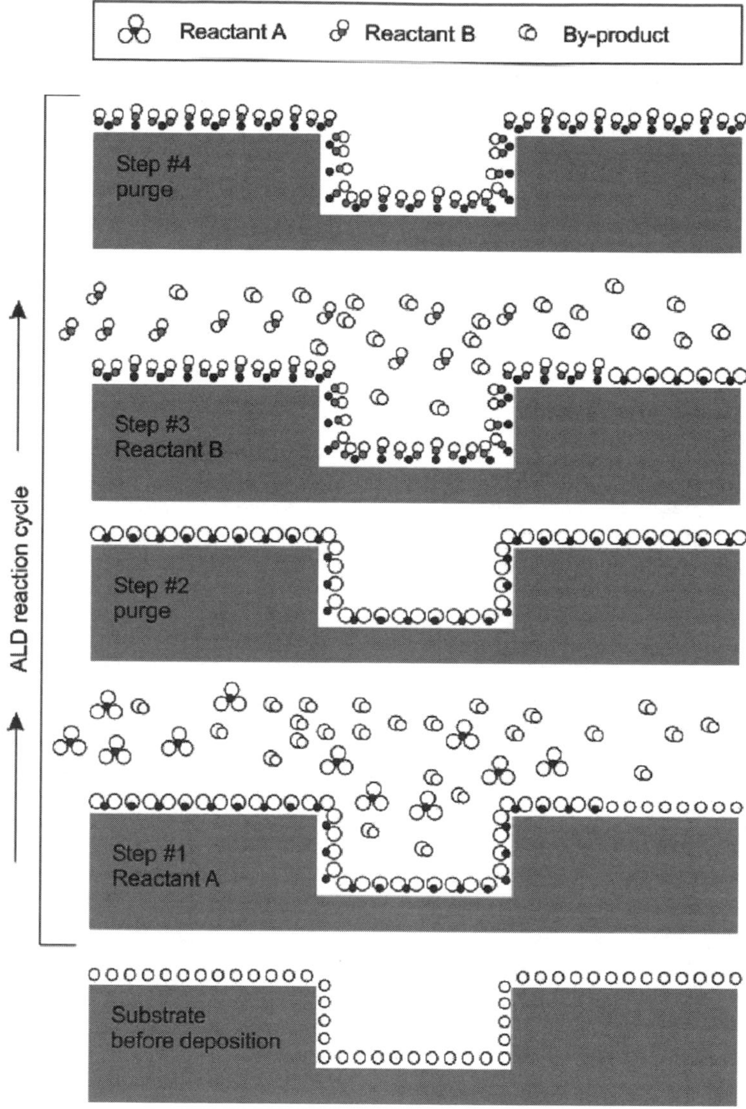

Fig. 4.8 Schematic diagram indicating the sequence of stages during atomic layer deposition [19]

Applications of uniform, high-quality thin films have included optical coatings for electroluminescent-displays, solar cells, microelectronic dielectric films, and coatings for catalysis and powders. More recent work has expanded the range of applications to include Micro Electro-Mechanical Systems (MEMS) and the coating of diatoms and nanowires. Excellent recent reviews of ALD processes are provided by Puurunen [19] and Ritala and Leskelä [17]. The growth of films relevant to semiconductor device applications has been addressed by Kim [20], and a comprehensive update on CVD is provided by Choy [21].

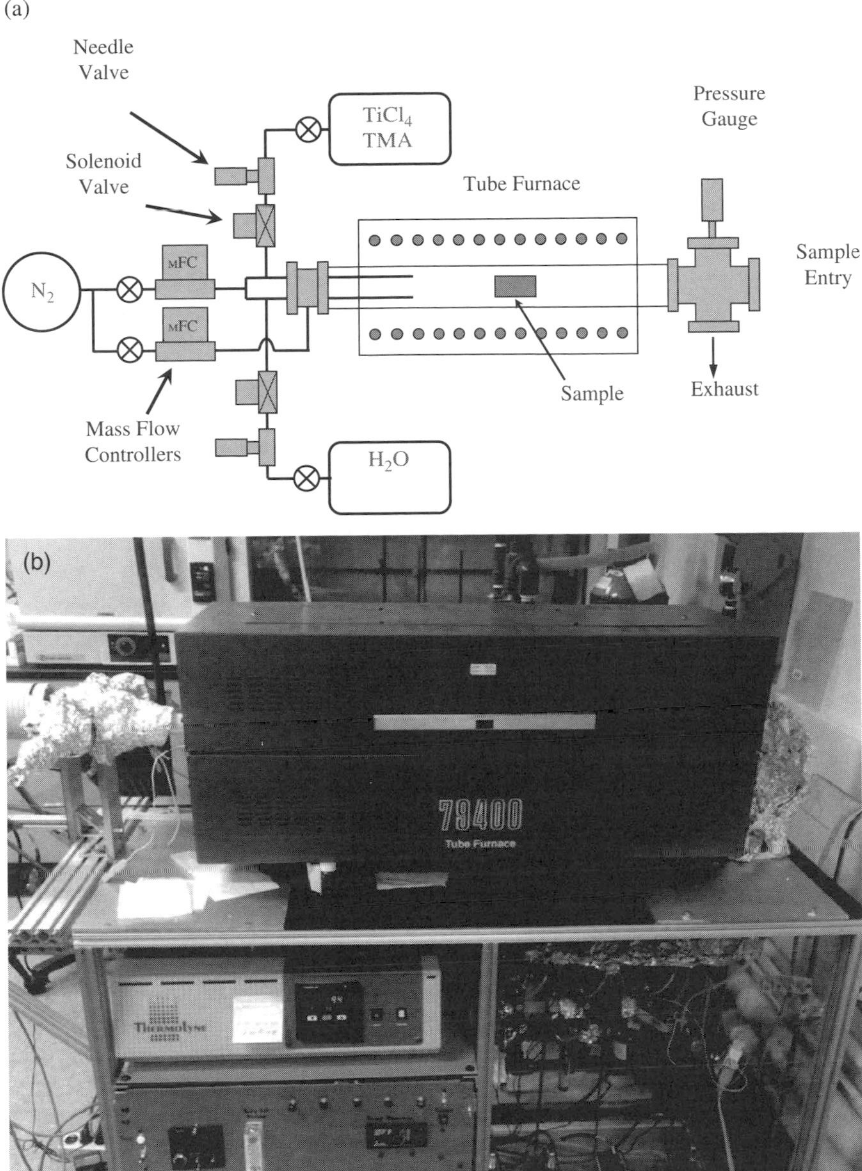

Fig. 4.9 (A) Schematic diagram and (B) photograph of ALD reactor for the deposition of alumina and titania. (courtesy of Prof. C. Summers, School of Materials Science and Engineering, Georgia Institute of Technology)

Early work on ALD was carried out by Soviet researchers, in which they refer to the process as "molecular layering" [22, 23]. The ALD (or ALE) process was originally developed for thin films of ZnS in electroluminescent displays, and has been extensively applied to semiconductors, metal oxides, and a range of other materials.

There are excellent reviews of early work authored by Niinisto et al. [24] and by Ritala and Leskelä [25].

Much has yet to be understood about the kinetics of the ALD process for different precursors and substrates. Among the important aspects are:

- How the processing conditions influence the self-limiting film growth process, and
- Whether or not a monolayer is defined.

The growth on different substrates, film morphologies, and surface chemistry will influence how a monolayer is defined in each growth cycle. A detailed review and study of aluFmina deposition from TMA and water reactants is discussed by Puurunen [19].

In general, a smaller molecular weight precursor is preferred, one that is volatile and thermally stable at the deposition temperature, but is also able to react with surface groups. For example, the metal halides are thermally stable and provide a reactive self limiting growth process with surface groups. Consider the growth process presented in Fig. 4.10. At the hydrated surface, $ZnCl_2$ reacts with the substrate, forming a self-limiting sub-monolayer film. The original glass substrate is terminated with hydroxides, and so the following mechanism is proposed:

$$(–OH)_{(s)} + ZnCl_{2(g)} \quad \rightarrow (–O–)ZnCl_{2\text{-}n} + nHCl_{(g)}$$

where n is 0 to 1, depending on the temperature and processing conditions. The exposure to H_2S results in the formation of the ZnS film:

$$(–O–)nZnCl_2 + +H_2S \rightarrow (–O–)nZnS + 2HCl_{(g)}$$

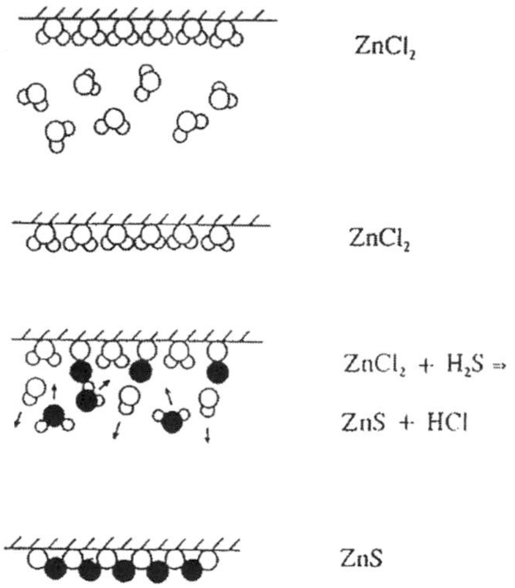

Fig. 4.10 Schematic diagram of cycle for ZnS growth from $ZnCl_2$ and H_2S [25]

This approach provides a reproducible layer thickness each cycle when the growth process is self-limiting.

Other precursors include metal organics such as alkyls, specifically trimethyl aluminium (TMA) for alumina growth. Other metal organics with a higher molecular weight include akloxides, β-diketonates, cycopentadienyls, and others.

When the precursor is a larger molecule, the steric hindrance may change the adsorption to sub-monolayer coverage. Additional precursors are required, beyond those currently available, to optimize the ALD processes. The reason is that the surface reaction must be self limiting so that the monolayer formation is independent of exposure time. This is the correct processing window when the temperature is sufficient to promote the surface reaction, but not sufficient for gas phase decomposition to occur. These considerations are illustrated schematically in Fig. 4.11.

In general terms, the different regimes potentially available include:

- L1 is where the condensation of reactants can take place.
- L2 is where there is insufficient thermal activation of the precursor.
- H1 is where decomposition can take place.
- H2 is where re-evaporation of the monolayer will occur.

Two factors affect the growth-per-cycle:

- The concentration of surface groups available for the reaction, and
- The saturation of the surface with the reactant.

When saturation occurs, sufficient time should be allowed to provide the opportunity for all of the surface area to be coated, independent of surface geometry. Then the self-limiting growth is independent of pressure and the ambient concentration of precursors.

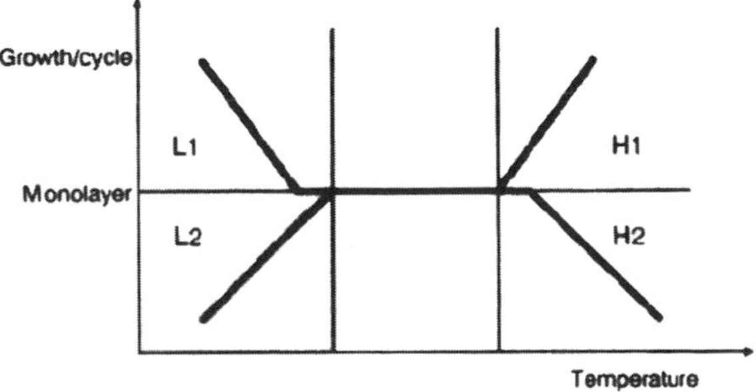

Fig. 4.11 Diagram indicating range of growth conditions available for ALD as a function of temperature. Region L1 condensation of reactants, L2 activation energy insufficient is limits complete monolayer formation, H1 decomposition of reactants which results in non-evaporating surface layer, H2 re-evaporation of monolayer occurs [18]

4.3.2 Semiconductors

The previous section discussed the pioneering work that focused on ALD deposition of ZnS formed from $ZnCl_2$and H_2S precursors. Because progress in the field of ZnS thin films for electroluminescent displays has led to a very successful and commercially available display technology, semiconductors have been extensively investigated. Work in this area includes GaAs [26,27], SiGe, silicon [28], and HgTe by electrochemical ALD by Venkatasamy et al. [29].

The detailed mechanisms for growth of GaAs is still controversial. Notably, there is a lack of complete agreement on the deposition mechanism [30]. The literature includes several reports of preliminary findings on other semiconductors, including:

- Silicon has been deposited from a silane precursor onto diamond substrates [28].
- Thin film solar cell performance is improved with ALD coatings [31].
- ZnOS films have been grown from dimethylzinc and hydrogen sulfide by Sanders [32] and indium sulfide for solar cells by Youshi et al. [33].

4.3.3 Dielectric Films

These sections discuss the use of dielectric films, focusing specifically on alumina films and on silica and mixed films.

4.3.3.1 Alumina Films

Growth of alumina is based on TMA and water precursors. Growth has been achieved at temperatures in the range of from under a hundred degrees to 400 °C [34]. When the growth process is self-limiting, the layer thickness and deposition rate are stable for each deposition cycle (see Fig. 4.12 for an illustration of the alumina thickness produced in each deposition cycle).

The reaction can be studied by monitoring the change in mass during the deposition cycle. Figure 4.13 shows the deposited mass measured with a quartz crystal microbalance, as a function of time, with the same precursors. The largest mass change occurs when the reaction of TMA—when compared to the smaller mass increase with exposure to water—is followed by a decrease that may indicate the desorption of surface hydroxyl groups or the desorption of molecular water.

A study of ALD coating of BN particles with the same precursors provided additional insight via FTIR spectra taken at different points in the cycle (shown in Fig. 4.14). Here the removal of the initial BO–H bond and BN–H_2 stretching vibration is replaced with AlC–H_3 vibrations, which substitution indicates the conversion of the surface groups. This uniform conformal growth is confirmed with a TEM image of the 9 nm thick coating of the BN particles (see Fig. 4.15).

The number of available surface groups will have an impact on the layer coating each cycle. This effect was indicated in a study of alumina coating of hydrogen terminated silicon and chemically treated silica surface, where the –OH surface

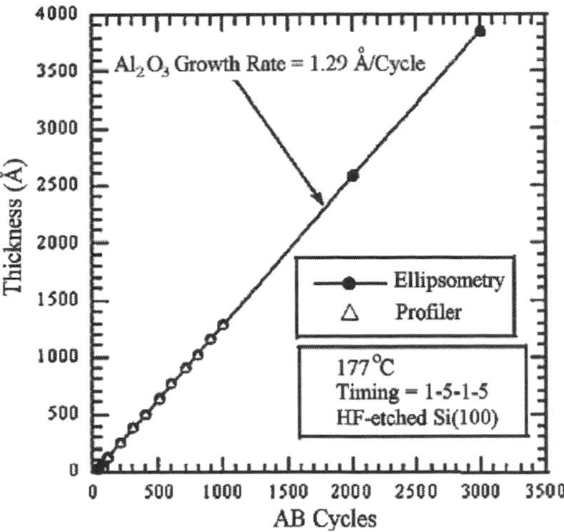

Fig. 4.12 Alumina deposition from TMA and water at 177 °C onto silicon (100) substrates. Growth per cycle measured with ellipsometer and profiler [137]

concentration was larger [19]. Figure 4.16 illustrates that the impact is significant: there is an increase in reaction rate when a higher surface concentration is present. There could then be some initial island growth, followed by layer growth for silicon; whereas, for the silica surface, layer by layer growth predominates throughout.

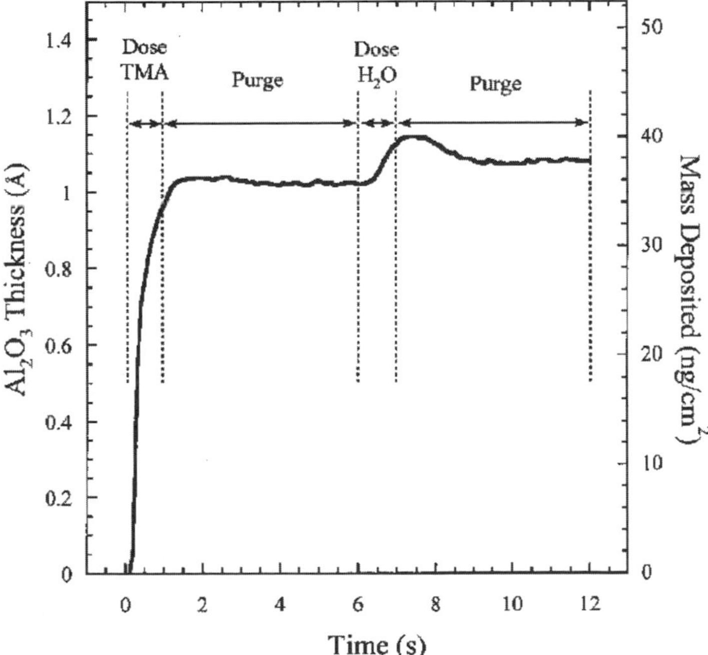

Fig. 4.13 Growth of alumina from TMA, measurement of mass increase as a function of time based upon quartz crystal microbalance. Each stage of the cycle is indicated over a period of 12 seconds [138]

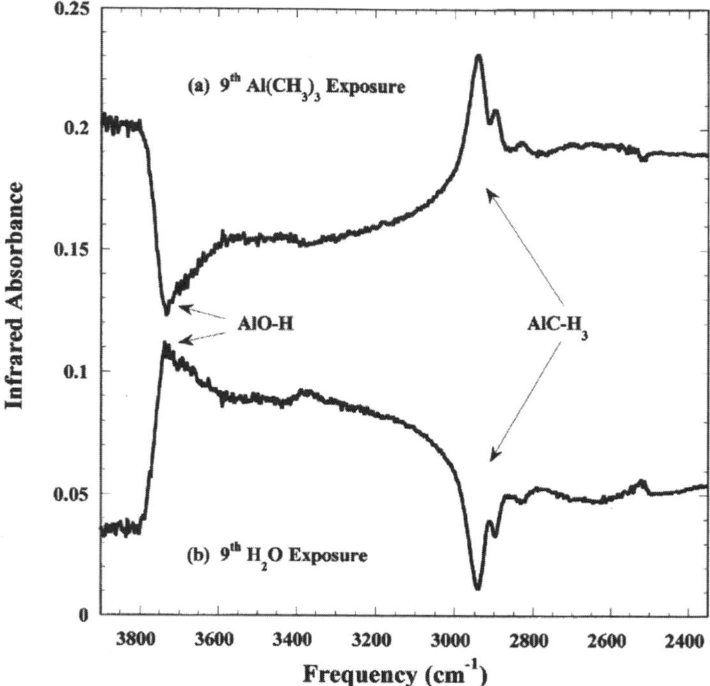

Fig. 4.14 FTIR difference spectra measured during growth of alumina on BN particles at 450 K indicate the various ligands present on the surface. (a) after the 9th cycle of exposure to TMA, and (b) after the 9the exposure to water. The reference spectra were recorded prior to any TMA or water exposure. [43]

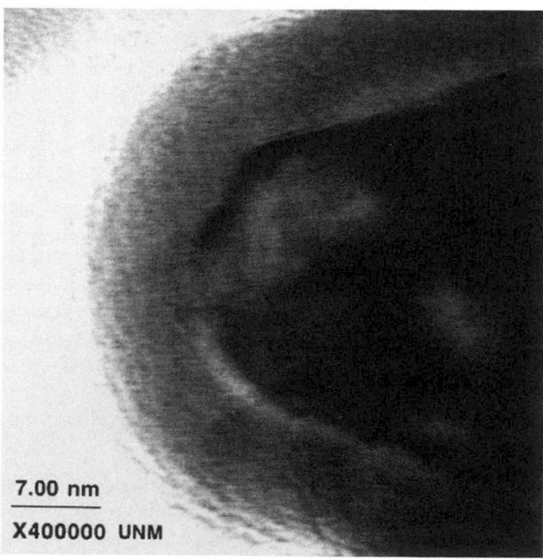

Fig. 4.15 TEM micrograph of 9 nm thick alumina coating formed after 50 cycles of ALD at 450 K on a BN particle [43]

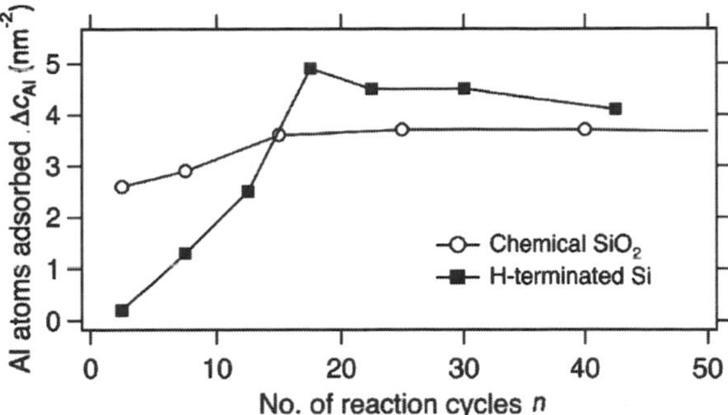

Fig. 4.16 Comparison of growth of alumina at 300 °C on silicon dioxide rich in –OH groups and a hydrogen terminated silicon substrate. Measurements indicate the number of aluminum atoms adsorbed as a function of number of reaction cycles [19]

The exact number of atoms in a monolayer is difficult to define. However, based upon the bulk density of the alumina, approx 10^{12} /cm^2 would comprise a monolayer. Here the slower growth, up to the 18th cycle, indicates island growth, followed by more rapid growth. This increased growth rate is followed in turn by planar 2-D growth when a constant per-cycle deposition is obtained.

A study of the properties of films grown at 350 to 400 °C and at 0.4 Torr (250 Pa) pressure was undertaken by Kim et al. [35] on silicon surfaces, employing ozone for the oxidation step. For semiconductor device applications, to maintain the high dielectric constant for the gate dielectrics, it is important the oxide layer thickness is minimized. Here the process uniformity was better than 2% at a per-cycle growth of 0.085 nm/cycle, which is similar when compared with the same process employing water. The growth per cycle and step coverage both increased with temperature.

Alternative growth precursors have been examined by Jeon et al. [36]: TMA and isopropyl alcohol at 250 °C, and 133 Pa pressure on silicon substrates. The resulting films had an average growth of 0.08 nm per cycle, and were essentially carbon free with no interfacial oxide layer. However, annealing after growth in argon or in oxygen, ambient at 800 °C for 5 minutes, produced an interfacial $SiAl_xO_y$ layer (see Fig. 4.17). The ALD of alumina has also been carried out at temperatures as low as 35 °C [34]. At 177 °C an average growth of 0.13 nm per cycle was achieved with precursor trimethyl-aluminum (TMA) and water. On (100) silicon and polysilicon surfaces, pinhole-free coatings were produced of stoichiometric Al_2O_3. This result was confirmed by Auger depth profiling and RBS measurements.

The use of alumina dielectric films on GaAs has shown promise in microelectronic devices. The advantage here is that high quality, thin gate oxides can be formed at a moderate temperature. For example, a 16 nm alumina layer demonstrated a leakage current of less than 10^{-4} A/cm^2 with a Ti/Au gate [37]. On the GaAs substrate, the breakdown electric fields were as high as 5 MV/cm. In fact, a low recombination velocity can be achieved with p-type silicon, passivated with alumina, with built in negative charge [38] Finally, alumina films provide what is potentially an excellent diffusion barrier for organic light emitting diodes [39]. Measurements indicate that, with a 25 nm film, the water diffusion rate was on the order of 6.5×10^{-5} g/m^2 day.

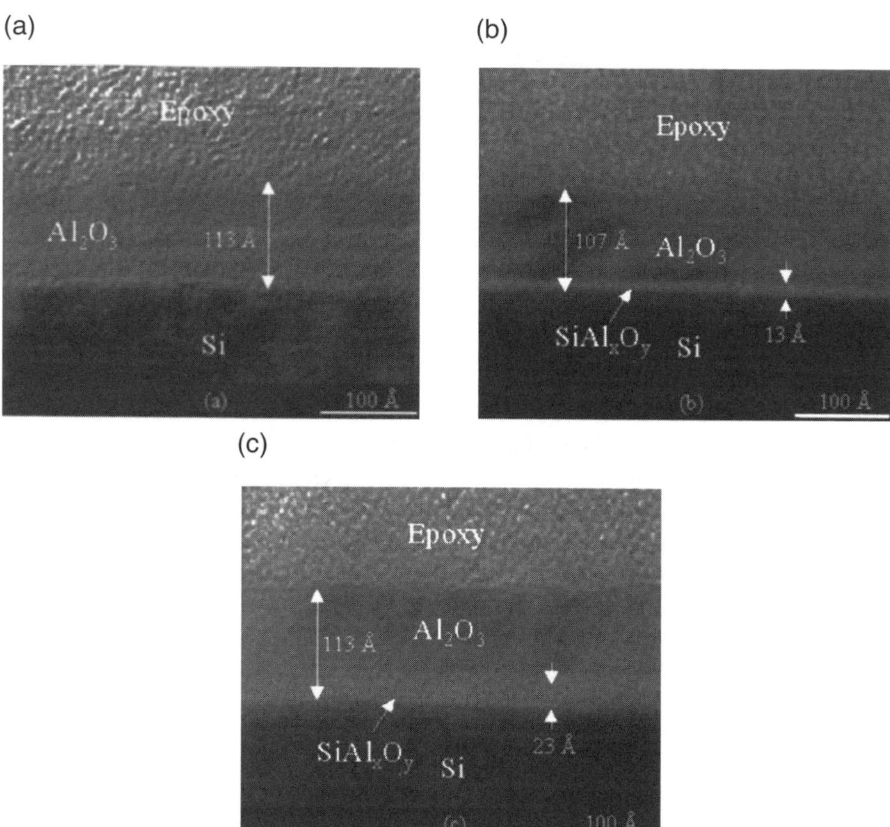

Fig. 4.17 TEM cross-section of alumina dielectric film grown with (a) TMA and isopropy alcohol at 250 °C on silicon (b) after annealing in argon at 800 °C for 5 minutes and (c) after annealing in oxygen at 800 °C for 5 minutes. The presence of silicon dioxide clearly visible in sublayers [36]

4.3.3.2 Silica and Mixed Dielectrics

A great deal of work has been undertaken toward the development of thin dielectric films for microelectronic gate dielectrics. Electronic devices with improved thin gate dielectric performance, with ALD films in combination with high-k dielectrics, can reduce the tunneling currents present in sub-100 nm gate dielectric films. While there is insufficient space to provide a comprehensive review of all this work, this section will discuss aspects of the efforts based on selected examples.

Microelectronic circuit compatibility with CMOS dictates certain requirements. For example, if the dielectric film is applied after metallization, the processing temperature must not reach 400 °C. Silica, alumina, titania, zirconia, hafium oxide, and others have been grown with the various precursors and water or oxygen, as the oxidizing agent.

For example, there have been investigations into oxide growth employing other sources of oxygen, including alkoxides [40], with an aim toward avoiding substrate oxidation:

$$M(OR)_a + MX_a \rightarrow 2MO_{a/2} + aRX$$
$$bM(OR)_a + aM'X_b \rightarrow M_bM'_aO_{(a+b)/2} + (a+b)RX$$

where *a* and *b* represent the integer number of ligands on the precursor.

In general, the deposition rates are higher than those with the metal halide/water or metal alkoxide/water precursor combinations. A number of materials have been investigated including Al, Ti, Ta, Si, and Ti. A review of precursors is provided by Leskelä and Ritala [16].

Silica can be deposited from silicon tetrachloride and water [41,42]. The surface reaction was found to be a function of the Si–OH groups, so that monolayer growth was achieved at a thickness of between 0.07 nm and 0.11 nm for each A/B cycle. Growth on BN particles has been investigated by Ferguson et al. [43] in a high vacuum system at 700 K with in situ FTIR. The purpose was to study the reaction mechanism and optimize the exposure time to reactants for both alumina and silica coatings.

As illustrated in Fig. 4.18, the change in species indicates a reaction from the SiO–H bonds to the Si–Cl_3 bonds, with the coverage a function of exposure time. The surface characterization method is critical for the further development of our understanding of the reaction mechanisms involved. In this work, the growth was found to be sub-monolayer each reaction cycle, and was a function of the exposed crystalline planes in the BN particle, reaching a steady-state after five deposition cycles. Such behavior is shared by other material systems. TEM imaging was able to confirm these effects by providing a measurement of the silica film thickness, which was not uniform on the basal plane of BN, with some areas having thicker coatings.

Silica is a material for which alternative precursors have been investigated including $SiH_2(N(CH_3)_2)_2$ and $SiH(N(CH_3)_2)_3$ by Kamiyama et al. [44,45], with processing at 133–655 Pa and 275 °C. Several other methods at lower temperatures have been studied, including Du et al. [46], in which a pyridine catalyst is added to lower the deposition temperature. The advantage of these precursors is a lower deposition temperature; however, the surface coverage of these larger molecules may be limited by steric hindrance. Growth per cycle was monitored by QCM, ellipsometry, and surface profilometry, as indicated by Fig. 4.19.

Nanolaminated dielectrics, as illustrated schematically in Fig. 4.20, have been studied extensively for the formation of improved dielectric properties, high *k* and low leakage currents, and high dielectric strength for microelectronic devices. Kukli et al. [47,48] demonstrate Ta_2O_3, Al_2O_3 and ZrO_2 stacks have improved charge storage over silica. In this work, the advantageous properties of one dielectric, such as a high dielectric constant, can be combined with the advantages of low leakage current to provide superior performance overall.

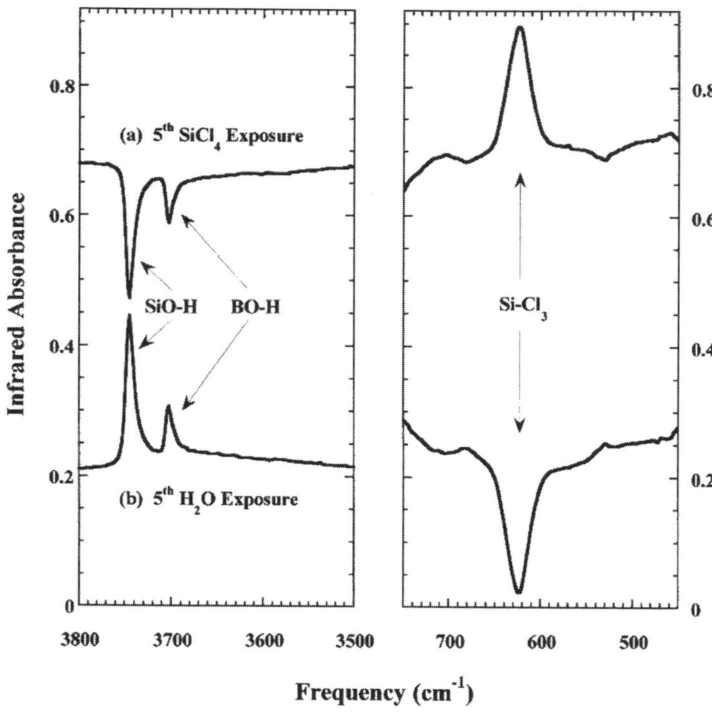

Fig. 4.18 Growth of silica on BN from silicon tetrachloride and water at 700 K. FTIR difference spectra indicate the O–H and Si–Cl stretching regions after (a) the 5th cycle of $SiCl_4$ exposure and (b) 5th cycle of water exposure indicating the presence of adsorbed Si–Cl_3 and Si–O–H and B–O–H groups [43]

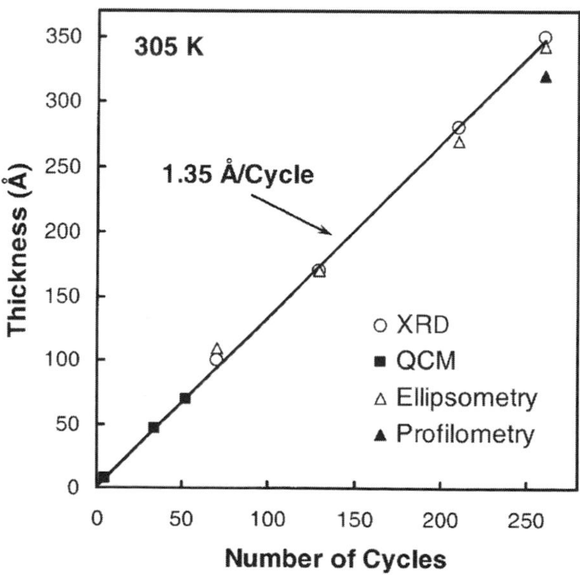

Fig. 4.19 Silicon dioxide thickness as a function of the number of cycles for growth from silicon tetrachloride and water with pyridine to catalyze the reaction at 305 K. Comparison of measurements made by quarts crystal microbalance, ellipsometry, profilometry and XRD analysis [46]

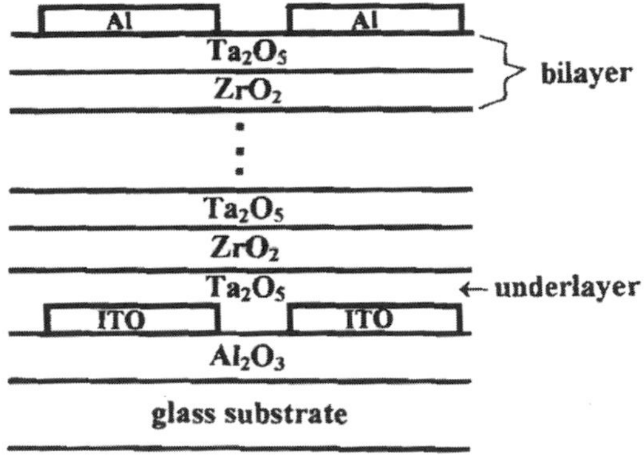

Fig. 4.20 Cross-sectional diagram of nanolaminate fabricated by ALD. Comprises 8–34 layers of Ta_2O_5, thickness 1.2–18.8 nm and HfO_2 so that the total thickness for each bilayer was 20, 10, or 5 nm. Total dielectric film of thickness 170 nm deposited at 325 °C onto an alumina coated glass substrate [48]

Titanium oxide has been studied extensively as an alternative dielectric for microelectronic devices. The surface roughness of TiO_2 was a function of the number of cycles, so that $TiCl_4$and water reactions are incomplete, resulting in an increase in roughness with film thickness. Ritala et al. [49] investigated different sources, such as titanium ethoxide. Additional applications for uniform, high quality, thin dielectric films include optical coatings for electroluminescent displays, solar cells, and coating catalysis and powders with Ta_2O_5 [50]. Thin films for high dielectric constant dielectric films and optical coatings include Ta_2O_3, ZrO_2 and HfO_2 [51]. The characteristics of hafnium oxide that make it attractive for microelectronic applications are the high dielectric constant and large breakdown voltage. Stacked gate structures can be fabricated by sequential processes. For example, work by Kukli et al. [52] demonstrated that pulsed tantalum ethoxide and tantalum chloride precursors allow the dielectric to be grown without the presence of water. The temperature range of 275–400 °C was investigated.

4.3.4 Other Metal Oxide and Nitride Films

TiN and TaN have been investigated as a diffusion barrier for microelectronic applications:

- Deposition by ALD with various precursors is described by Martensson et al. [53].
- Tiznado and Zaera have used $TiCl_4$and ammonia as precursors [54].
- In situ ellipsometry measurements during growth provided insight into the nucleation of the films [54] for alternative reactants.

Figure 4.21(a) shows the plasma assisted deposition system for growth from $TiCl_4$ and a N_2–H_2 100 W RF plasma created during each cycle [55]. One cycle consists of a:

- 5 s exposure to $TiCl_4$, followed by,
- 10 s argon purge,
- 15 s plasma exposure (10 mTorr–1.3 Pa), and
- 10 s pump down to base pressure (10^{-6} Torr, 0.13 mPa).

The film thicknesses is calculated, based upon in situ ellipsometric measurements, as a function of the growth temperature, as in Fig. 4.21(b). For this calculation, the assumptions of a Drude model was made for the optical properties of the TiN film. The growth in thickness per cycle demonstrates a reduced initial nucleation region, which is complete within 50 cycles. This is followed by a constant per-cycle growth, in which the rate of increase in growth per cycle is a function of temperature.

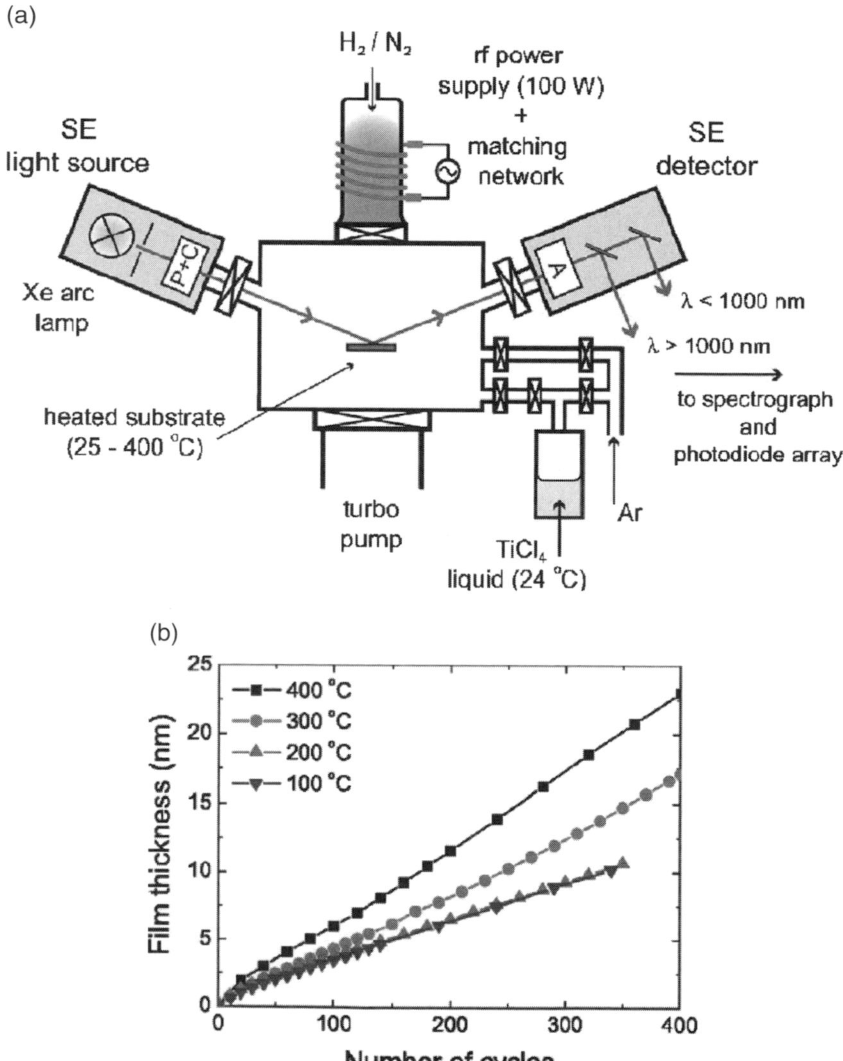

Fig. 4.21 (a) Schematic diagram of a plasma assisted atomic layer deposition for growth of TiN films, with in-situ thickness measurements using a spectroscopic ellipsometer. (b) Growth of TiN film from titanium tetrachloride and nitrogen on silicon, thickness as a function of number of cycles. Temperature is shown as a parameter between 100 and 400 °C. [54]

Plasma enhanced deposition with halogen precursors, studied by Rossnagel et al. [56], has the benefit of lowering the deposition temperature. Deposition of semiconducting metal oxides, such as SnO_2 for sensors and catalysis coatings, has also been investigated. Deposition from the $SnCl_4$, as compared to that from SnI_4, was investigated by Lu et al. [57] on alumina substrates. Growth at 350–750 °C, with nitrogen as carrier gas for the SnI_4vapor and oxygen at 1.3 kPa, produced a growth of 0.1–0.12 nm per cycle [58]. Processing below 475 °C resulted in very smooth surfaces. At a higher temperature, increased roughness was produced, with a higher growth rate due to CVD deposition. The orientation of SnO_2 was polycrystalline on SiO_2/Si wafers, but had a predominance of [110] on α-Al_2O_3 (012) substrates [59].

Growth of V_2O_5 films for rechargeable Li ion batteries was presented by Badot et al. [60]. The films were grown at room temperature 105 °C with a vanadyl triisopropoxide ($VO(OC_3H_7)$ precursor and water on a titanium substrate. A stable per-cycle growth was achieved, forming a final film thickness of 260 nm. After the films had been annealing in air at 500 °C for 2 hrs, a polycrystalline structure was obtained. Following deposition, the Li ions were inserted into the V_2O_5 film by an electrochemical method. Dielectric properties of the film were measured over the temperature range from 210 to 300 K.

Characterization of ALD films for the most part has been ex situ. Nevertheless, the integration of process-monitoring tools, including ellipsometry and quartz crystal microbalances, provide real-time feedback of the process conditions [61]. For a conductive film, the changes in conductance has been demonstrated for ZnO by Schuisky et al. [62].

4.3.5 Metals

Utiainen et al. [63] reports on the feasibility of depositing platinum, copper, and nickel by ALD. The precursors allow deposition at temperatures under 250 °C at a pressure of 100 Pa. The metallic nickel was also formed by deposition of NiO and conversion to Ni in a hydrogen ambient at 260 °C.

A great deal of research effort has focused on developing suitable precursors for copper deposition. This discussion provides a couple of examples of investigations into the process. However, these are only some examples, and are by no means intended as a comprehensive survey [64].

The scaling of copper interconnection down to small dimensions leads to work on copper seed layers, with copper(II)-1,1,1,5,5,5-hexafluoroacetylacetonate hydrate to promote reduction using methanol, ethanol, and formalin [65]. Layers deposited on a TiN or TaN seed layer at 300 °C were conformal and uniform, with resistivities of 4.25 $\mu\Omega$-cm for a 20 nm thickness. Earlier work on Cu deposition, which indicated multi-layer growth with a CuCl precursor at temperatures in the range of from 400 to 500 °C, is recorded by Juppo et al. [66]. The copper was produced in the growth chamber by a reaction in which the Zn removed chloride. However, some of the Zn dissolves into the Cu, producing an alloy, which in turn can have a reaction directly with the $CuCl_2$ in subsequent process steps. In other words, the Zn, when incorporated into the growing Cu film, prevents the self-limiting growth mechanism initially observed.

Tungsten is an attractive metal for electrical via formation in high-aspect ratio structures for DRAM. It is also attractive because the source $WF_6$is readily available. A study of sequential surface chemistry carried out Klaus et al. [67], over the temperature range of from 425 to 600 K, indicates the reaction with Si_2H_6 can be separated into two half reactions. And this sequence is self-limiting. The Si_2H_6 serves only to strip the fluorine from the layer without incorporating silicon into the film. In situ ellipsometry provides process film thickness and an average deposition rate of 0.25 nm per cycle (see Fig. 4.22). The structure of the tungsten film was amorphous, and the surface topography was smooth, with an RMS roughness of +/– 0.48 nm for a film thickness of 32 nm.

Nucleation and growth was studied by Elam et al. [68] with WF_6 and Si_2H_6, at 573 K, on silicon dioxide-coated silicon substrates. Following an initial nucleation process, the films grew by 0.25 nm per cycle. The effect of the substrate on nucleation was also studied by Grubbs et al. [69]. As shown in Fig. 4.23, the nucleation and growth rate for tungsten on alumina was very different from that obtained for alumina deposition on tungsten. Studies with WF_6 and B_2H_6 precursors have resulted in uniform nucleation and growth on a TiN seed layer [70] at 300° C. Pretreatment of the surface played an important role, as it provides improved step coverage. The 20 nm films are used as seed layers for tungsten CVD in the 70 nm DRAM process. Finally, rhodium [71] has been deposited by ALD from acetylacetonato-rhodium onto an ALD-deposited alumina layer at 250° C and 0.075 nm per cycle. The film thickness was proportional to the number of cycles between 15 and 145 nm. Resistivity measured 12 μΩ-cm for a film thickness of 20 nm.

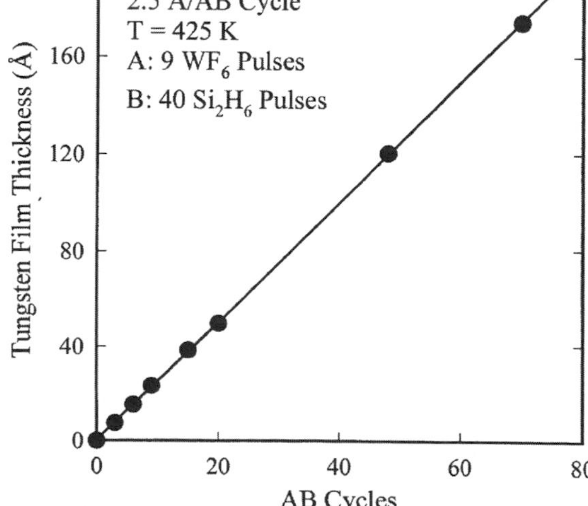

Fig. 4.22 Growth of tungsten onto silicon substrate with WF_6 and Si_2H_6 precursors at temperature of 425 K. Thickness measured with in-situ elliposmeter as a function of number of cycles [67]

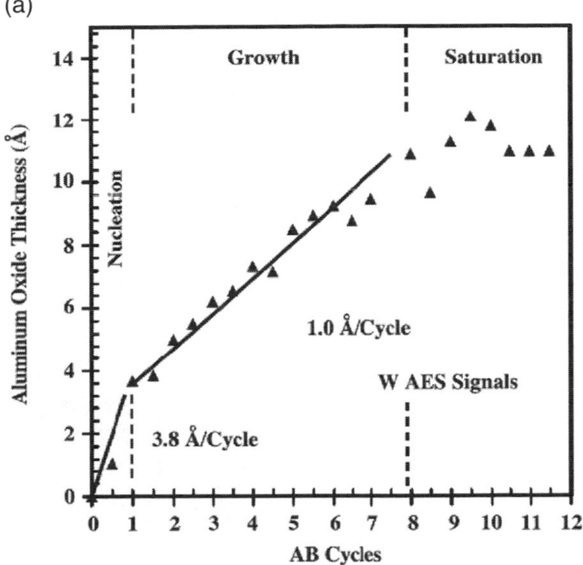

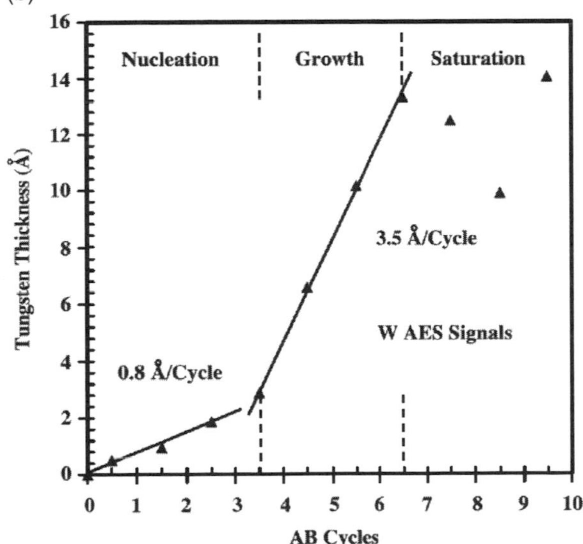

Fig. 4.23 (a) Growth of alumina from TMA/water at 450 K on tungsten. Film thickness as a function of number of cycles based upon decrease in intensity of AES signals for tungsten. (b) Growth of tungsten from WF_6 and Si_2H_6 on alumina at 473 K based upon the AES W signal intensity. Film thickness as a function of number of cycles demonstrates a nucleation region, increase in deposition per cycle, followed by saturation in deposition per cycle [69]

4.3.6 MEMS Applications of ALD

ALD coating is a particularly attractive method of coating MEMS devices because they are 3-dimensional, and achieving a conformal coating is a challenge for other deposition processes. There is also a need for protective hermetic encapsulation to prevent corrosion. The introduction of surface coatings in mechanical devices is also important, as it reduces surface friction and wear.

Cantilever beams, MEMS actuators, capacitors, and resonators have been successfully coated by ALD. Coated cantilever beam structures showed no change in their radius of curvature, indicating that the coating was uniform on both the top and bottom side of the beams. A small shift in resonant frequency was observed, which corresponds to the added mass of the ALD alumina coating (see Fig. 4.24). Another example is an ALD [72] of 10 nm-thick alumina, deposited at 168° C with a TMA precursor, at a pressure of 1 Torr (625 Pa), and coated on MEMS polysilicon micromotors for wear-resistance. Figure 4.25 shows the thickness of the layer deposited on the top surface (1). The hub (2) was 10 nm, compared with that produced under the gear (3), which was 10.5 nm. The surface roughness was 0.2 nm, indicating that adequate coating uniformity can be achieved with this process. An ALD of

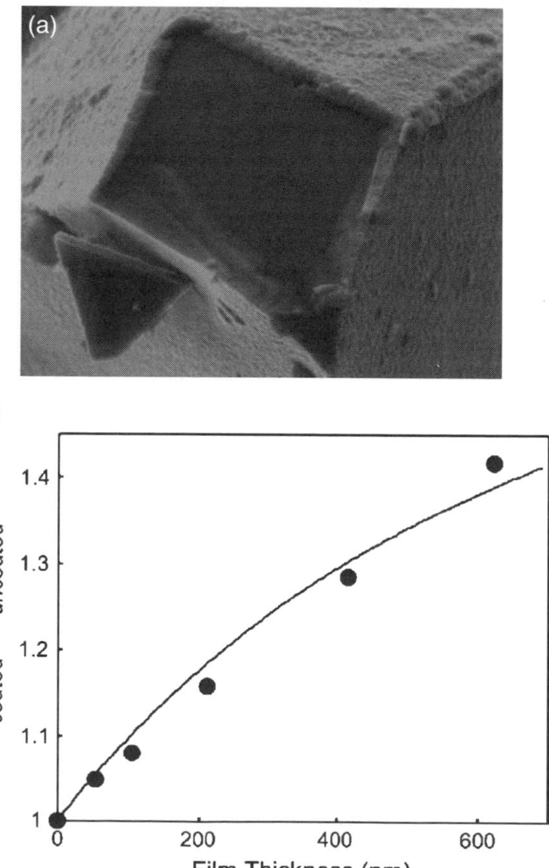

Fig. 4.24 Growth of silicon carbide film from disilabutane at 800°Con polysilicon, (a) SEM image of polysilicon microresonator coated with 210 nm of SiC. The cantilever tip has been removed to allow measurement of coating thickness. (b) Measured frequency shift for coated resonator compared to uncoated resonator as function of film thickness. The solid line is the calculated frequency ratio used to fit the elastic modulus of 360 GPa [74]

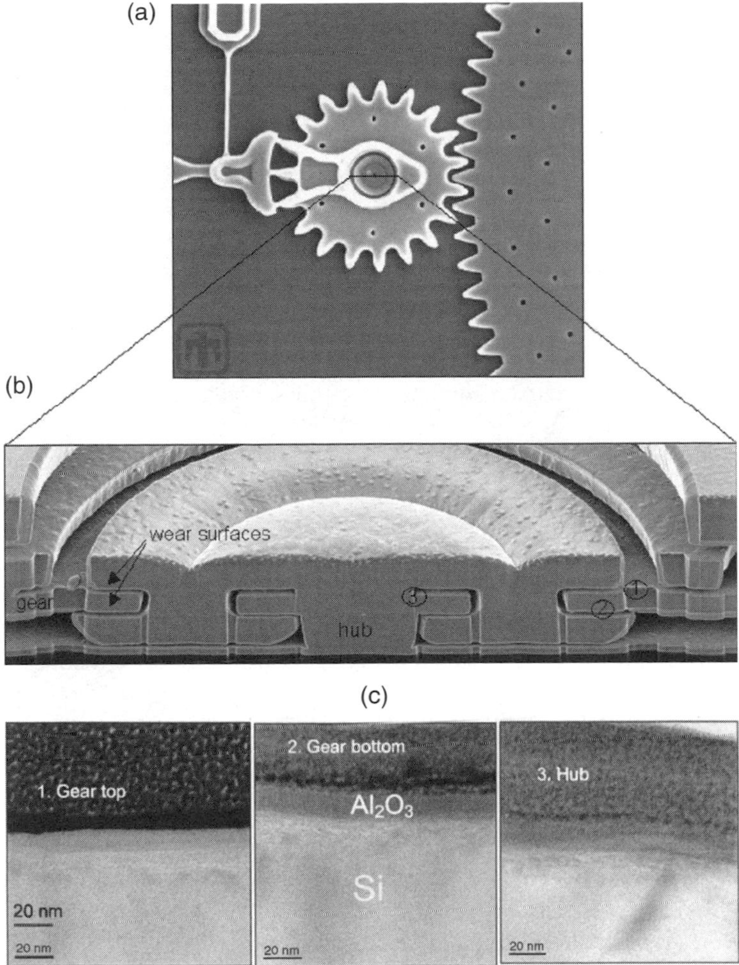

Fig. 4.25 Alumina coating, approximate thickness of 10 nm, on nominally 3 μm thick polysilicon MEMS microgears. The sacrificial silicon dioxide has been removed to release the mechanism from the substrate. (b) Cross-section of the hub regions showing the contact surfaces, and (c) TEM micrograph of regions labeled (1), (2) and (3) and shown in (b). Thickness ranges from 10 nm to 10.5 nm [72]

alumina has also been applied to optical MEMS [73]. MEMS device applications are reviewed by Stoldt and Bright [74]

Wear reduction via ALD tribological coatings in MEMS actuators is a highly desirable process. The process's suitability is due to its usability in the preparation of smooth and conformal coatings [75], which reduces friction levels when compared to those of silica surfaces. Static friction and wear were reduced in MEMS systems when they were coated with TiO_2 and ZrO_2 films. The ZrO_2 films were formed with precursors of zirconium tetra-tert-butoxide ($Zr(OC(CH_3)_3)_4$), and the TiO_2

films with ti-iso-propoxide ($Ti(OCH(CH_3)_2)_4$) and water. Furthermore, the ionic adsorption from a solution of MoS_2 and ZrO_2 films, when applied to test devices, also indicated that the coefficient of friction was reduced, as was the debris produced with wear in the presence of water vapor on silica surfaces.

The alumina layer provides a hydrophobic coating which dramatically changes the contact angle and reduces stiction forces [76]. Figure 4.26 demonstrates the dramatic change in contact angle for coated surfaces. Here the vapor phase treatment avoids the issues associated with solvents usage for surface micromaching sacrificial film release. The treatment also sidesteps the limitations of silane SAM films, which demonstrate instability when used to reduce stiction on polysilicon structures. The hydrophobic precursor consisted of tridecaflyoro-1, 1, 2, 2-tetrahydrootylmethyl-bis(dimethylamino) silane, which reacts selectively with the hydroxyl groups, providing a covalvent bond to the alumina. This film is stable up to 250° C in an ambient of air or nitrogen. Stiction measurements indicate that the coating applied to a polysilicon cantilever beam resulted in a 90% reduction in adhesion energy. The beams were cycled 125,000 times at 80% relative humidity without failure.

4.3.7 Integration with Porous Membranes and Templates

Coating of alumina porous membranes by ALD has been an effective means of reducing internal pore dimensions [77]. Alumina, deposited from TMA and water, demonstrated the self-limiting characteristic within the high aspect ratio pores for each half reaction. A reduction in pore diameter from 22 nm to 12 nm was achieved with 120 cycles. FTIR transmission measurements indicated the surface reactive groups present on the membrane and indicate that coverage was consistent with pore reduction size, as did ex situ gas permeation measurements with hydrogen and nitrogen. The processes carried out at a temperature of 500 K and at 0.5

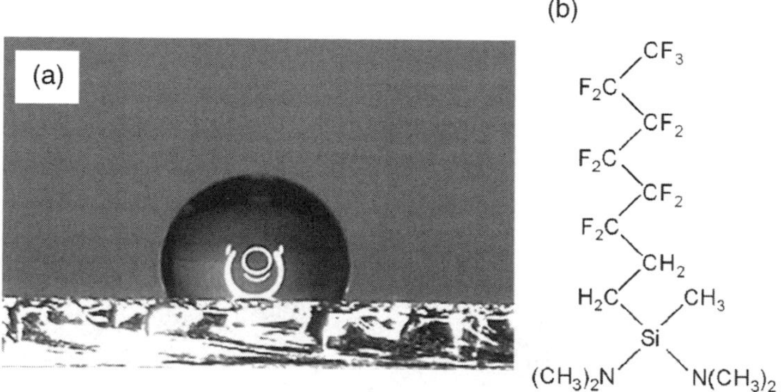

Fig. 4.26 Vapor phase conformal coating of ALD deposited alumina coated silicon surface produced a hydrophobic surfaces. Photograph of contract angle 108 ° +/–2 ° observed with coating of thin hydrophobic layer. The contact angle prior to coating, not shown in figure was 55 ° +/– 5 °. (b) Structural diagram of FOMB(DMA)S precursor [76]

Torr (312 Pa) pressure resulted in an average deposition of 0.25 nm per cycle. The small diameter pores in the membrane, 20 nm over a 20 μm length, were reduced to a size that was within the range of 5 nm to 1 nm. Figure 4.27 shows the size reduction as a function of the number of AB cycles. These permeable membranes can be used for conductance measurements for various gases. The ALD of tungsten on nanoporous aerogels was carried out by Elam et al. [138] with WF_6 and Si_2H_6, at 200° C, and separated by 5 min purges. Prior to growth, a 0.2 nm-thick film of alumina was deposited. The average deposition of tungsten was 0.72 nm per cycle, eventually forming a total thickness of 7 nm.

A template method based upon nano- or microspheres can use the spheres as a mask to expose regions of the substrate to direct film growth [78]. Nanobowl films of TiO_2, formed by ALD over a self-assembled monolayer of 505 nm-diameter polystyrene spheres, were removed in toluene following an ion beam etching of the spheres' top layer [79]. Figure 4.28 summarizes the fabrication process. High temperature processing was subsequently used to transform the amorphous material into polycrystalline titania.

Size-tunable metal nanostructures were also defined utilizing these periodic arrays. Here the process is modified to include a PMMA sacrificial film [80], as depicted in Fig. 4.29. The junction between the PMMA film and the polystyrene sphere provides a good contact. After ALD coating, the nanospheres and PMMA were removed to produce a free-standing sheet of nanobowls, each with a small hole at the center. This process provides a large area shadow mask, up to cm's in size, for the evaporation of gold. A gap between the mask and substrate of approximately 500 nm was produced by placing the nanobowl array so that the bowls were in contact with the wafer. Thus, by changing the angle of evaporation, a number of gold dots were defined close together, illustrated by Fig. 4.30.

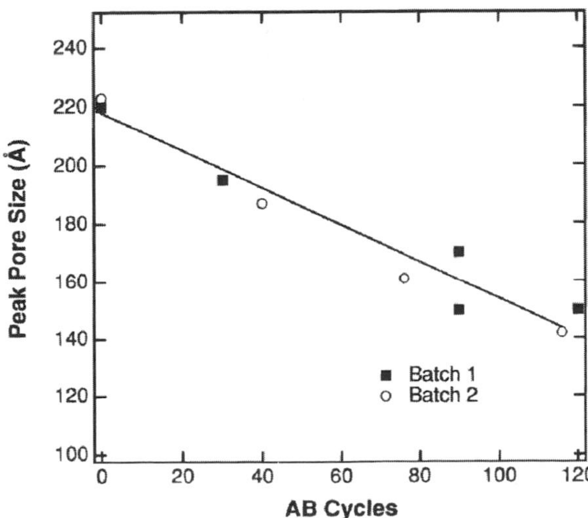

Fig. 4.27 Peak pore diameter versus number of alumina deposition AB cycles measured using liquid–liquid displacement permporometry [77]. Pore diameter decreases at 0.073 nm per cycle

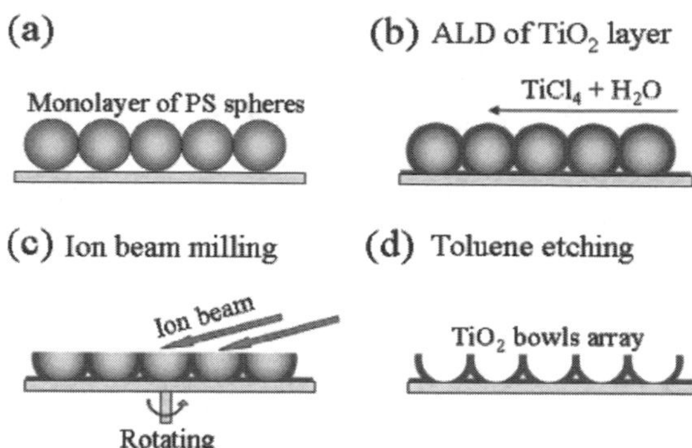

Fig. 4.28 Schematic diagrams indicated the experimental process for fabrication of TiO_2 nanobowl arrays. (a) arrangement of polystyrene spheres, nominally 505 nm in diameter sapphire substrate, (b) growth of 20 nm TiO_2 with 200 cycles, (c) ion beam milling to remove upper half of hemisphere, (d) dissolve polystyrene in toluene [79]

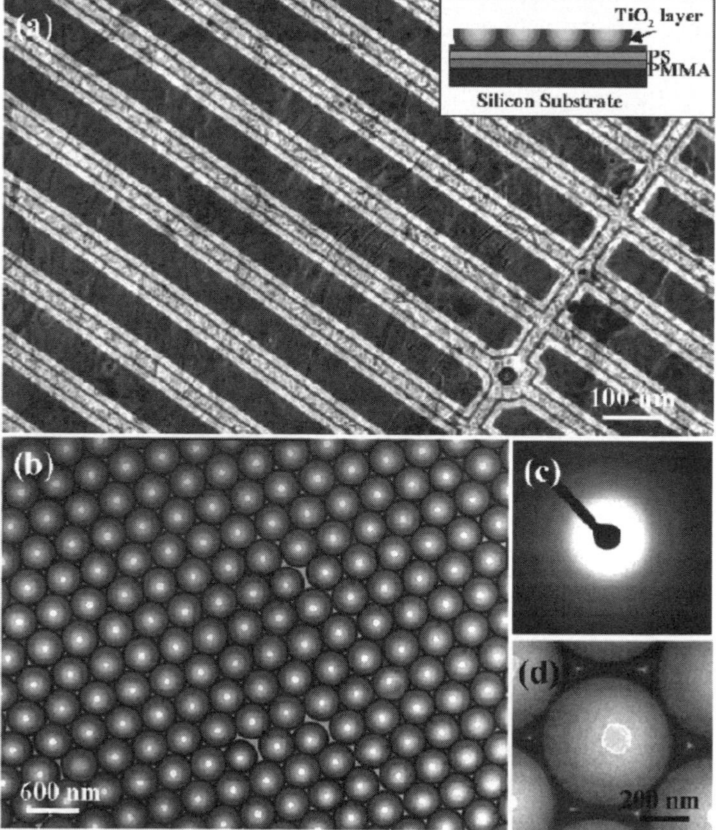

Fig. 4.29 Optical image of copper TEM grid covered by large-area nanobowl sheet. Inset: schematic of the modified configuration for fabrication of stable large-area nanobowl sheets. (b) A TEM image of the nanobowl sheet, (c) electron diffraction pattern of the amorphous TiO_2 nanobowl sheet, and (d) a higher magnification image of the nanobowls [79]

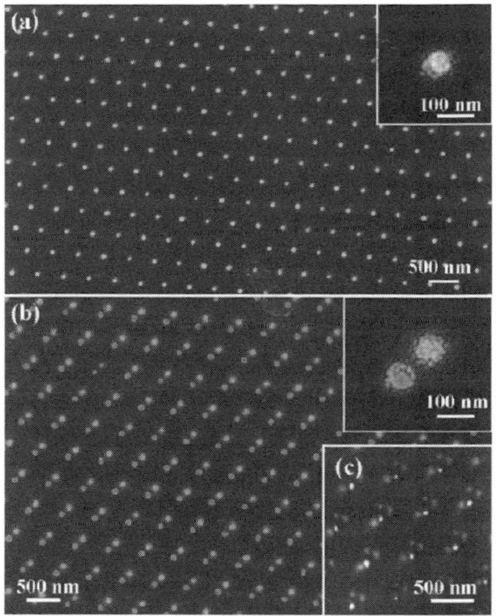

Fig. 4.30 High resolution SEM image of 20 nm thick gold dot pattern produced with nanobowl lithography: (a) triangular single-dot pattern, (b) triangular double-dot pattern, and (c) multi-dot pattern [80]

4.4 Focused Ion Beam Processing

4.4.1 Introduction

Ion beam usage has been ubiquitous in microfabrication for many years. Ion beams provide a source for lithography, a method for thin film deposition, and a process for etching. One of the advantages of a focused ion beam, as opposed to "blanket" area processing systems, is that the lithographic exposure, along with directed etching or localized deposition, can be carried out without the need for a mask. Commercial systems based upon liquid metal ion sources of high brightness have achieved a range of processing capabilities. In addition to lithography, etching, and deposition, surface modification can be carried out with ion beam irradiation and, at higher energies, ion implantation is possible.

This section includes a brief review of the physics of ion surface interactions. That background should make it easier to understand the processes under discussion. For more information on ion beam-based lithography, etching, and processing, refer to the detailed discussion of these subjects in texts by Brodie and Muray [2] and Giannuzzi and Stevie [81]. An introduction to nanolithography is also provided by Cui [1].

Commercial, focused ion beam systems include those made by FEI Company (Eindhoven, The Netherlands). Figure 4.31 is a cross-sectional diagram through a system. The illustration depicts the components of the column, ion-beam optics, and scanning system.

At the top of the column is the liquid metal ion source. A liquid metal (or alloy) field ionization source (in this case gallium) emits ions upon application of voltage between the needle-shaped tungsten outlet of the liquid metal reservoir and the extraction aperture. The ions are accelerated by the applied electric field into the ion column, which focuses the ions onto the sample surface by a set of electrostatic lenses. A mass separator (mass filter) that uses magnetic field deflectors is used to select one charge and isotope of gallium. The beam deflector—in this case an electrostatic deflector based on an octopole—deflects the beam sideways. The device can be turned on and off rapidly (beam blanking), which facilitates exposure control during the beam writing process. The ion beam can be focused to a diameter of < 50 nm at the substrate surface. The sample translation stage is usually based on a laser interferometer, which allows for precision positioning and for the rotation and tilting of the sample in the focal plane of the FIB system.

Characteristics of a Dual Beam FEI System:
Electron optics:
1.1 nm at 15 kV, 2.5 nm at 1 kV
Beam current < 20 nA
Voltage 0.2 kV to 30 kV.

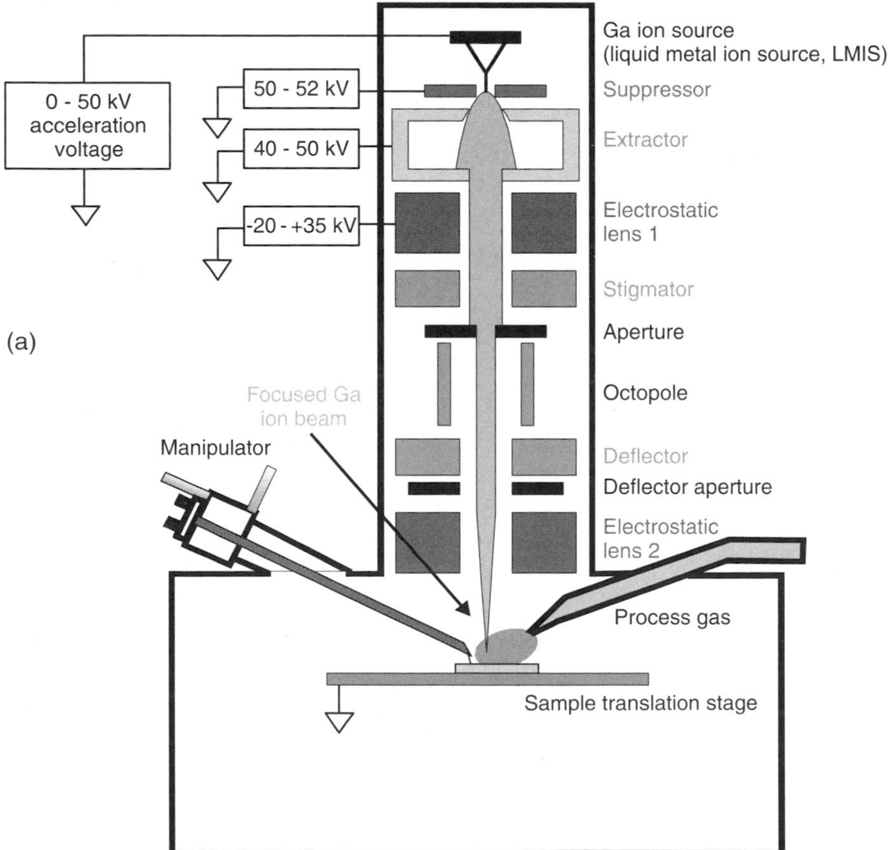

Fig. 4.31 (a) Schematic view of a focused ion beam system. (b) Schematic diagram of a liquid metal ion source

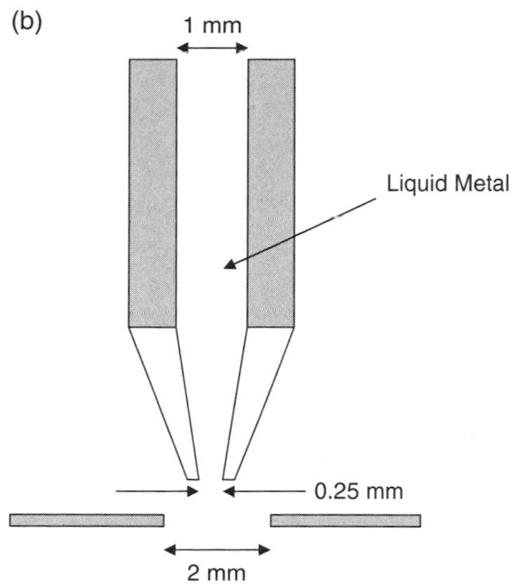

Fig. 4.31 (continued)

Ion optics:
7 nm at 30 kV,
Beam current 20 nA (range 2 nA–2 uA in commercial systems)
Energy 5 kV to 30 kV

4.4.1.1 Ion Source Operation

Liquid metal ion sources have provided bright, stable sources for focused ion beam systems, as portrayed in Fig. 4.31(b) [82]. The sharp tungsten needle is filled with the liquid metal, such as gallium. Gallium is a good choice because it has both a low melting point and low vapor pressure, as well as excellent mechanical, electrical, and vacuum properties. The surface tension brings the liquid to the front of the source, and a high-bias voltage is applied to extract ions from the liquid meniscus. The diameter of the source is only 5–10 um. A voltage of typically 2–10 kV is used to extract ions into a cone. At the sharp radius of the molten metal meniscus, an electric field strength of $\sim 10^{10}$ V/m develops. This field strength is sufficient to produce a current of 1–3 uA, corresponding to a current density as high as 10^6 A/cm^2 through the aperture plate. In addition to single charged ions, multiple charged ions can also be emitted. These are removed from the beam by the mass-selective quadrupole, as illustrated in Fig. 4.31. There is also a spread in the ion beam, a result of space charge effects, which results in coulombic repulsion between the charged Ga^+ ions, effectively enlarging the focus of the ion beam at the substrate.

4.4.1.2 Ion Surface Interactions and Energy Loss

The interaction of the ions with the surface is determined principally by their kinetic energy. For example, with gallium (atomic mass 69.72) and a source operation at 50 keV, the velocity can be calculated as:

$$\text{Energy} = 50,000 \times 1.6 \times 10^{-19} = 6 \times 10^{-15}\text{J/ion}$$

$$\nu = \sqrt{\frac{2Ee}{m}} = \sqrt{\frac{2x50x1.6x10^{-19}}{69.72x10^{-3}/6.602x10^{-23}}} = 12,309m/s$$

The rate of energy loss is a function of the mass of the ion and the properties of the substrate. The stopping is defined as the rate of energy loss per unit depth of penetration into the substrate. In the case of lighter ions, the energy loss is due primarily to nuclear interactions, and can be estimated using a model based upon collisions with substrate atoms. However, for heavier ions there is also significant energy loss through the electron cloud, which gives rise to a more gradual loss of energy. Definition of Range is the mean depth of the ion when it comes to rest. The Straggle is the standard deviation when a Gaussian distribution is fit to the Range data. The Lindhard-Scharff-Schiott (LSS) theory allows the *Range* and *Straggle* to be calculated [83]. The energy loss due to the scattering of atoms and electrons within multiple films on the substrate can be modeled with the commercially available program "TRIM" [84].

The ion penetration is a function of the ion energy. Figure 4.32 shows a typical Ga ion depth profile in silicon. In the case of a resist film coating on the substrate,

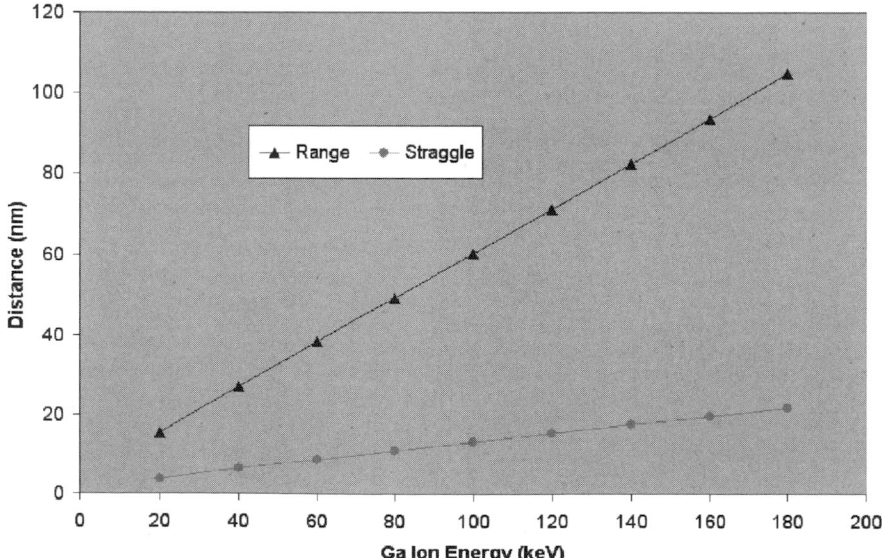

Fig. 4.32 Range and straggle for Ga ions in silicon as a function of the incident ion energy for normal incidence [139]

which is present with lithographic processing, the range affects the exposure dose of the resist. For single crystal substrates, including silicon, there is an additional bias based upon the alignment of the crystal axis with the trajectory of the incident ions. This channeling is manifest through a deeper than expected penetration, because the substrate atoms are in crystallographic alignment. In such an alignment, the open spaces between the atoms reduce the stopping, and hence contribute to the increased penetration depth.

4.4.2 Applications of FIB

4.4.2.1 Ion Sputter Etching

TEM sample preparation is one of the largest applications of FIB milling techniques. TEM sample preparation with FIB is a simple and versatile process, to prepare thin sections a two-step process is employed. First, trenches are milled into the sample at a higher beam ion flux density, followed by cutting at lower beam flux densities to obtain a clean sample, free of re-deposited material produced by the initial milling process. An example is shown in Fig. 4.33(a). Also examples of etching of silicon for form grooves with a Ga^+ ion beam at different ion fluxes are shown in Fig. 4.33 for (b) widths of nominally 50 nm and (c) nominally 300 nm width.

In fact, the FIB system provides a precise method for etching into a variety of materials. The sputter yield is a function of the atomic mass and energy of the incident ion. The sputter yield is the number of sputtered substrate atoms per incident ion.

Figure 4.34(a) shows contour plots for etching silicon with 50 keV Ga^+ ions. The beam is cylindrically symmetric with a Gaussian profile, with a nominal beam diameter of 68 nm and a beam current density of 0.8 A/cm^2. The yield of sputtered ions is

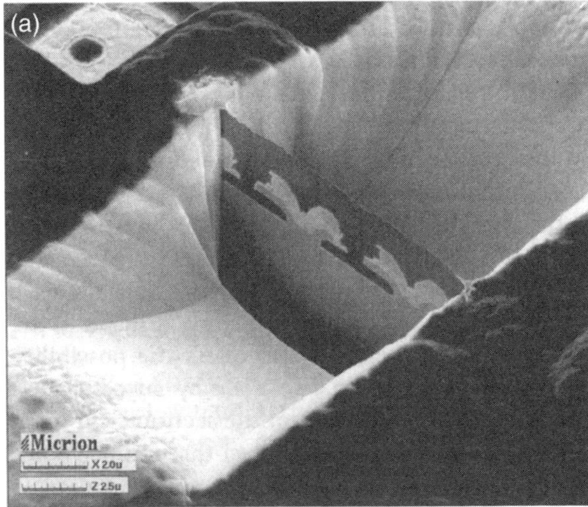

Fig. 4.33 Examples of structures etched by FIB milling. (a) SEM image of substrate prepared for TEM analysis. Two pits are milled at either side of the area of interest and then the remaining wall is trimmed to a thin membrane using a smaller beam current in the FIB (b) FIB image of FIB etched nominally 50 nm wide trenches in Si with doses from 0.5 to 3 nC/μm^2 in steps of 0.5 nC/μm^2, (c) same as (b) for nominally 300 nm wide trenches in silicon [85]

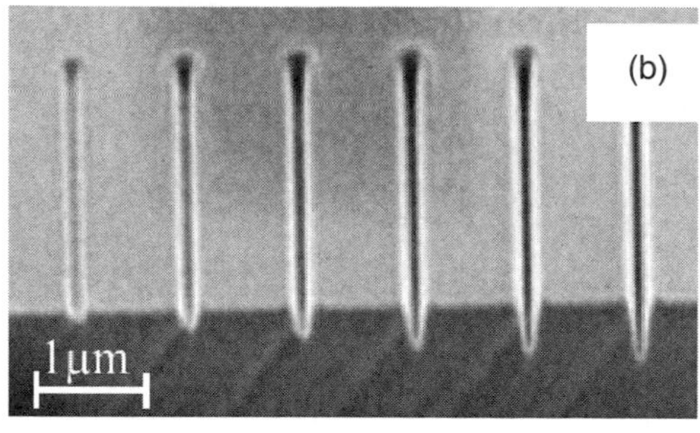

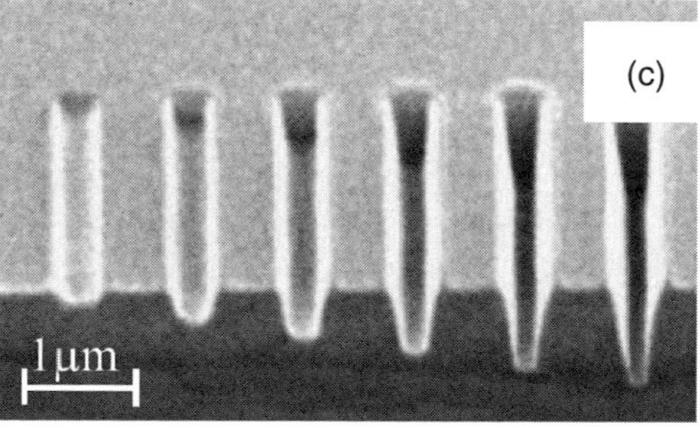

Fig. 4.33 (continued)

a function of the surface geometry and dose. Local sputtering and amorphorization of the silicon in the low dose regions results in a swelling at the edges, which diminishes at larger doses. Figure 4.33(b) and (c) illustrates the influence of scanning width on trench depth, and renders the profile for silicon as a function of dose between 0.5 and 3 nC/mm^2 [85].

The differential sputter yield when milling silicon to various dimensions at a fixed ion flux is shown in Fig. 4.34(b) and (c). The yield per incident ion is also a function of depth. For narrow trenches some yield increase was observed, due to ion focusing into the center of the trench, which becomes more significant for high aspect ratio structures, as graphically portrayed in Fig. 4.34(b). The sputtering yield

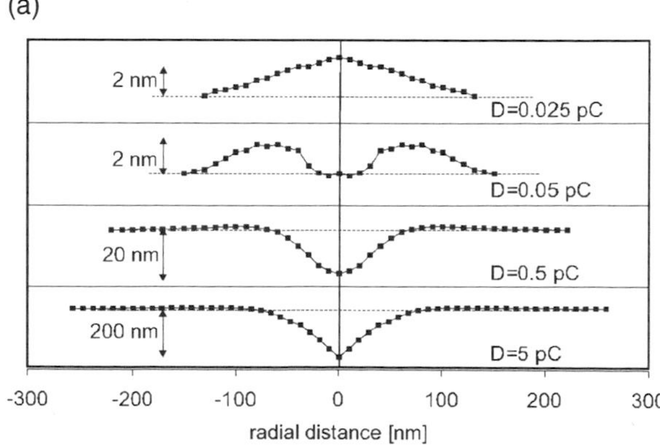

Fig. 4.34 (a) Surface contour plots of silicon after etching with incident Ga$^+$ ions of 50 keV energy and current density of 0.8 A/cm^2 beam diameter 68 nm. The dashed line indicates the level of the silicon surface prior to etching. (b) Differential sputter yield for silicon as a function of milling depth and trench width for milling a 5×5 μm^2 sized area; and (c) yield for 50, 100, 300 and 500 nm wide trenches. [85]

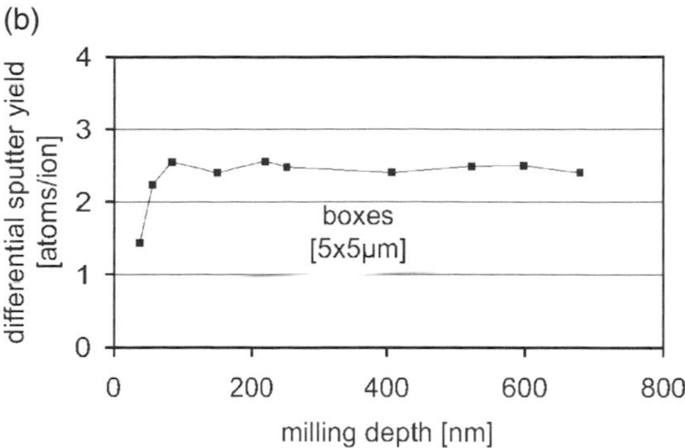

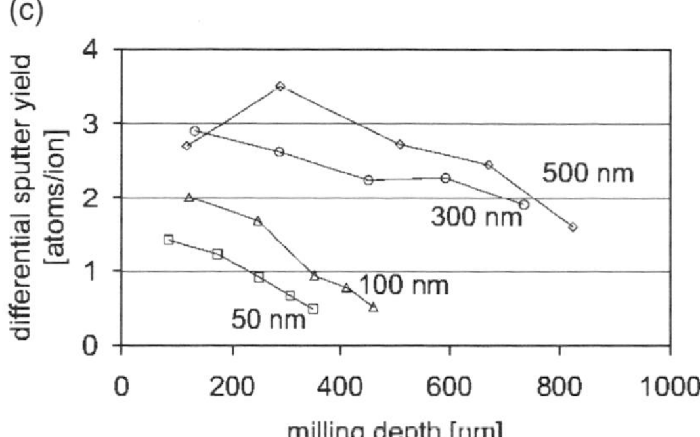

Fig. 4.34 (continued)

is also a function of incident angle, as ions incident on the surface at a more oblique angle generally have a higher sputter yield. At higher energies the sputter yield drops as the incident Ga^+ ions penetrate to a greater depth; hence more of them remain implanted in the substrate.

Re-deposition of sputtered material has to be considered when carrying out an FIB machining process. The concern is that the current density and writing speed should be adjusted to minimize re-deposition in the region where greater precision is needed, or where the feature sizes are critical. Figure 4.35 shows an example of a nanoscale, comb-shaped, electrode structure fabricated for an electrochemical sensor. An overview of the fabrication process is depicted in Fig. 4.35(a). After formation of a trench via a deep reactive ion etching (DRIE) process, a thermal silicon dioxide film of 50 nm was grown. A sputtered layer of 10 nm Ti and 50 nm platinum was deposited to cover the structure. Focused ion beam milling with Ga^+ at 30 keV and 100 pA was used to define the remaining platinum layer. The result, illustrated in Fig. 4.35(b), is what remains after removal of platinum from the base of the trench and the top of each channel mesa. Due to the directional etching characteristic of the FIB process, platinum-on the sidewalls of each trench remains, and this lining forms the comb electrodes [86].

The formation of high resolution stencils for vapor deposition of materials has also been developed in low stress silicon nitride. This formulation process achieves a resolution of 100 nm [87]. A shadow mask is formed, such that the vapor deposition

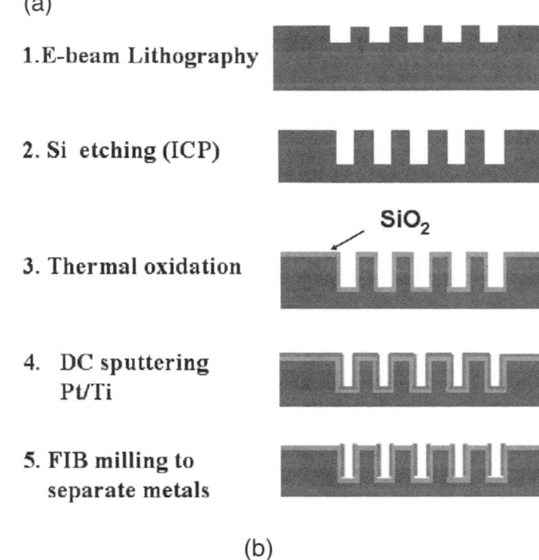

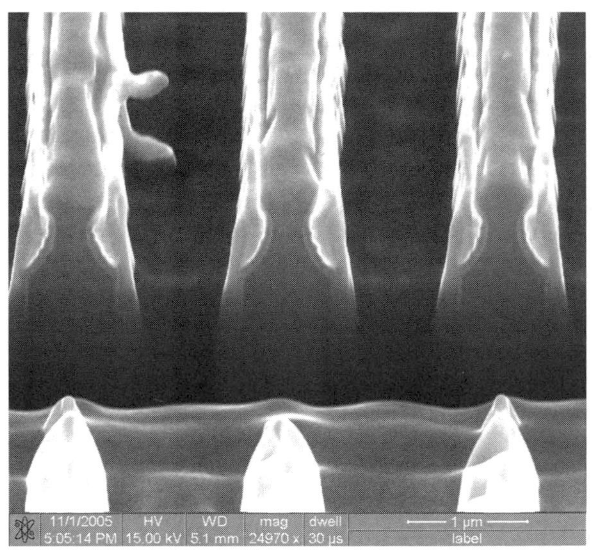

Fig. 4.35 Example of FIB defined pattern for a biosensor comb electrode array (A) schematic diagram showing the fabrication process, and (B) SEM image of comb structure with platinum electrodes defined on opposing walls

of aluminum is defined through the mask. The advantage in this instance is that no photoresist processing is required for the procedure. The presence of a reactive halogen gas can greatly enhance the etch rate and selectivity, as indicated in Table 4.2, where the etch rates for several materials are listed. The halogen the containing gas adsorbs onto the surface, so that with ion impact a volatile product is produced. Chloride significantly increases the etch rate of silicon and aluminum, where as XeF_2 does not etch SiO_2. Advantages to this process are that a lower ion dose is required to achieve etching, and higher aspect ratio structures can be produced.

Table 4.2 Ion assisted etch rate enhancement factors [134]

	Substrate material			
Gas	Silicon	Aluminum	Tungsten	Silicon Dioxide
Chlorine	7–10	1–10	None	None
Xenon Difluoride	7–12	None	7–10	7–10

4.4.2.2 Examples of FIB Structures Built

FIB is a versatile tool for micro- and nanofabrication and can be utilized in conjunction with MEMS processing. For example, a MEMS microaccelerometer has been demonstrated on a Silicon on Insulator (SOI) wafer with capacitive sensing. The sensing electrode was obliquely cut to define an electrode gap by FIB milling [88]. Figure 4.36 shows the top view of the accelerometer proof mass and a higher magnification image of the metallized capacitive sensor. After milling by FIB, platinum was deposited with FIB deposition (see the next section).

Another example is mask-less fabrication of electrode contacts for a Junction Field Effect Transistor (JFET). This fabrication procedure provides a convenient means for shallow implantation and contact formation by FIB processes [89]. Finally, AFM probe tips can be modified by FIB etching to integrate Scanning Electrochemical Microscope (SECM) functionality into the AFM probes. Microfabricated SECM can be tuned more accurately to the desired operation by FIB machining of the insulation film.

Figure 4.37 shows the process developed by Kranz et al. [90], in which the FIB process etched eight cycles to define an SECM electrode. The process employed produced a square, narrow, noble metal electrode defined a fixed distance away from the tip via a single beam FIB system. With current state-of-the-art dual beam FIB systems allowing bitmap-assisted FIB milling, this method can be implemented

Fig. 4.36 (A) SEM micrograph of a microaccelerometer built with SOI silicon with proof mass thickness 7.5 μm, and suspension 3 μ above the substrate. The readout gap has been cut at 45° angle to surface by FIB milling. (B) SEM cross-section of the capacitive sensing gap after conductive platinum metallization has been applied by FIB-CVD from organometallic precursor [88]

Fig. 4.36 (continued)

in a semi-automated fashion, thus providing a manufacturing process for AFM-SECM probes. The ring/square electrode provides electrochemical data simultaneously in a single scan, independent from the topography. Successful imaging of redox processes at gold electrodes has been achieved [91]. Successful tapping mode imaging, with low damage to soft sample, and measurement of enzyme activity on a silica surface was demonstrated by Kranz et al. [92]. FIB has also been used to deposit platinum on AFM tips to improve resolution for electric field imaging [93].

4.4.3 FIB CVD

A wide variety of three dimensional structures can be produced by FIB-assisted CVD. Here the precursor is absorbed onto the solid surface, and the local enhanced reaction rate facilitates localized deposition of material, as portrayed schematically in Fig. 4.38. Various materials have been deposited, including carbon, diamond-like carbon, silica, tungsten, and platinum.

Diamond-like carbon nanopillars were deposited by Fujita et al. [94] with phenanthrene $C_{14}H_{10}$ as the source gas and a Ga^+ ion beam, as shown in Fig. 4.39. The nanopillars are set into vibration in the SEM microscope with a piezoactuator, resulting in a resonance at 1.21 MHz and a Q of 1200. The Young's Modulus of the beam was estimated to be the order of 100 GPa. Further work on the growth of amorphous carbon nanopillars by Ishida et al. [95] from the same precursor indicated mechanical properties with an initial density of between 2.5 and 4.2 g/cm^3. In that procedure, annealing at 600° C removed the gallium incorporated during film growth, and that step produced a more uniform density. The spring constants were evaluated in the range 0.07–0.1 N/m. Additionally, Morita et al. [96] demonstrated the formation of free space structures in amorphous carbon and ion beam deposition of diamond-like carbon. The electrical resistivity of the structures measured with

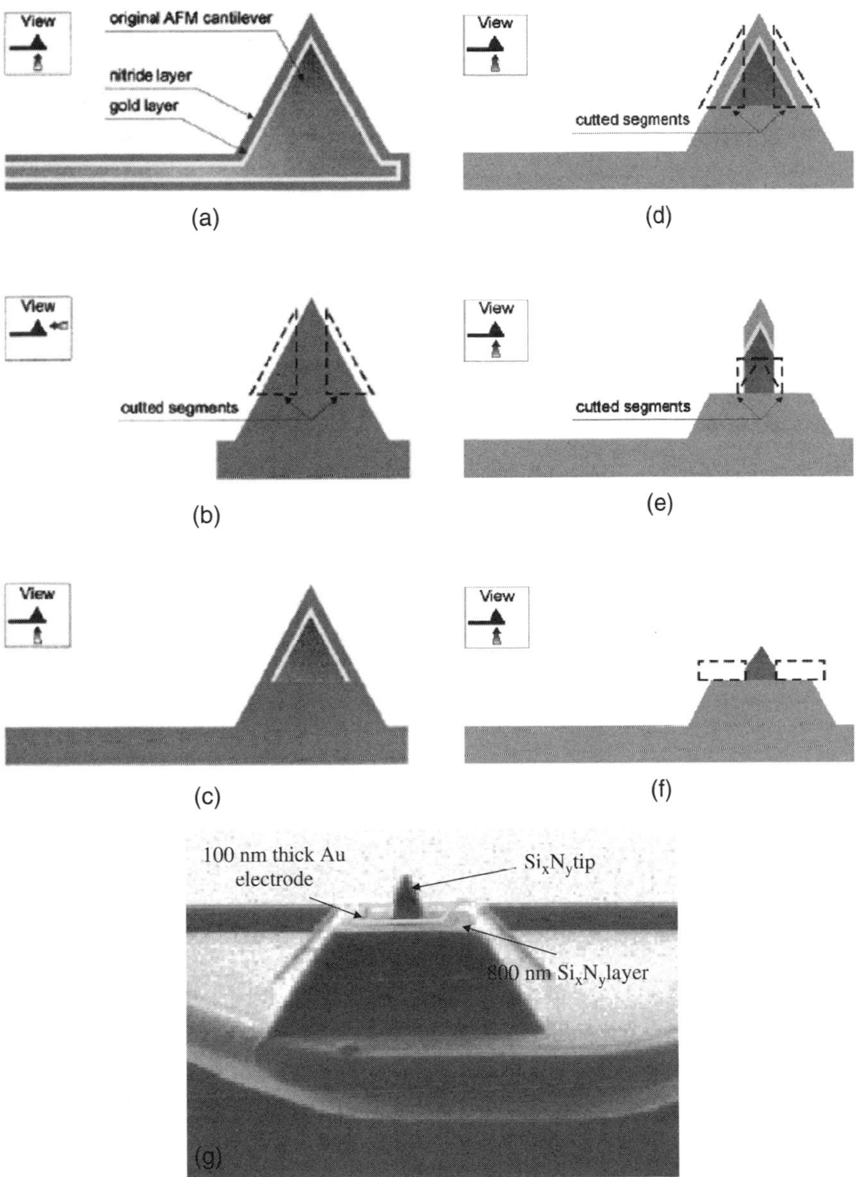

Fig. 4.37 A SECM probe for integrated electrochemical measurements, (edge length, 2.2 μm). Schematic diagram summarizing the fabrication process sequence for a single beam FIB system: (a) AFM cantilever after coating with the gold layer and the silicon nitride insulation; (b) diametrically opposed FIB cuttings along the dotted lines; and (c) side view after step, (d) repetition of the diametrically opposed FIB cuttings of step (b) along the dotted lines after turning the cantilever by 90°, creating a free-standing square pillar; (e) remodeling of the nonconductive AFM tip by FIB cuttings along the dotted lines on all four sides of the square pillar; and (f) after "single pass milling" along the dotted lines for removal of redeposited material from the electroactive surface (g) FIB image of the final integrated frame microelectrode and AFM tip. [90] reprinted with permission of ACS

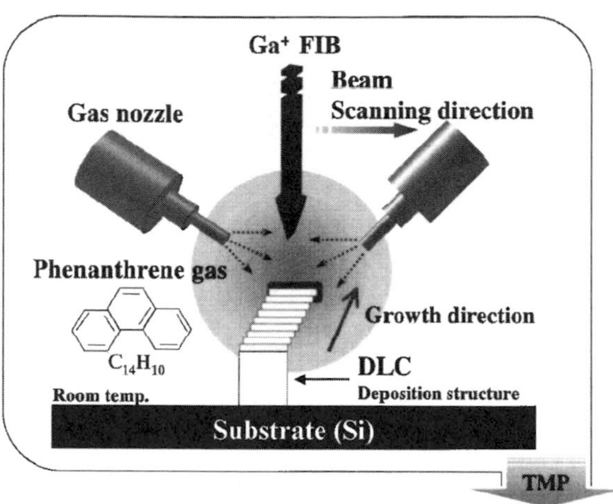

Fig. 4.38 Schematic diagram of FIB chemical vapor deposition system for the deposition of a diamond like carbon film from phenanthrene gas (Pressure $\sim 10^{-4}$ Pa) with 30 keV Ga^{+} ions on a silicon substrate at room temperature [96]

gold electrodes was 100 ohms at room temperature. Figure 4.40 shows the typical structures grown at a beam current of 0.4 pA, with a 7 nm spot size in a 30 keV Ga^{+} beam and a background pressure of 1×10^{-4}Pa. A three dimensional rotor produced by FIB CVD in diamond-like carbon by Igaki et al. [97] utilized a Ga^{+} beam in phenanthrene vapor and a current ranging from 5 to 200 pA at 30 keV. Figure 4.40 shows a nanosheet of thickness 100 nm and a flat rotor of 5.5 um diameter. The fabrication took 50 minutes for each one.

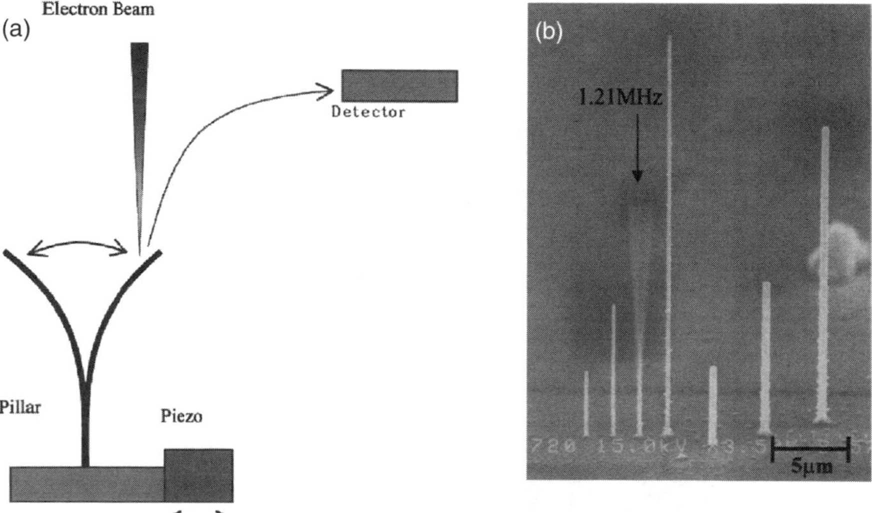

Fig. 4.39 (a) Schematic diagram of the vibration monitoring system set up in the SEM microscope, and (b) SEM image of the vibration induced by piezodevice at the resonance frequency of 1.21 MHz [94]

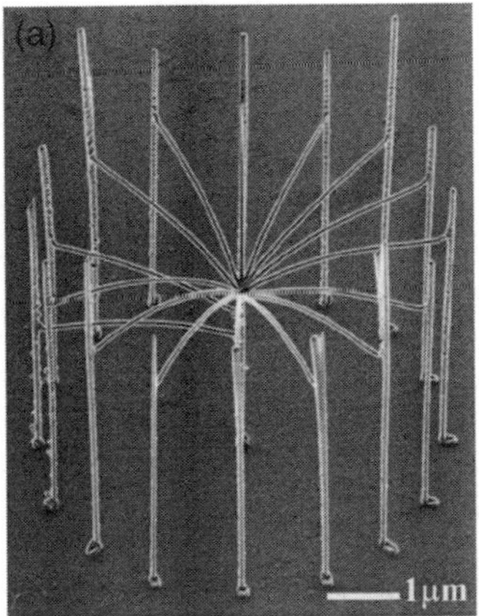

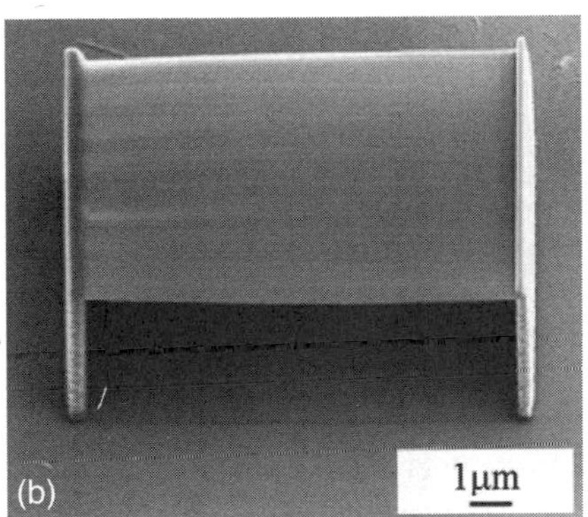

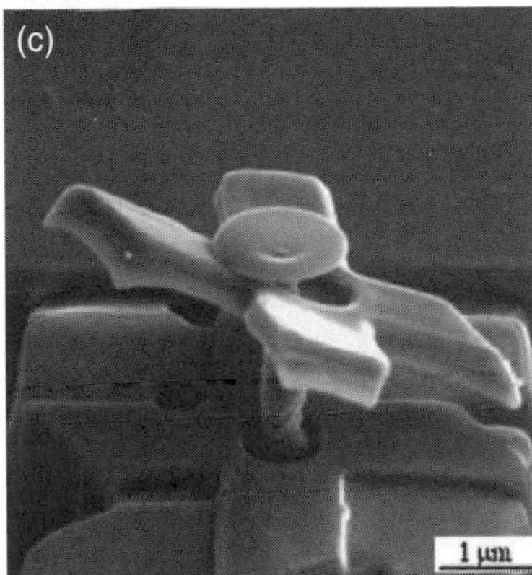

Fig. 4.40 Diamond like carbon structures produced by FIBCUD: (a) radial free-space wiring grown in 16 directions from the center [96], (b) Nano-sheet, 12 μm × 12 μm with 100 nm thickness [97], and (c) Flat rotor with 5.5 μm diameter, 1.2 μm wing-width and 0.57 μm wing-thickness, [97]

Deposition of silica films with tetramethylcyclotetrasilozane (TMCTS) as the precursor has been achieved by Puers and Reyntjens [98]. This process produced dense films that were suitable for the construction of enclosed silica chambers. In tests these cavities provided hermetic encapsulation for over 54 days. The additive fabrication process covered an area of 14 μm × 14 μm. The process employed repeated depositions, such that overhanging structures completed the top of a chamber. This chamber formation sealed the piezoresistive pressure sensor inside (see Fig. 4.41). The sensor exhibited no drift during testing, indicating that a leak-free seal was formed during the deposition process.

Tungsten CVD has been demonstrated by Ishida et al. [99]. That process employed a tungsten hexacarbonyl ($W(CO)_6$) precursor at a pressure of 3×10^{-3} Pa, and used a 9 pA Ga^+ beam. The pillars formed were mechanically characterized, demonstrated a maximum density of 13 g/cm^3, and a Young's Modulus of 300 GPa. When growth is undertaken in a mixed ambient of tungsten hexacarbonyl

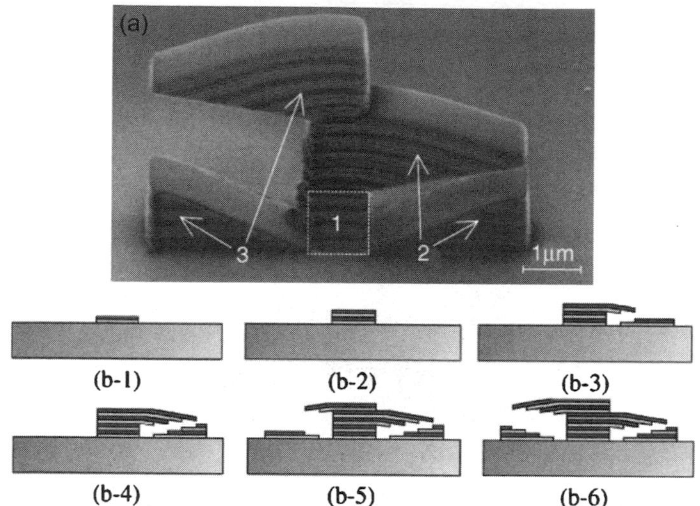

Fig. 4.41 Deposition of silica in an overhanging structure (A) SEM image of completed structure, and (B) schematic diagram indicating different stages in the fabrication process. [98]

and phenanthrene, unique characteristics are produced [100]. Figure 4.42 shows nanosprings formed with a beam current of 1 pA and a spot size of 7 nm, with 30 keV Ga^+ at a pressure of 1×10^{-3} Pa. The basis for the spring size diameter control was the scan rate of the ion beam. Mechanical tensile testing of nanosprings demonstrated an extension of up to 4.9 μm. Nonosprings that had a 3% higher tungsten content resulted in a Young's modulus of over 200 GPa.

Local CVD can also be carried out with electron beams (EB). Igaki et al. [101] compared the CVD method to that of FIB for amorphous carbon. The FIB process is

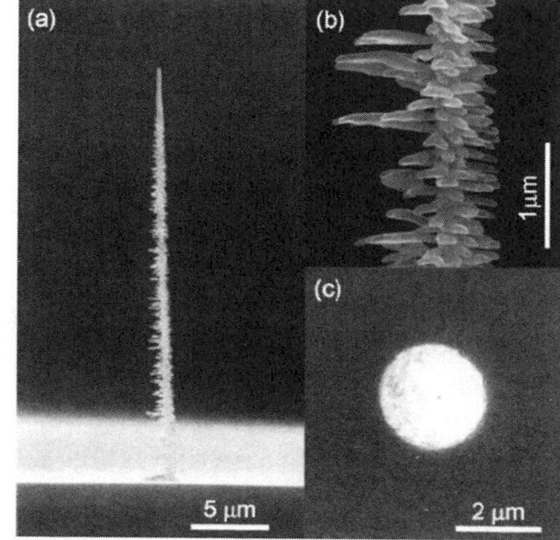

Fig. 4.42 SEM micrographs of a tungsten pillar (a) as grown side view, (b) magnified view, (c) top view, (d) side view after cleaning and (e) top view after cleaning with FIB milling [99]

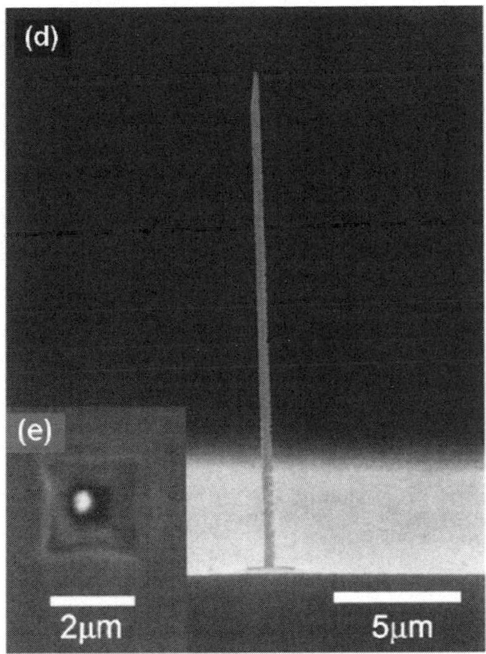

carried out at 5 to 30 keV at 1 pA, and the electron beam at 5 keV at 160 pA to 2 nA, with both systems demonstrating a resolution of 5 nm. Figure 4.43 shows a pillar of carbon produced by each method for comparison. The FIB method produced a pillar with a diameter of 290 nm and length 7.96 μm. The EB-fabricated pillar is 160 nm in diameter with length 4.15 μm. Branch-like growth is also seen in tungsten

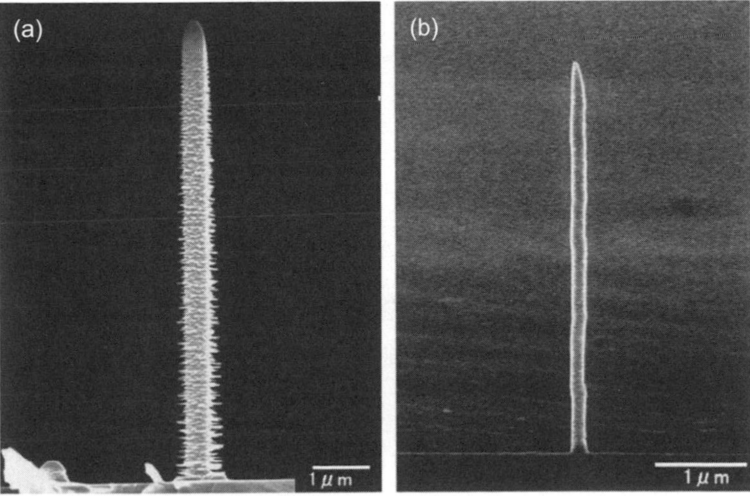

Fig. 4.43 SEM micrographs of (a) FIB CVD pillar and (b) electron beam CVD deposited pillar [101]

pillar formation, and is attributed to the redeposition of sputtered material nucleating whiskers, which accumulate on the perimeter. This effect was not present for the EB-deposited structures.

In general, for the same deposition current, the volumetric growth rate of the FIB system is approximately 100 times faster than the EB growth method. The structures produced are amorphous. However, the film grown via FIB contained 97% C and 3% Ga, and was electrically conductive, while the structure growth via the EB process was insulating and composed of carbon only.

Platinum films, grown from the precursor trimethyl methyl cyclopentadienyl platinum ($C_9H_{10}Pt$), with a focused ion beam of Ga^+, were deposited to provide a microelectronic electrical interconnect repair. Deposition on heated silicon substrates with areas of between $50\,\mu m \times 50\,\mu m$ to $200\,\mu m \times 200\,\mu m$ produced a film thickness of 100 nm. The resistivity of the conductive film produced was a function of the carbon content in the film [102]. Also, Teng and Prewett [103] have defined platinum conductors on microfabricated silicon nitride cantilevers with a focused ion beam CVD of platinum with sub-micron dimension. A section of the silicon nitride cantilever was pre-thinned to 15 nm with a beam current of 100 pA. The platinum was deposited from methyl cyclopentadientyl trimethyl platinum $(CH_3)_3Pt(CpCH_3)$ using a beam current of 1 pA, 0.2 μs dwell time. The process used a 200% overlap during the scan based upon a beam current of 2–6 pA/um^2.

Lapicki et al. [104] demonstrated the deposition of cobalt with FIB CVD on glassy carbon substrates from the precursor octacarbonyl dicobalt ($Co(CO)_8$). This was done at a pressure of 4×10^{-5} Pa with a beam current in the range of 4–10 pA. The 100–150 nm diameter flat islands of cobalt were characterized by an AFM and magnetic force microscopy (MFM). The magnetic properties of arrays of dots were characterized with a vibrating magnetometer, demonstrating an Hc $\sim$ 50–200 Oe and Ms $\sim$ 1200 (+/–300) emu/cm^3.

4.5 Electroplating of Nanostructures

There has been a great deal of interest in electroplating nanostructures and electroplating into nanometer-dimension templates for the purpose of forming nanostructures. Much of this interest was piqued following the pioneering work of Martin et al. [105]. The reader is referred to the excellent introductory texts on electrochemical methods by Bard and Faulkner [106], to texts on electroplating by Paunovic and Schlesinger [107], and to the comprehensive *Handbook of Electroplating* [108]. This section presents a brief review of the electroplating process.

4.5.1 Electrochemical Cells for Electroplating

Electroplating is a process that makes use of an electrolytic cell. Electroplating of metals and alloys involves the reduction of metal ions from the electrolyte. The electrolyte can be aqueous, an organic solvent, or a molten salt at high temperature. The reduction reaction can be achieved with either an external power supply or another chemical reaction: a process known as electroless plating.

The electrons from the reduction of the metal ions at the electrode result in the Faradaic current. The more the reduction and oxidation reaction occurs, the greater is the Faradaic current. The direction of the electron flow is shown in the Fig. 4.44. (However, due to a historical convention, thc conventional electrical current flows counter to the electron flow.)

In a three-terminal electrochemical cell, plating takes place at a working electrode (WE). The other electrode is known as the counter electrode (CE). The reference electrode (RE) is used to measure the potential at the working electrode. In the electroplating process, the cathode is the part to be plated. The ion in solution is transported to the electrode by convection, and, therefore, the mixing of the bath is important for uniform plating processes. There it undergoes a reduction reaction, forming a metal layer through several stcps at the nanoscale. These steps include the nucleation of islands, growth of islands, and the formation of a continuous metal film adherent to the surface.

Chemical additives can be added to the electroplating bath to improve the surface finish and reduce surface roughness. These additives are known as brighteners and levelers. The current density controls the film growth. The process of employing a constant current is known as DC plating. A time-dependant application of the potential with a sine wave is known as AC plating; with time-dependant pulses, it is pulsed plating (PP). The current flow, I, for an electrode of area, A, is given by the following expression:

$$I = nA\,\mathrm{F}\eta N'$$

Where n is the number of electrons involved in the reduction ration, η is the electroplating efficiency, F is Faraday constant, and N' is the rate of molar deposition per

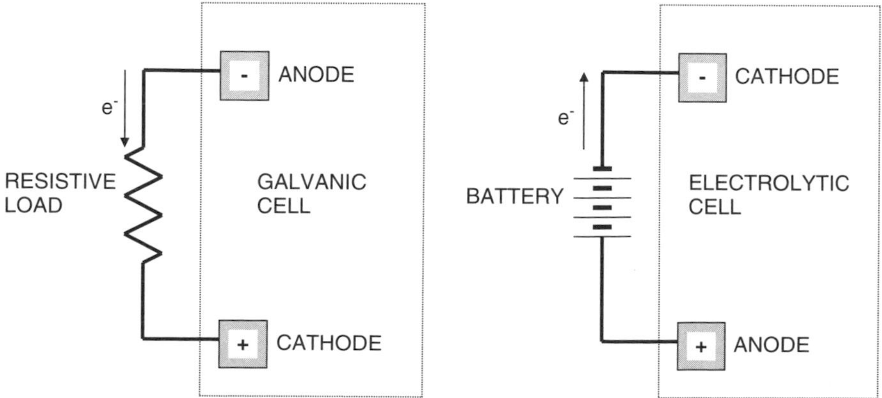

Fig. 4.44 Electrical circuit representation of (a) Galvanic cell and (b) Electrolytic cell

unit area. The film thickness can be calculated based upon the atomic mass, m, and film density, ρ, and as follows:

$$r = mN'/\rho$$

Where r is the film thickness growth rate.

The distance between the anode and the cathode is set, an important consideration for uniform growth and control at a given current density. Pulses of current can also be used to improve the uniformity of growth into a photoresist mold and to achieve the plating of metal alloys. Applied pulses are depicted schematically in Fig. 4.45. The pulse can be at the rest potential; or, as in reverse pulsed plating, the pulse can be more anodic, allowing for some dissolution of the plated film.

The application of time-dependant plating allows the concentration of the ion in the electrolyte to recover from diffusion limitations. Thus this method is much more effective in filling vias [109]. Hydrogen evolution can also take place at cathodic potentials. Thus care should be taken in the selection of additives to the bath. The additives that reduce the probability of the formation of gas bubbles are of benefit. A gas bubble would block electrodeposition from taking place at the interface, at least until the bubble is dislodged from the surface.

4.5.2 Templates

Several types of membranes are available for plating through templates (per Hulteen and Martin [110]). Polycarbonate membranes are produced by nuclear tracks and are commercially available (Whatman, Nuclepore Membranes). Alumina templates with defined dimensions are commercially available from Whatman Inc. (Florham Park, New Jersey). These templates have pore sizes of 20–200 nm and a pore density of up to 10^{20} /cm^2 (depicted schematically in Fig. 4.46). The templates are formed by the anodization of aluminum with controllable channel lengths of up to 10's of micrometers [111].

The pioneering work on the formation of membranes for selective molecular transport, purification and filtering by Martin et al. [113]. This excellent review also

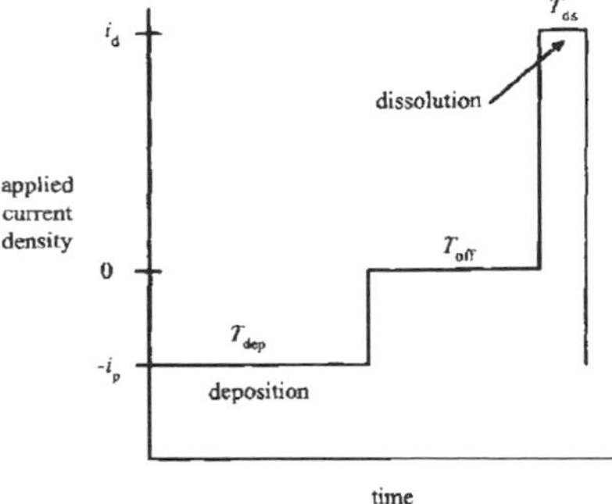

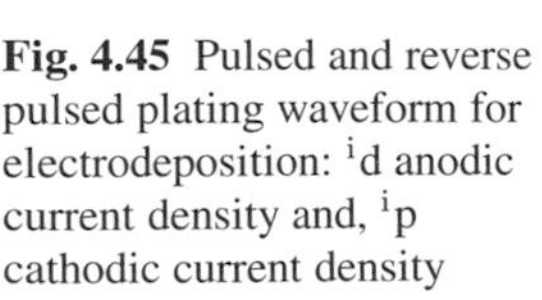

Fig. 4.45 Pulsed and reverse pulsed plating waveform for electrodeposition: ${}^{i}d$ anodic current density and, ${}^{i}p$ cathodic current density

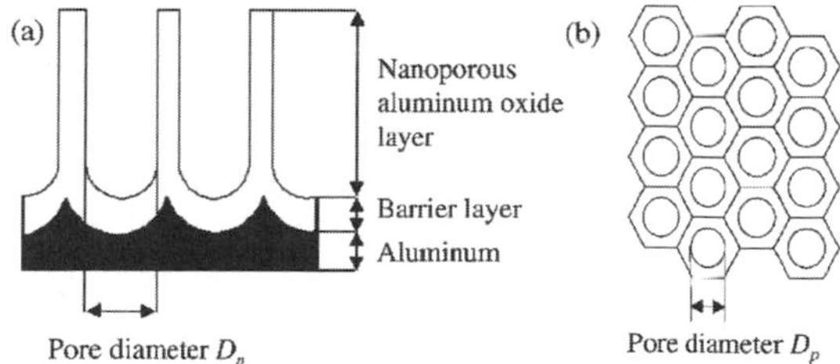

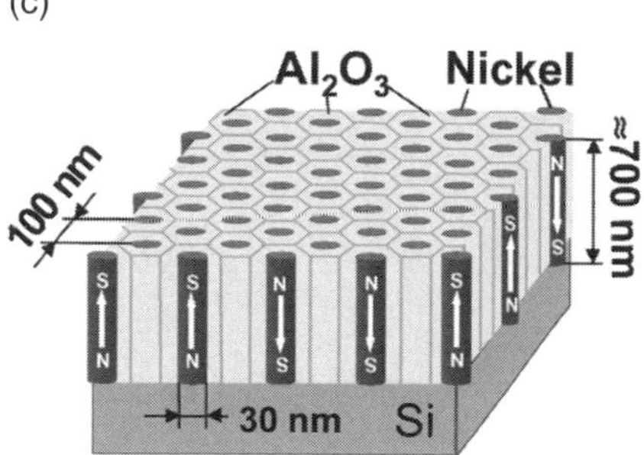

Fig. 4.46 Schematic diagram of alumina templates formed by anodization of aluminium (a) cross-section and (b) plan view of [126]. (c) Alumina template on a silicon wafer with schematic representation of Ni filled pores with magnetic polarizations indicated in each. [140]

includes work on membrane formation and filling. The membranes show ion selective transport properties. Notably, data demonstrates cation permselective behavior over a range of concentrations. The internal surface of the nanotube membrane is coated with gold. The gold has thiols absorbed with a defined surface charge. The charge is a function of the terminal ligand. Electrolyte composition and pH influence the ion transport through the membrane. The selective transport of other anions and cations were also analyzed, with separation based upon size exclusion. Future studies could apply such membranes to mimic selective transport through natural ion channels.

Template-based formation, using track-etched membranes for the formation of superconducting nanowires, has been demonstrated in work by Dubois [112]. This work included the formation of electroplated nanowires, nanorods, and nanotubes. Formation of small diameter indium, tin, and zinc nanowires was pioneered by Possin [114]. Work with a alumina template and for iron and cadmium is presented by Al-Mawlawi et al. [115]. That work involved a sulfate bath with boric acid, with an AC voltage at a frequency of 200 Hz, 16AC(RMS), and used a graphite electrode. The nanowires were released from the alumina template by dissolving it in a mixture of phosphoric acid and chromic acidic solution. The initial pore geometry can be widened by etching of the alumina in oxalic acid, so that pore diameters in the range of between 10 and 200 nm were accessible for wire formation.

Fabrication of individual copper nanowires was undertaken with a scanning probe by Suryavanshi and Yu [116]. The electrolyte was carried in a pipette probe, eliminating the need for a bath, and creating a local solution for the formation of the vertical solid polycrystalline nanowire (see Fig. 4.47).

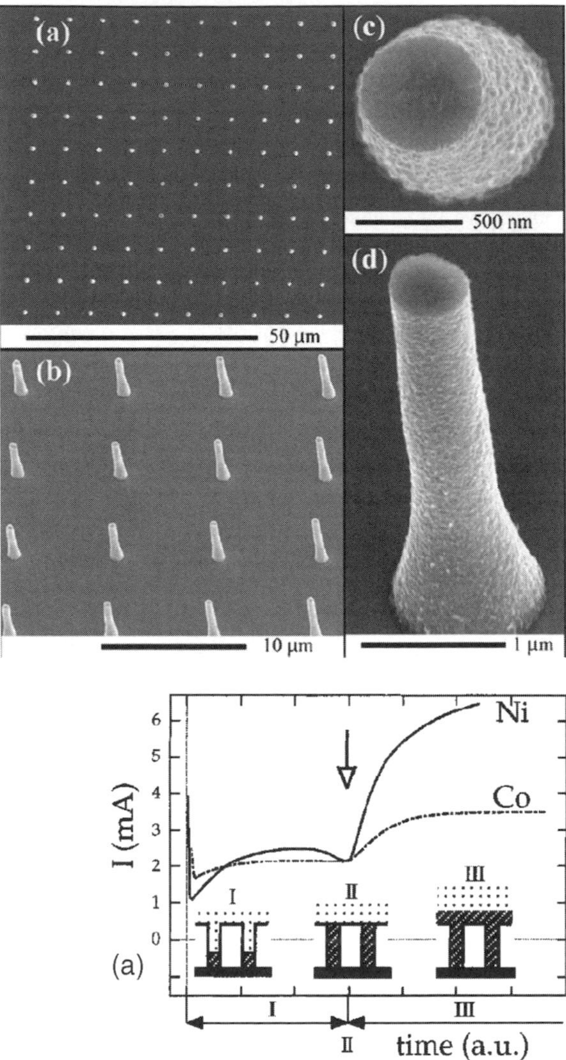

Fig. 4.47 (a) Scanning electron microscope image of electroplated copper nanowire array with grid spacing of 7 μm formed on a Au/Si substrate. (b) view of 4×4 array of wires, (c) single Cu nanowire, and (d) higher magnification images [116]

4.5.3 Ferromagnetic Nanowire Materials

Nickel and copper multilayers were electroplated by Wang et al. [117] using a polycarbonate membrane. TEM measurements of the 80 nm diameter wires indicated a preferential <111> crystalline direction [113] growth with DC plating; and up to 220 alternating layers were formed in a wire. It was found that the layer thickness increased from the initial 2 nm to 7 nm after 100 layers had been formed. This result is consistent with the current density, which increases, indicating a more rapid growth was taking place.

Similar effects were observed by Schonenberger et al. [118] for cobalt, copper, gold and nickel wires when using alumina templates. Figure 4.48(a) shows the

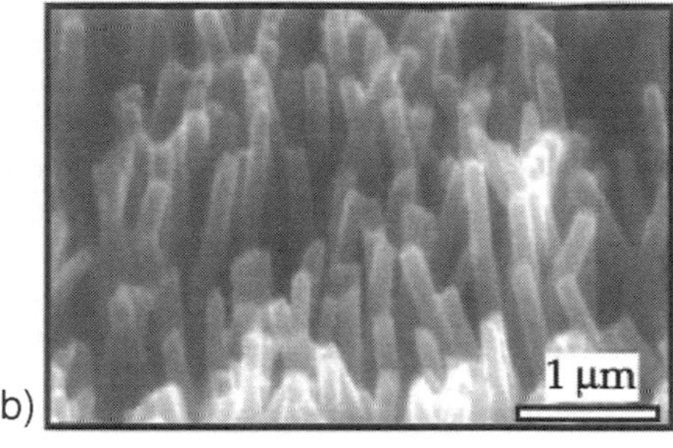

Fig. 4.48 Electroplating into porous alumina template (a) stages of current versus time for potentiostatic plating of Ni and Co (at $-1.1V_{SCE}$ for Ni and $-1.2V_{SCE}$ for Co) in pores of polycarbonate membrane with minimal pore diameter of 80 nm. Schematic represents the three stages of process I filling pores, II pores just filled at transition region, and III plating over whole membrane surface; (b) SEM micrograph of Ni nanowires electroplated in region I, after the template ! has been removed [118]

current versus time during the plating, and Fig. 4.48(b) shows an array of nickel nanowires. Cobalt and copper multilayer nanowires with diameters of 8–10 nm were fabricated in polycarbonate membranes by Blondel et al. [119], by Liu et al. [120], and by Voegeli et al. [121]. DC current and a sulfate with boric acid and DC plating conditions was used. Wires of diameter 8–10 nm and length 10–100 nm were formed in a polycarbonate membrane. The magnetoresitive properties of these structures was measured at low temperatures Alternating layers of 5 nm cobalt and 2 nm copper formed structures with magnetoresistance properties at 5 K. When contamination of the cobalt layer with copper occurred, then the magnetoresistance effect was no longer present.

Composite wires were observed to have interesting optical properties by Mock et al. [122]. Figure 4.49 depicts gold, silver and nickel composite wires of 30 nm diameter made with polycarbonate templates. A combination of electroless plating and electroplating was used to form the wires. The selection of size was based on that size falling within the active range for surface Plasmon resonance.

Kroll et al. [123] deposited 12 nm (mean) diameter Fe, Ni, and Co nanowires in porous alumina substrates using an AC plating procedure. Wires of up to 100 μm in length can be fabricated with these processes. The magnetic properties were measured as a function of temperature for each material. Strong anisotropy with squareness values of 0.95 were observed, even at room temperature. The wire diameter is less than a typical magnetic domain size of 50 nm, indicating that the wires should exhibit one-dimensional magnetic properties. Measurements on the Co nanowires indicate a competition between the shape-induced magnetic anisotropy and the magnetic cryatalline properties at 314 and 5 K (as depicted in Fig. 4.50). The microstructure of the cobalt nanowires indicates a combination of HCP and FCC structure.

Cobalt nanowires of up to 22 μm in length have also been formed by electroplating into polycarbonate membranes [124]. The electrolyte solution used was cobalt

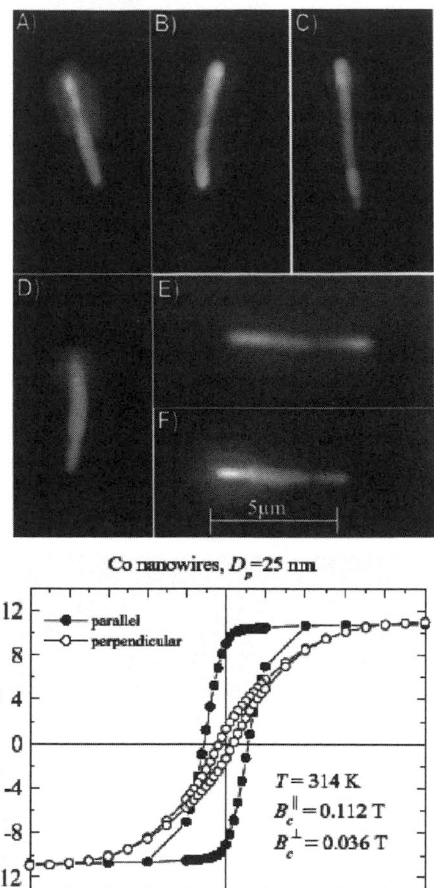

Fig. 4.49 Digital camera color images of coded nanowires illuminated with white light polarized along the short axis of each wire. (A) silver/gold, (B) gold/silver/gold, (C) gold/nickel/gold/nickel, (D) silver/gold/nickel, (E) silver/gold/nickel illuminated with light polarized along the long axis, and (F) silver/gold/nickel illuminated with light polarized along the short axis. The emission wavelength correspond to composition of nanowires [122]

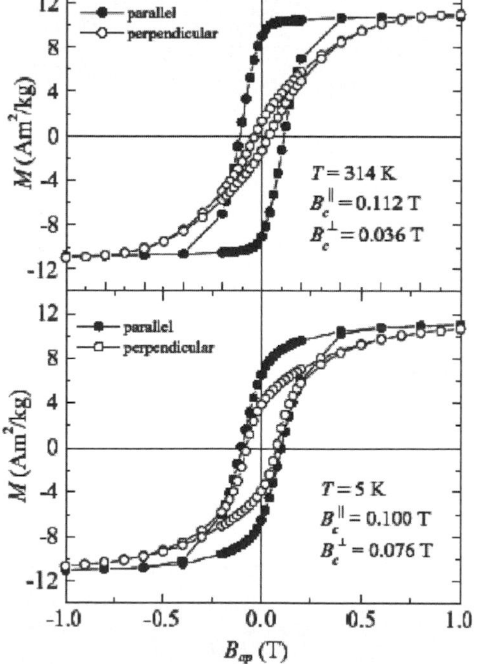

Fig. 4.50 Magnetic properties of cobalt nanowires at 314 and 5 K [123]

sulfate with H_3BO_3 at a pH level of between 2 and 4. Deposition at room temperature was in the potential range: 0.95 V versus the Ag/AgCl reference electrode. The pH of the bath has a strong influence on the structural properties and hence an influence on the magnetic hysteresis.

Arrays of nickel nanowires were formed in alumina membranes with typical diameters of 18–35 nm and a length of a few micrometers [125]. Pulsed AC electroplating with 16VAC was used to fill the nanopores [126]. Magnetic properties were characterized by VFM as a function of the wire diameter and plating bath composition.

NiFeCu composite wires were formed by coating 20 μm copper wires with NiFe—via electroplating in a Watts-type bath (Seet et al. [127])—achieving extremely high permeability materials for sensors. Also studied was the effect of the current density on the composition, grain size, and magnetic properties.

Finally, multilayers of cobalt and copper to form nanowires were electroplated by Pirota and Vazaquez [128] with a pulsed plating method at 50 °C. This process used a cobalt sulfate, cobalt chloride, and H_2BO_4 bath (see Table 4.3). A pulse of 2 ms with a current of 30 mA was followed by a voltage pulse of 5 V with a 3 s duration, followed by a resting pulse of 1 s at zero current. X-ray diffraction (XRD) indicates that more HCP structure is formed in the layered wire with a total plating time of 120 minutes, with 3 minutes per layer of cobalt or copper. In addition, multilayered nanowire had an increase in the coercivity, from 850 Oe to 1230 Oe, and the remanence to magnetic saturation ratio increased from 0.25 to 0.62.

4.5.3.1 Electroless Plating of Ferromagnetic Nanodots

Selective area formation of nanodots of CoNiP was demonstrated by Kawaji et al. [129]. Electroless plating, into patterned pores defined on oxidized silicon wafers, with diameters of less than 100 nm, resulted in an aspect ratio of greater than five. A chemical cleaning of the silicon—to obtain a hydrogen terminated surface prior to plating in the CoNiP bath—was required. The resulting film's magnetic properties were characterized and found to have a higher perpendicular squareness ratio than a continuous CoNiP film. The effect of the chemical activation process prior to deposition, with a palladium complex preferentially adsorbed to the exposed silicon surface, was examined to enhance nucleation of the plated film.

4.5.4 Noble Metal Nanowires

Platinum nanowires were formed by Yoo and Lee [130] in a Technic Platinum AP (Cranston. Rhode Island) bath with DC current. They found that the formation of the platinum and paladium wires were a function of the current density. At a low current density of 0.1–1 mA/cm^2, smooth, compact nanowires were produced. However a higher currently density of 20–200 mA/cm^2 produced a hollow morphology. After plating the alumina, it was removed in 1.8 wt% chromic acid and 6 wt% phosphoric acid over 24 hrs. Figure 4.51 shows the different nanowire and nanotube structures that can be formed. Figure 4.52 is a schematic diagram that suggests how the enhanced film growth gives rise to the hollow structure.

Silver nanowires were formed by Cheng and Cheng [131] with electrolyte silver nitrate and silver sulfite, as listed in Table 4.3. The DC pulse plating was carried

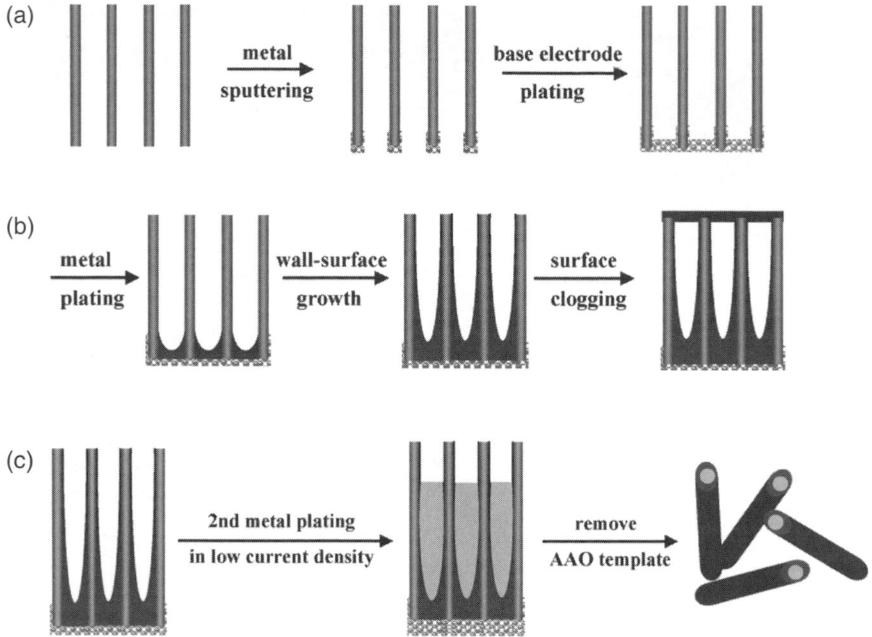

Fig. 4.51 Schematic diagram indicating the procedure of electroplating metal nanotubes and core-shell tubular metallic wires. (a) contact metal sputtering onto porous alumina template, (b) initial high-current density electroplating, (c) second plating was at a lower current density to ensure filling of pores, and removal of template [130]

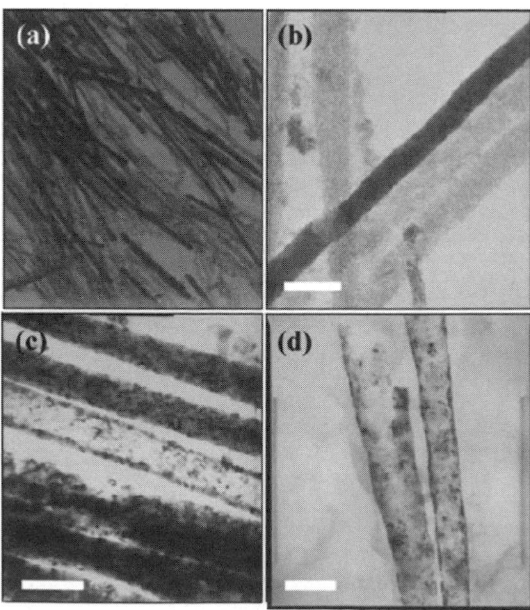

Fig. 4.52 Platinum naowires and nanotubes obtained by high-current density electroplated: (a) TEM image at low magnification. Three different kinds of naostructures were produced: (b) close-packed nanowire, (c) thick-wall nanotube, and (d) hollow nantubes [130]

Table 4.3 Composition of plating baths

Material	Composition	Conditions	Reference
Electroplating			
Ag	$AgNO_3$ 10 g/l Na_2SO_3 10 g/l CH_3COONH_4 20 g/l		[131]
Bi	48.57 g $Bi(NO_3)_3.H_2O$	RT	[132]
Co	248g/l $CoSO_4.7H_2O$ 40 g/l H_3BO_3 Add H_2SO_4 to achieve pH (citric acid 25 g/l)	RT	[124] ([135])
Cu	45 g/l $CuSO_4$, and 4 g/l $HsBO_4$		[128]
NiFe	125 g/l $NiSO_4.6H_2O$ 40 g/l $H_3BO_3$20 g/l $NiCl_2.6H_2O$ 8–18 g/l $FeSO_4.7H_2O$ 4 g/l saacharin	55°C	[127]
Electroless			
CoNiP	$NaH_2PO_2.H_2O$ 0.2M/l $(NH_4)_2SO_4$ 0.5M/l $Na_2C_3H_2O_4.H_2O$ 0.75 M/l $Na_2C_3H_4O_6.2H_2O$ 0.2 M/l $Na_2C_4H_4O_5.1/2H_2O$ 0.375 M/l $NiSO_4$. $6H_2O$ 0.16 M/l $CoSO_4$. $7H_2O$ 0.06 M/l	80 °C adjusted to pH 9.6	[129]

RT– Room Temperature M/l – Moles per liter

out typically over the range 1–4 V which had a pronounced effect on the various nanostructure morphologies that were created, as shown in Fig. 4.53. The silver wires appear also to have a high density of silver particles of approximately 19 nm in diameter comprising their morphology. In addition, varying the amplitude of the pulsed plating of 50 ms, with a duty cycle of 100 ms, resulted in the nucleation of silver particles on the nanowires, and achieved a wire growth length of up to 400 nm with a 40 nm diameter. Here, the faster growth rate corresponds to a more rapid reduction of silver taking place at the tips of the structures, and thus an increase in length.

4.5.5 Metal Oxide Nanowires

Plating of bismuth for the formation of δ-Bi_2O_3 nanowires was carried out by Huang et al. [132]. First, bismuth nanowires were formed in a bismuth nitrate electrolyte over a period of 10 hrs. This step was followed by annealing at temperatures as high as 550° C to oxidize the bismuth. The structure of the film was determined by XRD. The electrode was attached to a gold-coated silicon substrate and the alumina

Fig. 4.53 Field emission SEM micrographs of silver nanowires before [(a), (b), (c)] and after [(d), (e), (f)] pulse electroplating with 1V, 3V and 4V electroplated nanowire [131]

dissolved in 3M NaOH over 2 hrs. Figure 4.54 shows the δ-Bi_2O_3 nanowires formed, although normallyα-Bi_2O_3 is the stable phase below 723° C.

Metal and metal oxide nanowries fabricated by Tresback et al. [133] evidence interesting properties. Au–SnO_2–Au and Au–NiO–Au were electroplated with alumina membranes, with 220 nm pores at a constant current of 0.5 mA. After the alumina was removed in 3 M NaOH, the wires were annealed in a two-step process to form the contact and metal oxide. For Au–Sn–Au, 197° C for 0.5 h, followed by 650° C for 0.5 h, was sufficient. For Au–Ni–Au wire, there was a one-step process: heating to 600° C for 2 hrs. The resultant wires are shown in Fig. 4.55. The electron diffraction pattern confirms that the dark part is gold and the lighter region is polycrystalline tetragonal SnO_2. Although a jagged interface is present, the junction in the wire forms a Schottky contact, the electrical properties of which can be measured.

Fig. 4.54 SEM micrograph of electroplated bismuth nanowrics converted to Bi_2O_3 by oxidation in air. Inset shows bismuth nanowires, average diameter 47 nm, after removal of the porous alumina template [132]

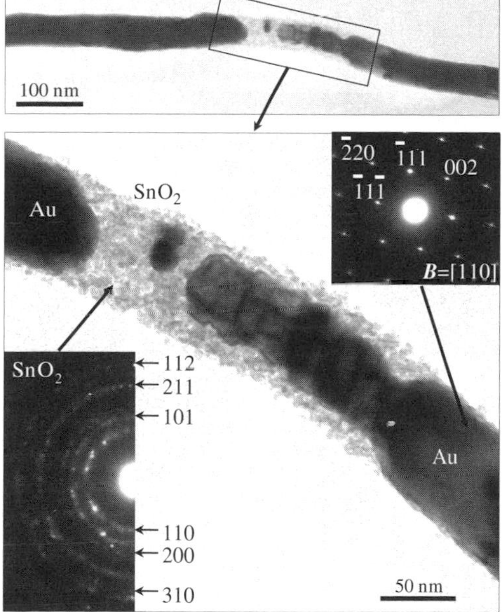

Fig. 4.55 TEM bright field micrograph showing –Au–SnO_2–Au nanowires synthesized inside the alumina template [133]

4.6 The Future

Physical and chemical deposition processes are inherently suitable for nanoscale fabrication because of their atomic level growth mechanisms. However, process control is an important element, as the objective is to reproducibly achieve the desired nanostructures. Given the range of nanofabrication processes available for the construction of nanodevices, the optimization of process control is a fertile area for research.

Another subject of importance is that of nanostructure inspection methods. Accurate and formalized inspection methodologies need to be developed for nanomanufacturing processes to become routine and commercially viable. The extent to which the available materials can be utilized in the production of novel devices is inextricably dependant upon the development of low-cost, mass production methods.

Currently, one area in which nanoscale material applications are coming into prominence is medical diagnostics, specifically cell tagging. The use of quantum dots directed to detect cancer cell surface antigens is only one example. Nanoscale devices can be implemented in other biomedical devices, in which the device size is on the same scale as that of the cell's internal structure.

The success of this and similar applications promises to lead to intriguing advances. One such promise is the development of nanodevices that can interface with the internal mechanisms of a living cell. Such an application can potentially give rise to new types of biomedical testing tools and even to methods for the growth of artificial organs.

This brief summation of some of the important research to date should give rise to optimism for the future of this exciting field. As nanofabrication techniques are enhanced and refined, commercially viable nanomanufacturing of novel devices—with important applications to any number of fields—seems the inevitable outcome.

References

1. Cui Z. Micro-Nanofabrication: Technologies and Applications. Springer, 2005.
2. Brodie I., J.J. Muray. The Physics of Micro/Nano-Fabrication. Plenum Press: New York, 1992.
3. Madou M.J. Fundamentals of Microfabrication: The Science of Miniaturization, CRC Press, 2002.
4. Campbell S.A. The Science and Engineering of Microelectronic Fabrication. Oxford University Press: Oxford, 2001.
5. Mahan J.E. Physical Vapor Deposition of Thin Films. John Wiley: New York, 2000.
6. Wolf S., R.N. Tauber. Silicon Processing for the VLSI Era. Lattice Press, 2000.
7. Chase M.W., C.A. Davies, J.R. Downey, D.J. Frurip, R.A. McDonald, A.N. Syverud. JANAF Thermochemical Tables. American Institute of Physics: New York, 1986.
8. Kittel C, H. Kroemer. Thermal Physics. W.H. Freeman and Co.: New York, 1980.
9. Freund L.B., S. Suresh. Thin Film Materials: Stress, Defect Formation and Surface Evolution. Cambridge University Press: Cambridge, 2003.
10. Adamson A.W. Physical Chemistry of Surfaces. John Wiley & Sons, Inc.: New York, 1990.
11. Hirth J.P., G.M. Pound. Condensation and Evaporation. Macmillan Co.: New York, 1963.
12. Michely T, J. Krug. Islands, Mounds and Atoms. Springer-Verlag: New York, 2004.
13. Pak S.C. Thin Film Impedance Based Immunosensor. Bioengineering. University of Illinois at Chicago: Chicago, 1999.
14. Pak S.C., P.J. Hesketh, W.R. Penrose. An ultrathin platinum film sensor to measure biomolecular binding. Biosensors and Bioelectronics 2001;16:371–379.
15. Ritala M. Advanced ALE processes of amorphous and polycrystalline films. Applied Surface Science 1997;112:223–230.
16. Leskelä M, M. Ritala. Atomic layer deposition (ALD): from precursors to thin film structures. Thin Solid Films 2002;409:138–146.
17. Ritala M, M. Leskelä. In: Nalwa HS, editor. Handbook of Thin Film Materials. Academic Press: San Diego, 2002: 103–159.
18. Suntola T. Atomic layer epitaxy. Materials Science Reports 1989;4:261–312.
19. Puurunen R.L. Surface chemistry of atomic layer deposition: A case study for the trimethylaluminum/water process. Journal of Applied Physics 2005;97:121301–121352.

20. Kim H. Atomic layer deposition of metal and nitride thin films: Current research efforts and applications for semiconductor device processing. Journal of Vacuum Science and Technology, B 2003;21:2231–2261.
21. Choy K.L. Chemical vapor deposition of coatings. Progress in Materials Science 2003;48:57–170.
22. Aleskovskii V.B. Zh Prikl Khim 1974;47:2145.
23. Malygin A.A. Zh Obshch Khim 2002;72:617.
24. Niinisto L., M. Ritala, M. Leskelä. Synthesis of oxide thin films and overlayers by atomic layer epitaxy for advanced applications. Materials Science and Engineering B 1996;41: 23–29.
25. Ritala M, M. Leskelä. Atomic layer epitaxy – a valuable tool for nanotechnology. Nanotechnology 1999;10:19–24.
26. Heitzinger J.M., J.M. White, J.G. Ekerdt. Mechanisms of GaAs atomic layer epitaxy: A review of progress. Surface Science 1994;299–300:892–908.
27. Isshiki H, Y. Aoyagi, T. Sugano. (GaAs)m(GaP)n low dimensional short-period superlattice fabricated by atomic layer epitaxy. Microelectronic Engineering 1998;43–44:301–307.
28. Wang Y.H.Z., J.B. Lu, J. Qi, X.H. Characterization of silicon films grown by atomic layer deposition on nanocrystalline diamond. Diamond and Related Materials 2006;15: 1434–1437.
29. Venkatasamya V., N. Jayarajua, S.M. Coxb, C. Thambiduraia, M. Mathea, J.L. Stickney. Deposition of HgTe by electrochemical atomic layer epitaxy (EC-ALE). Journal of Electroanalytical Chemistry 2006;589:195–202.
30. George S.M., A.W. Ott, J.W. Klaus. Surface chemistry for atomic layer growth. Journal of Physical Chemistry 1996;100:13,121–113,131.
31. Platzer-Bjorkman C, T. Torndahl, D. Abou-Ras, J. Malmstrom, J. Kessler, L. Stolt. Zn(O,S) buffer layers by atomic layer deposition in Cu(In,Ga)Se_2 based thin film solar cells: band alignment and sulfur gradient. Journal of Applied Physics 2006;100:445060–445069.
32. Sanders BW. Zinc oxysulfide thin films grown by atomic layer deposition. Chemical Materials 1992;4:1005–1011.
33. Yousfi EB, T. Asikainen, V. Pietu, P. Cowashi, M. Powalla, D. Lincot. Cadmuim-free buffer layers deposited by atomic layer epitaxy for copper indium dieselenide solar cells. Thin Solid Films 2000;361–362:183–186.
34. Hoivik N.D., J.W. Elam, R.J. Linderman, V.M. Bright, S.M. George, Y.C. Lee. Atomic layer deposition of protective coatings for released micro-electromechanical systems. Sensors and Actuators A 2003;103:100–108.
35. Kim J, K. Chakrabarti, J. Lee, K.-Y. Oh, C. Lee. Effects of ozone as an oxygen source on the propertie of the Al_2O_3 thin films prepared by atomic layer deposition. Materials Chemistry and Physics 2003;78:733–738.
36. Jeon W.-S., S. Yang, C.-S. Lee, S.-W. Kang. Atomic layer deposition of Al_2O_3 thin films using trimethyaluminum and isopropyl alchol. Journal of Electrochemical Society 2002;149:C306–C310.
37. Ye P.D., G.D. Wilk, J. Kwo, B.H.-J. Yang, L. Goossmann, M. Frei, S.N.G. Chu, J.P. Mannaerts, M. Sergent, M. Hong, K.K. Ng, J. Bude. GaAs MOSFET with oxide gate dielectric grown by atomic layer deposition. IEEE Electron Device Letters 2003;24:209–211.
38. Agostinelli G., A. Delabie, P. Vitanov, Z. Alexieva, H.F.W. Dekkers, S. De Wolf, G. Beaucarne. Very low surface recombination velocities on p-type silicon wafers passivated with a dielectric with fixed negative charge. Solar Energy Materials and Solar Cells 2006;90:3438–3443.
39. Carcia P.F., R.S. McLean., M.H. Reilly, M.D. Groner, S.M. George. Ca test of Al_2O_3 gas diffusion barriers grown by atomic layer deposition on polymers. Applied Physics Letters 2006;89:31915–31913.
40. Ritala M., K. Kukli, A. Rahtu, P.I. Raisanen, M. Leskelä, T. Sajavaara, J. Keinonen. Atomic layer deposition of oxide thin films with metal alkoxides as oxygen sources. Science 2000;288:319–321.

41. Klaus J.W.O.S., S.M. George. Growth of SiO_2 at room temperature with the use of catalyzed sequential half-reactions. Science 1997;278:1934–1936.
42. Klaus J.W., A.W. Ott, J.M. Johnson, S.M. George. Atomic layer controlled growth of SiO_2 films using binary reactioin sequence chemistry. Applied Physics Letter 1997;70:1092–1094.
43. Ferguson J.D., A.W. Weimer, S.M. George. Atomic layer deposition of $Al_2O_3$and SiO_2 on BN particles using sequential surface reactions. Applied Surface Science 2000;162–163:280–292.
44. Kamiyama S., T. Miura, Y. Nara. Comparison between Hf-silicate films deposited by ALD with BDMAS [$SiH_2(N(CH_3)_2)_2$] and TDMAS [$SiH(N(CH_3)_2)_3$] precursors. Electrochemical and Solid-State Letters 2006;8:F37–F39.
45. Kamiyama S., T. Miura, Y. Nara. Comparison between SiO_2 films deposited by atomic layer deposition with $SiH_2[N(CH_3)_2]_2$ and $SiH[N(CH_3)_2]_3$ precursors. Thin Solid Films 2006;515:1517–1521.
46. Du Y., X. Du, S.M. George. SiO_2 film growth at low temperatures by catalyzed atomic layer deposition in a viscous flow reactor. Thin Solid Films 2005;491:43–45.
47. Kukli K., J. Ihanus, M. RItala, M. Leskelä. Properties of Ta_2O_5-baded dielectric nanolaminates deposited by atomic layer epitaxy. Journal of Electrochemical Society 1997;144:300–306.
48. Kukli K., J. Ihanus, M. Ritala, M. Leskelä. Tailoring the dielectric properties of HfO_2-Ta_2O_5 nanolaminates. Applied Physics Letter 1996;68:3737–3739.
49. Ritala M., M. Leskelä, E. Rauhala. Atomic layer epitaxy growth of titanium dioxide thin films from titanium ethoxide. Chemical Materials 1994;6:556–561.
50. Pessa M., R. Makela, T. Suntola. Characterization of surface exchange reactions used to grow compound films. Applied Physics Letters 1981;38:131–132.
51. Becker J., E. Kim, R.G. Gordon. Atomic layer deposition of insulating hafnium and tirconium nitrides. Chemical Materials 2004;16:3497–3501.
52. Kukli K., M. Ritala, M. Leskelä. Atomic layer deposition and chemical vapor deposition of tantalum oxide by successive and simultaneous pulsing of tantalum ethoxide and tantalum chloride. Chemical Materials 2000;12:1914–1920.
53. Martensson P., M. Juppo, M. Titala, M. Leskelä, J.-O. Carlsson. Use of atomic layer epitaxy for fabrication of Si/TiN/Cu structures. Journal of Vacuum Science and Technology, B 1999;17:2122–2128.
54. Langereis E., S.B.S. Heil, M.C.M. Van De Sanden, W.M.M Kessels. In situ spectroscopic ellipsometry study on the growth of ultrathin TiN films by plasma-assisted atomic layer deposition. Journal of Applied Physics 2006;100:235341–235349.
55. Heil S.B.S., E. Langereis, A. Kemmeren, F. Roozeboom, M.C.M., van de Sanden, W.W.M. Kessels. Plasma-assisted atomic layer deposition of TiN moniored by in situ spectroscopic ellipsometry. Journal of Vacuum Science and Technology, A 2005;23:L5–L8.
56. Rossnagel S.M., A. Sherman, F. Turner. Plasma-enhanced atomic layer deposition of Ta and Ti for interconnect diffusion barriers. Journal of Vacuum Science and Technology, B 2000;18:2016–2020.
57. Lu J.J.S., M. Ottosson, A. Tarre, A. Rosental, J. Aarik, A. Harsta. Microstructure characterization of ALD-grown epitaxial SnO_2 thin films. Journal of Crystal Growth 2004;260:191–200.
58. Sundqvist J., J. Lu, M. Ottosson, A. Harsta. Growth of SnO_2 thin films by atomic layer deposition and chemical vapour deposition: A comparative study. Thin Solid Films 2006;514:63–68.
59. Sundqvist J., A. Tarre, A. Rosental, A. Harsta. Atomic layer deposition of epitaxial and polycrystalline SnO_2 films from the SnI_4/O_2 precursor combination. Chemical Vapor Deposition 2003;9:21–25.
60. Badot J.C., A. Mantoux, N. Baffier, O. Dubrunfaut, D. Lincot. Submicro- and nanostructural effects on electrical properties of $Li0._2V_2O_5$ thin films obtained by atomic layer deposition (ALD). Journal of Physical and Chemistry of Solids 2006;67:1270–1274.

61. Rahtu A., T. Alaranta, M. Ritala. Insitu quartz crystal microvalance and qudrupole mass spectrometry studies of atomic layer deposition of aluminum oxide from trimethylaluminum and water. Langmuir 2001;17:6506–6509.
62. Schuisky M.J.W.E., S.M. George. In situ resistivity measurements during the atomic layer deposition of ZnO and W thin films. Applied Physics Letters 2002;81:180–182.
63. Utriainen M., M. Kroger-Laukkanen, L.-S. Johansson, L. Niinisto. Studies of metallic thin film growth in an atomic layer epitaxy reactor M(acac)2 (M = Ni, Cu, Pt) precursors. Applied Surface Science 2000;157:151–158.
64. Norman J.A.T. Advanced in copper CVD for the semiconductor industry. Journal of Physics IV, Part 3 2001;11:497–503.
65. Solanki R., B. Pathangey. Atomic layer deposition of copper seed layers. Electrochemical and Solid-State Letters 2000;3:479–480.
66. Juppo M., M. Ritala, M. Leskelä. Deposition of copper films by an alternate supply of CuCl and Zn. Journal of Vacuum Science and Technology A, 1997;15:2330–2333.
67. Klaus J.W., S.J. Ferro, S.M. George. Atomic layer deposition of tungsten using sequential surface chemistry with a sacrificial stripping reaction. Thin Solid Films 2000;360: 145–153.
68. Elam J.W., C.E. Nelson, R.K. Grubbs, S.M. George. Nucleation and growth during tungsten atomic layer deposition on SiO_2 surfaces. Thin Solid Films 2001;386:41–52.
69. Grubbs R.K., C.E. Nelson, N.J. Steinmetz, S.M. George. Nucleation and growth during the atomic layer deposition of W on Al_2O_3 and Al_2O_3 on W. Thin Solid Films 2004;467: 16–27.
70. Kim S.-H., N. Kwak, J. Kim, H. Sohn. A comparative study of the atomic-layer-deposited tungsten thin films as nucleation layers for W-plug deposition. Journal of the Electrochemical Society 2006;153:G887–G893.
71. Aaltonen T., M. Ritala, M. Leskelä. ALD of Rhodium thin films from $Rh(acac)_3$ and oxygen. Electrochemical and Solid-State Letters 2005;8:C99–C101.
72. Mayer T.M., J.W. Elam, S.M. George, P.G. Kotula, R.S. Goeke. Atomic-layer deposition of wear-resistant coatings for microelectromechanical devices. Source: Applied Physics Letters 2003; 82:2883–2885.
73. Lin Y.-C., J.-C. Chiou, W.-T. Lin, Y.-J. Lin, S.-D. Wu. The design and assembly of surface-micromachined optical switch for optical add/drop multiplexer application. IEEE Transactions on Advanced Packaging 2003;26:261–267.
74. Stoldt C.R., V.M. Bright. Ultra-thin film encapsulation processes for micro-electro-mechanical devices and systems. Journal of Physics D: Applied Physics 2006;39:R163–R170.
75. Nistorica C., J.-F. Liu., I. Gory, G.D. Skidmore, F.M. Mantiziba, B.E. Gnade, J. Kim. Tribological and wear studies of coatings fabricated by atomic layer deposition and by successive ionic layer adsorption and reaction for microelectromechanical devices. Journal of Vacuum Science and Technology A: Vacuum, Surfaces and Films 2005;23: 836–840.
76. Herrmann C.F.D., W. Frank, V.M. Bright, S.M. George. Conformal hydrophobic coatings prepared using atomic layer deposition seed layers and non-chlorinated hydrophobic precursors. Journal of Micromechanics and Microengineering, 2005;15:984–992.
77. Ott A.W., J.W. Klaus, J.M. Johnson, S.M. George, K.C. McCarley, J.D. Way. Modification of porous alumina membranes using Al_2O_3atomic layer controlled deposition. Chemical Materials 1997;9:707–714.
78. Haynes C.L., R.P. Van Duyne. Nanosphere lithography: A versatile nanofactricaiton tool for studies of size-dependent nanoparticle optics. Journal of Physical Chemistry B 2001;105:5599–5611.
79. Wang X.D., E. Graugnard, J.S. King, Z.L. Wang, C.J. Summers. Large-scale fabrication of ordered nanobowl arrays. NanoLetters 2004;4:2223–2226.
80. Wang X., C. Lao, E. Graugnard, C.J. Summers, Z.L. Wang. Large-size liftable inverted-nanobowl sheets as reusable masks for nanolithiography. Namo Letters 2005;5:1784–1788.

81. Giannuzzi L.A., F.A. Stevie. Introduction to Focused Ion Beams. Springer: New York, 2005.
82. Swanson L.W., G.A. Schwind, A.E. Bell, J.E. Brady. Emission characteristics of gallium and bismuth liquid metal field-ion sources. Journal of Vacuum Science and Technology 1979;16:1864–1867.
83. Lindhard L., M. Scharff, H. Schiott. Atomic collisions II: Range concepts and heavy ion ranges. K Dan Vidensk, Selsk, Mat Fys Medd 1963;33:1.
84. Biersack J.P., L.A. Haggmark. Monte Carlo computer program for the transport of energetic ions in amorphous targets. Nuclear Instrumentation and Methods 1980;174:257–269.
85. Lugstein A., B. Basnar, J. Smoliner, E. Bertagnolli. FIB processing of silicon in the nanoscale regime. Applied Physics A 2003;76:545–548.
86. Peng Z., P. Hesketh. Nanoparticle/Microfluid Based Electrochemical Biosensor System. 209th ECS Meeting. The Electrochemical Society: Denver, Colorado, 2006: #1275.
87. Kim G.M., M.A.F. van den Boogaart, J. Brugger. Fabrication and application of a full wafer size micro/nanostencil for multiple length-scale surface patterning. Microelectronic Engineering 2003;67–68:609–614.
88. Daniel J.H., D.F. Moore. A microaccelerometer structure fabricated in silicon-on-insulator using a focused ion beam process. Sensors and Actuators 1999;73:201–209.
89. DeMarco A.J., J. Melngailis. Maskless fabrication of JFETs via focused ion beams. Solid-State Electronics 2004;48:1833–1836.
90. Kranz C., G. Friedbacher, B. Mizaikoff, A. Lugstein, J. Smoliner, E. Bertagnolli. Integrating an ultramicroelectrode in an AFM cantilever. Combined technology for enhanced information. Analytical Chemistry, 2001;73:2491–2500.
91. Kueng A., C. Kranz, B. Mizaikoff, A. Lugstein, E. Bertagnolli. Combined scanning electrochemical atomic force microscopy for tapping mode imaging. Applied Physics Letter 2003;82:1592–1594.
92. Kranz C., A. Kueng, A. Lugstein, E. Bertagnolli, B. Mizaikoff. Mapping of enzyme activity by detection of enzymatic products during AFM imaging with integrated SECM-AFM probes. Ultramicroscopy 2004;100:127–134.
93. Menozzi C., G.C. Gazzadi, A. Alessandrini, P. Facci. Focused ion beam-nanomachined probes for improved electric force microscopy. Ultramicroscopy 2005;104:220–225.
94. Fujita J., M. Ishida, T. Sakamoto, Y. Ochiai, T. Kaito, S. Matsui. Observation and characteristics of mechanical vibration in threedimensional nanostructures and pillars grown by focused ion beam chemical vapor deposition. Journal of Vacuum Science and Technology B 2001;19:2834–2837.
95. Ishida M., J. Fujita, Y. Ochiai. Density estimation for amorphous carbon nanopillars grown by focused ion beam assisted chemical vapor deposition. Journal of Vacuum Science and Technology B 2002;20:2784–2787.
96. Morita T., R. Kometani, K. Watanabe, K. Kanda, T. Hoshino, K. Kondo, T. Kaito, T. Ichihashi, J. Fujita, M. Ishida, Y. Ochiai, T. Tajima, S. Matsui. Free-space-wiring fabrication in nano-space by focused-ion-beam chemical vapor deposition. Journal of Vacuum Science and Technology B 2003;21:2737–2741.
97. Igaki J., R. Kometani, K. Nakamatsu, K. Kanda, Y. Haruyama, Y. Ochiai, U. Fujita, T. Kaisto, S. Matsui. Three-dimensional rotor fabrication by focused-ion-beam chemical-vapor-deposition. Microelectronic Engineering 2006;83:1221–1224.
98. Puers R., S. Reyntjens. Fabrication and testing of custom vacuum encapsulations deposited by focused ion beam direct-write CVD. Sensors and Actuators A 2001;92:249–256.
99. Ishida M., J. Fujita, T. Ichihashi, Y. Ochiai. Focused ion beam-induced fabrication of tungsten structures. Journal of Vacuum Science and Technology B 2003;21:2728–2731.
100. Nakamatsu K., J. Igaki, M. Nagase, T. Ichihashi, S. Matsui. Mechanical characteristics of tungsten-containing carbon nanosprings grown by FIB-CVD. Microelectronic Engineering 2006;83:808–810.
101. Igaki J., K. Kanda, Y. Haruyama, M. Ishida, Y. Ochiai, U. Fujita, T. Kiato, S. Matsui. Comparison of FIB-CVD and EB-CVD growth characteristics. Microelectronic Engineering 2006;83:1225–1228.

102. Telari K.A., B.R. Rogers, H. Fang, L. Shen, R.A. Weller, D.N. Braski. Characterization of platinium films deposited by forcused ion beam-assisted chemical vapor deposition. Journal of Vacuum Science and Technology B 2002;20: 590–595.
103. Teng J., P.D. Prewett. Focused ion beam fabrication of thermally actuated bimorph cantilevers. Sensors and Actuators A 2005;123–124:608–613.
104. Lapicki A., K. Kang, T. Suzuki. Fabrication of magnetic dot arrays by ion beam induced chemical vapor deposition (IBICVD). IEEE Transactions on Magnetics 2002;38: 2589–2591.
105. Martin C.R. Membrane-based synthesis of nanomaterials. Chemistry of Materials 1996;9:1739–1746.
106. Bard A.J., L.R. Faulkner. Electrochemical Methods – Fundamentals and Applications. Wiley Inc.: New York, 1980.
107. Paunovic M., M. Schlesinger. Fundamentals of Electrochemical Deposition. Wiley Inc.: New York, 1998.
108. Schlesinger M., M. Paunovic. Modern Electroplating. Wiley Inc.: New York, 2000.
109. West A.C., C.-C. Cheng, B.C. Baker. Pulse reverse copper electrodeposition in high aspect ratio trenches and vias. Journal of Electrochemical Society 1998;145:3070–3074.
110. Hulteen J.C., C.R. Martin. A general template-based method for the preparation of nanomaterials. Journal of Materials Chemistry 1997;7:1075–1087.
111. Hornyak G.L., K.L.N. Phani, D.L. Kunkel, V.P. Menon, C.R. Martin. Fabrication, characterization and optical theory of aluminum nanometal/nanoporous membrane thin film composites. NanoStructured Materials 1995;6:839–842.
112. Dubois S.A.M., J.P. Eymery, J.L. Duvail, L. Piraux. Fabrication and properties of arrays of superconducting nanowires. Journal of Materials Research 1999;14:665–671.
113. Martin C.R., M. Nishizawa, K. Jirage, M. Kang, S.B. Lee. Controlling ion-transport selectivity in gold nanotube membranes. Advanced Materials 2001;13:1351–1362.
114. Possin G. A method for forming very small diameter wires. Review of Scientific Instruments 1970;41:772–774.
115. Al-Mawlawi D., C.Z. Liu, M. Moskovits. Nanowires formed in anodic oxide nanotemplates. Journal of Materials Research 1994;9:1014–1018.
116. Suryavanshi APaMFY. Probe-based electrochemical fabrication of freestanding Cu nanowire array. Applied Physics Letter 2006;88:831031–831033.
117. Wang L., K.Y. Zhang, A. Metrot, P. Bonhomme, M. Troyon. TEM study of electrodeposited Ni/Cu multilayers in the form of nanowires. Thin Solid Films 1996;288:86–89.
118. Schonenberger C., B.M.I. van der Zande, L.G.J. Fokkink, M. Henny, C. Schmid, M. Kruger, A. Bachtold, R. Huber, H. Birk, U. Staufer. Template synthesis of nanowires in porous polycarbonate membranes: electrochemistry and morphology. Journal of Physical Chemistry B 1997;101:5497–5505.
119. Blondel A., B. Doudin, J. Ph. Ansermet. Comparative study of the magnetoresistance of electrodeposited Co/Cu multilayered nanowires made by single and dual bath techniques. Journal of Magnetism and Magnetic Materials 1997;165:34–37.
120. Liu K., K. Nagodawithana, P.C. Searson, C.L. Chien. Perpendicular giant magnetoresistance of multilayered Co/Cu nanowires. Physical Review B 1995;51:7381–7385.
121. Voegeli B., A. Blondel, B. Doudin, J. Ph. Ansermet. Electron transport in multilayered Co/Cu nanowires. Journal of Magnetism and Magnetic Materials 1997;151:388–395.
122. Mock J.J., S.J. Oldenburg, D.R. Smith, D.A. Schultz, S. Schultz. Composite plasmon resonant nanowires. NanoLetters 2002;2:465–469.
123. Kroll M.W.J.B., D. Grandjean, R.E. Benfield, F. Luis, P.M. Paulus, L.J. de Jongh. Magnetic properties of ferromagnetic nanowires embedded in nanoporous alumina membranes. Journal of Magnetism and Magnetic Materials 2002;249:241–245.
124. Encinas A., M. Demand, J.-M. George, L. Piraux. Effect of pH on the microstructure and magnetic properties of electrodeposited cobalt nanowires. IEEE Transactions on Magnetics 2002;38:2574–2576.

125. Vazquez M., M. Hernandez-Velex, K. Pirota, A. Asenjo, D. Navas, J. Velazquez, P. Vargas, C. Ramos. Arrays of Ni nanowires in alumina membranes: magnetic properties and spatial ordering. European Journal of Physics B 2004;40:489–497.
126. Paulus P.M., F. Luis, M. Kroll, G. Schmid, L.J. de Jongh. Low-temperature study of the magnetization reversal and magnetic anisotropy of Fe, Ni and Co nanowires. Journal of Magnetism and Magnetic Materials 2001;224:180–196.
127. Seet H.L., X.P. Li, Z.J. Zhao, L.C. Wong, H.M. Zheng, K.S. Lee. Current density effect on magnetic properties of nanocrystalline electroplated Ni_80Fe_20/Cu composite wires. Journal of Magnetism and Magnetic Materials 2006;302:113–117.
128. Pirota K.R., M. Vazquez. Arrays of electroplated multilayered Co/Cu nanowires with controlled magnetic anisotropy. Advanced Engineering Materials 2005;7:1111–1113.
129. Kawaji J., F. Kitaizumi, H. Oikawa, D. Niwa, T. Homma, T. Osaka. Area selective formation of magnetic nanodot arrays on Si wafer by electroless deposition. Journal of Magnetism and Magnetic Materials 2005;287:245–249.
130. Yoo W.C., J.K. Lee. Field-dependent growth patterns of metals electroplated in nanoporous alumina membranes. Advanced Materials 2004;16:1097–1101.
131. Cheng Y.H., S.Y. Cheng. Nanostructures formed by Ag nanowires. Nanotechnology 2004;15:171–175.
132. Huang C.C., I.C. Leu, K.Z. Fung. Fabrication of δ-Bi_2O_3 nanowires. Electrochemical and Solid-State Letters 2005;8:A204–A206.
133. Tresback J.S., A.L. Vasiliev, N.P. Padture. Engineered metal-oxide-metal heterojunction nanowires. Journal of Materials Research 2005;20:2613–2617.
134. Casey J.D., A.F. Doyle, R.G. Lee, D.K. Stewart. Gas-assisted etching with focused ion beam technology. Microelectronic Engineering 1994;24:43–50.
135. Cohen-Hyams T, W.D. Kaplan, J. Yahalom. Structure of electrodeposited cobalt. Electrochemical and Solid-State Letters 2002;5:C75–C78.
136. Ming Y. Characterization of Evaporated Ultrathin Metal Films for Immunobiosensors. Electrical Engineering and Computer Science. University of Illinois at Chicago: Chicago, 1996: MS Thesis.
137. Hoivik N.D., J.W. Elam, R.J. Linderman, V.M. Bright, S.M. George, Y.C. Lee. Atomic layer deposited protective coatings for micro-electromechanical systems. Sensors and Actuators A 2003;103:100–108.
138. Elam J.W., M.D. Groner, S.M. George. Viscous flow reactor with quartz crystal microbalance for thin film growth by atomic layer deposition. Review of Scientific Instruments 2002;73:2981–2987.
139. Gibbons J.F. Ion implantation in semiconductors – Part 1: Range distribution theory and experiments. Proceedings of the IEEE 1968;56:295–319.
140. Nielsch K., R.B. Wehrspohn, J. Barthel, J. Kirschner, S.F. Fischer, H. Kronmuller, T. Schweinbock, D. Weiss, U. Gosele. High density hexagonal nickel nanowire array. Journal of Magnetism and Magnetic Materials 2002;249:234–240.

List of Symbols and Abbreviations

Symbol	Designation
δ**A**	Area of effusion cell
A	Electrode area
F	Faraday constant
Hc	Magnetic coersive field
$\Delta_{\mathrm{vap}} H_A^o$	Standard enthalpy of vaporization
I	Current
$\mathbf{J}_\Omega$	Flux from an ideal effusion cell
J	Flux of atoms per unit area
$\mathbf{j}_S$	Flux of atoms arriving a substrate
K	Boltzmann's constant
m	Atomic mass
Ms	Magnetic saturation field
n	Number of electrons in redox reaction
N'	Number of atoms deposited per unit area per second
$\mathbf{P}_{eq}$	Equilibrium vapor pressure
$\mathbf{P}^0$	Standard Pressure
R	Standard gas constant
r	Film growth rate
$\Delta_{\mathrm{vap}} S_A^o$	Standard entropy of vaporization
T	Temperature
θ	Polar angle
ϕ	Angle between evaporation source and surface normal at the substrate
η	Efficiency of electroplating process

Abbreviation	Explanation
AFM	Atomic Force Microscope
ALD	Atomic Layer Deposition
ALE	Atomic Layer Epitaxy
CE	Counter Electrode
CVD	Chemical Vapor Deposition
DRAM	Dynamic Random Access Memory
DRIE	Deep Reactive Ion Etching

EB	Electron Beam(s)
FCC	Face Center Cubic
FIB	Focused Ion Beam
FTIR	Fourier Transform Infrared
HCC	Hexagonal Close Packed
ICP	Inductively Coupled Plasma
JFET	Junction Field Effect Transistor
LCR	Left – Center – Right (mnemonic for Inductance, Capacitance & Resistance), thus an LCR meter is an Inductance, Capacitance, Resistance meter
MEMS	Micro Electro-Mechanical Systems
MFM	Magnetic Force Microscopy
MTS	mercapto trimethyl-ethyl-silazane
NEMS	Nano Electro-Mechanical Systems
PP	Pulsed Plating
PVD	Physical Vapor Deposition
QCM	Quartz Crystal Microbalance
RE	Reference Electrode
SCE	Standard Calomel Electrode
SEM	Scanning Electron Microscope
SECM	Scanning Electrochemical Microscopy
SOI	Silicon on Insulator
TEM	Transmission Electron Microscope
TMA	trimethly aluminum
TMCTS	tetramethylcyclotetrasilozane
VFM	Vibrating Flux Magnetometer
WE	Working Electrode
XRD	X-ray Diffraction

Chapter 5
Micro- and Nanomanufacturing via Molding

Harry D. Rowland and William P. King

Abstract Molding is a simple manufacturing process that enables fabrication of feature sizes ranging from 1 nm to 1 m at high volume and low cost. This chapter introduces micro- and nanomolding applications, processes, and design rules. Micro-and nanomolding processes have created high resolution lithographic patterns and fabricated functional applications in microfluidics, optics, and other areas. Analysis of polymer flow during local cavity filling and nonuniform long range polymer transport makes it possible to predict the physical driving mechanism governing flow and develop guidelines for optimized processing via micro- and nanomolding. Micro- and nanomolding processes offer a low cost, scalable alternative to silicon based microfabrication that capitalize on the high resolution, ease of processing, and wide range of mechanical, optical, or chemical properties of polymers. Successful high resolution, high yield micro- and nanomolding processes can enable widespread fabrication of nanotechnology-related products.

Keywords: Micromolding · Nanomolding · Hot embossing · Nanoimprint lithography · Polymer deformation · Single peak flow · Squeeze flow · Viscous flow · Capillary flow · Process design

5.1 Introduction

Molding is a simple manufacturing process whereby fluid fills a master tool and then solidifies in the shape of the tool cavity. Molding has long allowed fabrication of plastic components with feature sizes typically ranging from 1 mm to 1 m. Molding can be performed quickly and can be parallelized, which enables manufacturing at high volume and low cost. Fabrication of microelectromechanical systems (MEMS) and integrated circuit (IC) components having feature sizes ranging from 100 nm

---

W. P. King
Department of Mechanical Science and Engineering, University of Illinois Urbana-Champaign, Urbana, IL 61801, USA, +1 (217) 244–3864, +1 (217) 244–6534
e-mail: wpk@uiuc.edu

P. J. Hesketh (ed.), *BioNanoFluidic MEMS.*

to 100 μm has traditionally followed complex multi-step processing with stringent requirements on lithography for critical dimension control, pattern placement, and defect control. As desired feature sizes shrink to 10 nm, the demands on resists, masks, and processing equipment may render conventional micro- and nanofabrication methods cost prohibitive. Extending molding and related processes to fabrication of microstructures and nanostructures with feature sizes ranging from 1 nm to 100 μm could facilitate the widespread development and use of nanotechnology-related products.

The goal of this chapter is to serve as an introduction to micro- and nanomolding of polymers, highlighting current applications of molded components with feature sizes < 1 mm, reviewing polymer flow in features of size 100 nm – 1 mm and surrounding areas, and discussing design methodologies to guide intelligent processing via micro- and nanomolding.

5.2 Review of Molding Processes

Traditional plastic molding processes typically form soft polymers above the glass transition temperature T_g to create components of intricate three dimensional shape with little or no waste. The manner of material flow, heat transfer, and subsequent solidification distinguishes various molding processes, a few of which are detailed for micromolding applications in [1]. Compression molding or hot embossing, in this chapter referred to simply as molding, applies mechanical force to press heated soft polymer into the shape of a master tool or mold. Variations to compression molding reduce cycle times and material stresses that can develop during processing. Injection molding reduces high cycle times caused by mechanical motion and thermal history of the mold during compression molding, instead forcing hot polymer to flow into a fixed mold where the polymer quickly cools into the conformal shape of the master tool. Reaction injection molding reduces residual stresses in the molded component by molding a highly formable low molecular weight M_W polymer. A post-processing step cross-links or cures the polymer chemically, thermally, or via ultraviolet radiation to increase the mechanical strength and toughness of the final molded component.

For most practical applications, micro- and nanomolding processes are essentially surface forming operations based on macroscale compression molding or reaction compression molding, a combination of compression and reaction injection molding. Nanoimprint lithography (NIL) [2, 3] uses compression molding processes to form microstructures and nanostructures. Figure 5.1 shows the molding process during NIL whereby heat and force transfer the pattern of the mold to a thermoplastic polymer layer supported on a substrate. UV-NIL or step and flash imprint lithography (SFIL) [4] is a similar process to NIL but instead uses reaction compression molding to form microstructures and nanostructures. In UV-NIL, transparent molds press a liquid of low M_W polymer at room temperature and UV radiation cross-links the polymer in the shape of the master tool. UV molding can

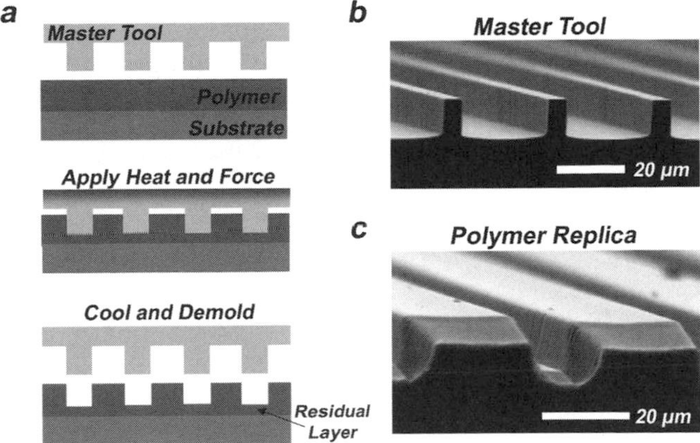

Fig. 5.1 (a) Illustration of micro-/nanomolding process in nanoimprint lithography. (b,c) Scanning electron micrographs of microfabricated silicon mold and corresponding polymer replica formed by molding. Reprinted from [58] with permission from IOP Publishing

allow for optimized, rapid fabrication of high resolution nanostructures but requires more advanced tooling and is more limited in materials selection than compression molding.

Material selection for master tools of micro- and nanomolding processes depends on the type of molding process, desired process conditions, and the material target. UV molding requires use of optically transparent molds composed of quartz or cured polymer such as PDMS. Molding of high M_W materials at high pressures requires structurally robust molds composed of silicon, quartz, or metal. All master tools regardless of material are initially fabricated in silicon or quartz by electron beam lithography or photolithography followed by chemical etching processes. The patterned silicon or quartz can then be used directly as master tools or as templates for cured polymer master tools fabricated by casting or imprinting [5–8] or metal molds fabricated by electroplating [9, 10]. Successive imprinting and casting or electroplating steps can create multiple polymer or metal master tool replicas from a single patterned silicon master template. Fabricated molds are then coated with an anti-adhesion release layer, designed for specific mold-polymer surface chemistry, to reduce stresses during demolding [11–14]. Demolding stresses can also be reduced by optimizing demolding temperatures [15, 16]. Molds fabricated in this manner have manufactured microstructures and nanostructures with a variety of applications.

5.3 Applications of Micro- and Nanomolding

Micro- and nanomolding processes offer a low cost, scalable alternative to silicon based microfabrication that capitalize on the formability of polymers to fabricate features for direct function or lithographic patterning. Direct forming of polymers through molding can reduce manufacturing steps and eliminate the need for expensive optical enhancement tooling required by typical silicon-specific microfabrication of feature sizes < 1 μm. The available library of deformable polymers with a wide range of material properties enables fabrication of components with tailored mechanical, optical, or chemical properties. A recent review [17] introduces a few functional and patterning applications of micro- and nanomolding.

5.3.1 Functional Micro- and Nanomolded Applications

Polymer components < 1 mm in size have been molded for direct functional use in a variety of nanoelectronics and nanoelectromechanical systems (NEMS)/MEMS. Figure 5.2 shows a few applications of micro- and nanomolded polymer in microfluidics [10, 18, 19], biomedical engineering [20], optics [21], and high density data storage [22]. Microfluidic lab-on-a-chip applications use micromolding to capitalize on the ease of processing, mechanical toughness, and the range of chemical properties of polymers [10, 19, 23, 24]. Tissue engineering applications use molded microstructures and nanostructures to study alignment, growth, and differentiation of a wide range of cell types [20, 25–27]. Polymers with specific index of

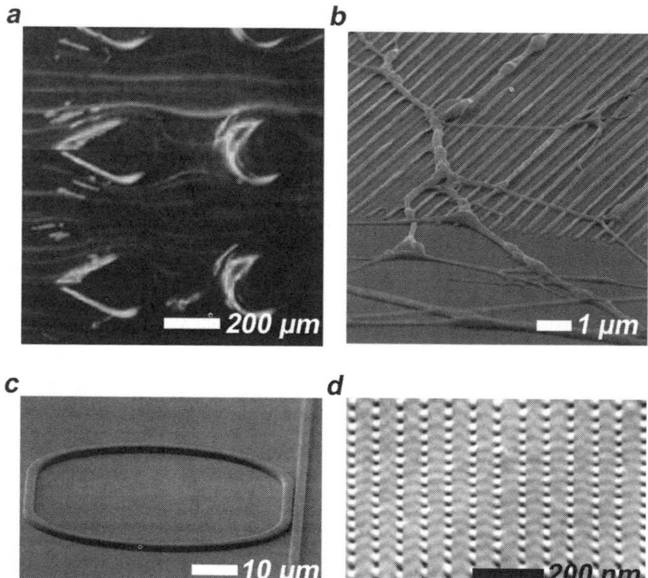

Fig. 5.2 Example applications formed by micro- and nanomolding processes where the molded material directly provides function. (a) Micromolded fluidic components for lab on a chip applications [10, 18, 19]. Reprinted from [18] with permission from SPIE. (b) Nanomolded cell engineering scaffolds for study of nerve cell growth. Reprinted from [20] with permission from Elsevier © 2006. (c) Micromolded ring resonator for optical waveguides. Reused with permission from [21] © 2002, AVS The Science & Technology Society. (d) Nanomolded indentations from parallel arrays of atomic force microscope probes for high density data storage. Reprinted with permission from [22] © 2002 IEEE

refraction are readily micromolded to produce micro-ring optical resonators [21], microlenses [28], and optical waveguides integrated into microfluidic devices [29]. Micromolding also shapes structural wells for conductive inks of electronic paper displays [30]. Nanomolding applications have also been developed. Scanning probe data storage based on AFM nanoindentation with 10,000 cantilevers operating in parallel has demonstrated data density as high as 1 Tbit/in^2 [22].

5.3.2 Lithographic Patterning via Micro- and Nanomolding

An even greater number of applications use micro- and nanomolding processes for lithographic patterning. After plasma removal of the residual layer shown in Fig. 5.1, the molded polymer is used as an etch mask to selectively pattern the features of the master tool into a substrate, typically silicon. Silicon field effect transistors (FETs) [31], flexible silicon nanowire FETs on plastic substrates [32], organic thin film transistors [33], surface acoustic wave filters [34], interdigitated cantilever arrays [35], Fresnel zone plates [36], and a variety of optical applications [37] have used micro- and nanomolding for lithography.

Perhaps the most promising applications of micro- and nanomolding are in high resolution NIL or UV-NIL for patterned media data storage and next generation lithography for IC fabrication with feature sizes < 20 nm. Current extensions of optical projection lithography at 193 nm wavelength include immersion lithography where high index of refraction lenses and fluids increase pattern resolution or extreme ultraviolet lithography where use of 13.5 nm wavelength increases pattern resolution. Both techniques introduce significant new technical challenges of contamination and material development for resists, masks, and lenses, while still requiring tooling and process costs to be less expensive than alternative technologies. With proper mold fabrication, alignment and overlay capabilities, and defect control during molding and demolding, NIL or UV-NIL could pattern high resolution features at high throughput without expensive optical resolution enhancement tooling. NIL or UV-NIL offers scalable manufacture of microstructures and nanostructures with resolution better than 10 nm [38] over areas > 100 mm^2 [39]. Recent studies have shown use of Moire fringes to nanoposition molds during processing to achieve alignment < 20 nm between two-layer molding processes [40].

Figure 5.3 shows the promise of nanomolding for high resolution patterning for data storage and IC fabrication. Figure 5.3a shows UV-NIL patterned cross-bar platinum nanowire arrays for electronically addressable memory that could enable data density > 500 Gbit/in^2 [41]. Nanomolding has also patterned magnetic media that could increase magnetic recording data density toward and above 100 Gbit/in^2 [42,43]. Figure 5.3b shows nanomolded lines of width 7 nm and spacing 7 nm for patterning metal gates in IC fabrication [44]. Nanomolding allows true molecular scale replication, as Fig. 5.3c,d show successful replication of single walled carbon nanotubes by UV-NIL [45,46] and replication of a single atom crack tip by UV casting processes similar to UV-NIL [47]. The high resolution and processing scalability

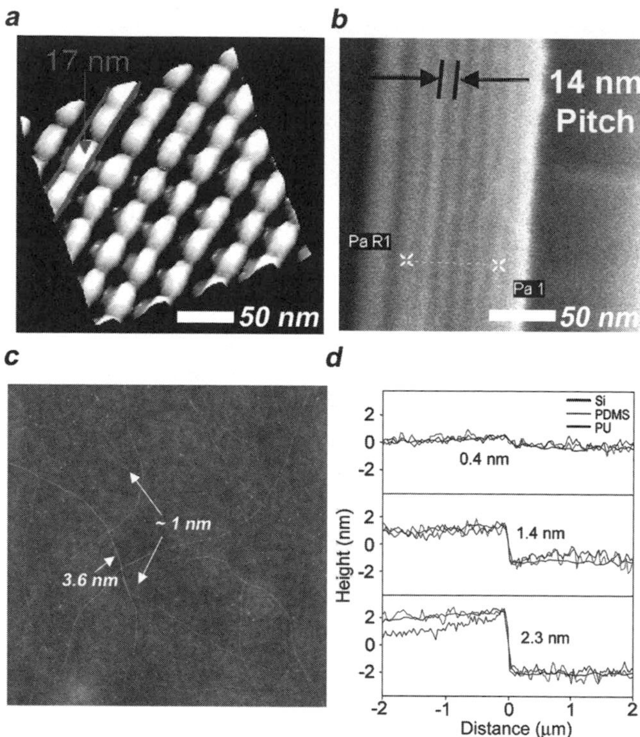

Fig. 5.3 Nanomolding processes for high resolution lithographic patterning. (a) UV-NIL produced electronically addressable memory at 17 nm half pitch. Reprinted with permission from [41]. (b) UV-NIL patterned lines of 7 nm width spaced 7 nm. Reused with permission from [44] © 2004, American Institute of Physics. (c) Molded replica of single walled carbon nanotube via UV-NIL. Reprinted with permission from [46] © 2006 IEEE. (d) UV-cast replica of single atom at the tip of a crack. Reprinted with permission from [47]

of nanomolding could enable widespread production of molecular electronics and other nanotechnology-related products.

5.4 Polymer Flow During Molding

Critical for widespread use of micro- and nanomolding applications is a fundamental understanding of polymer flow during processing to allow intelligent design of molds and molding processes. This problem is particularly acute for nanomolding polymer films at thickness below 100 nm, where polymer relaxation processes deviate from expected bulk response [48, 49]. The observed behavior of polymers at critical length scales < 100 nm suggests the existence of additional modes of polymer mobility [50] and enhanced segmental cooperativity [51] not observed in bulk materials. A deep understanding of molding at length scales < 100 nm requires development of new measurements of polymer nanorheological properties and

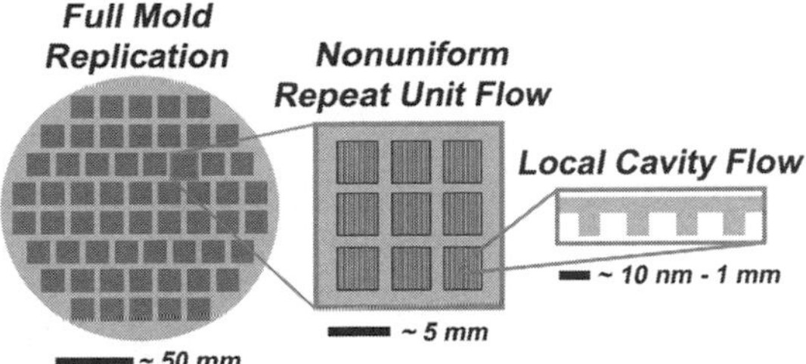

Fig. 5.4 For high yield of many applications, micro- and nanomolding must successfully replicate features over a range of length scales in local cavity flow, nonuniform repeat unit flow, and full mold replication

molecular-level understanding of polymers [52], an active area of research which is beyond the scope of this chapter. This review focuses on flow of polymer during micromolding at film thickness and length scales > 100 nm where polymer properties are not expected to deviate from bulk behavior. The principles presented in the following sections can be applied to nanomolding at length scales < 100 nm when proper, reliable polymer nanorheological properties can be defined.

Successful micromolding must allow for simultaneous polymer flow over a range of disparate length scales. Figure 5.4 shows the different flow geometries that must be considered for high yield of micromolding processes. Applications demand controlled molding for single features of size 1 nm–1 mm in local cavity flow as well as full replication of 100–300 mm diameter wafer-scale molds, or simultaneous replication of a range of 10^6 in length scales. For an extensive overview of massively parallel nanostructure formation, the reader is referred to a comprehensive review article [53]. This chapter focuses on understanding polymer flow during filling of single local cavities and then filling of adjacent areas of the mold within a nonuniform repeat unit. The insight provided by analysis of local and long range polymer flow can guide process design for high yield wafer scale replication.

5.4.1 Local Cavity Flow

Polymer flow during local cavity filling has been investigated in several single pitch architectures with feature sizes and film thickness ranging from 100 nm to 10 mm. Figure 5.5 shows measurements of single and dual peak polymer flow during local cavity filling of feature sizes from 100 nm to 500 μm. Experiments over the range of length scales measured similar deformation response despite differing length scales and differing process conditions governed by nonlinear elasticity [54], residual stress [55–57], linear and nonlinear viscous flow [58], compressive stress and elongational flow [59], and capillary force [60,61]. Simulations of rubber elastic [54,62] and viscous flow [63] also observed both single and dual peak polymer deformation.

The similar measured and simulated polymer flow despite competing physical effects arises from a geometric dependence of stress distribution governing polymer

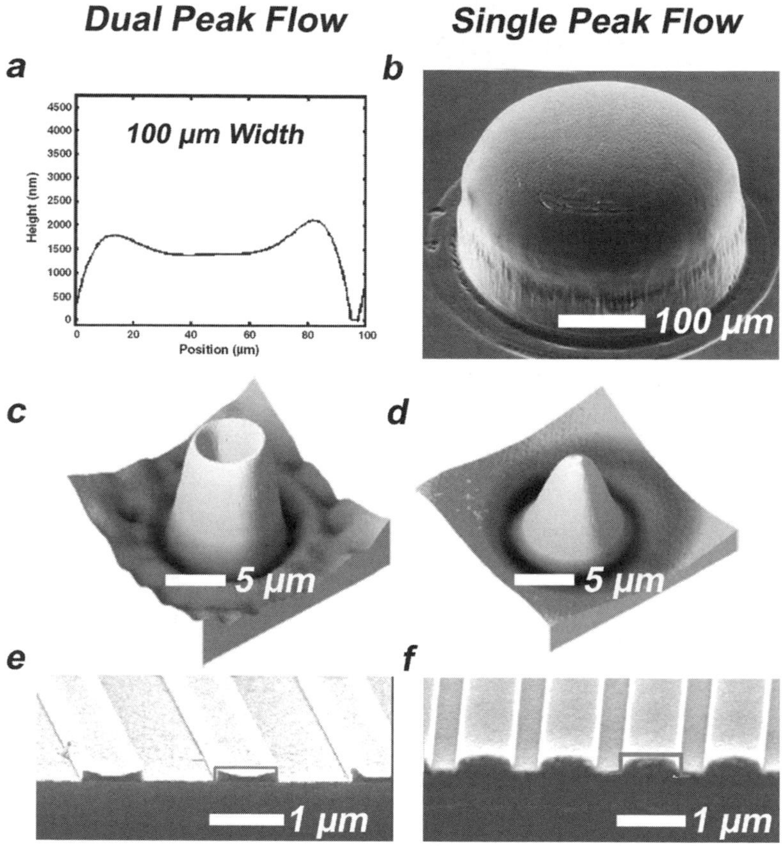

Fig. 5.5 Single and dual peak polymer deformation from 100 nm to 500 μm. (a,b) Cavity width > 100 μm [58, 80]. (a) is reprinted from [58] with permission from IOP Publishing. (b) is reprinted from [80] with permission from Elsevier © 2002. (c,d) Cavity width 10 μm. Reprinted with permission from [55] © 2003 IEEE. (e,f) Cavity width < 1 μm. Reused with permission from [54] © 2004, AVS The Science & Technology Society

flow during local cavity filling, *independent of length scale* [63]. During micromolding operations, the master tool applies a stress field to the polymer fluid where the shear stress gradients and resulting direction of flow depend on the specific geometry of the mold and film thickness. Figure 5.6 shows the labeled geometry of local cavity filling and shear stress contours resulting from three distinct flow modes defined by geometry and viscous flow theory. Simulations showed the difference between single and dual peak deformation is governed by the range of shearing and can be predicted by a dimensionless directional flow ratio w/h_o, where w is the local cavity half width and h_o is the initial film thickness [63]. Shear stress localizes near the indenting master tool, distributing radially into the cavity over a distance equivalent to the film thickness. Thus for mold cavities with $w/h_o < 1$, single peak deformation occurs while for mold cavities with $w/h_o > 1$, dual peak deformation occurs.

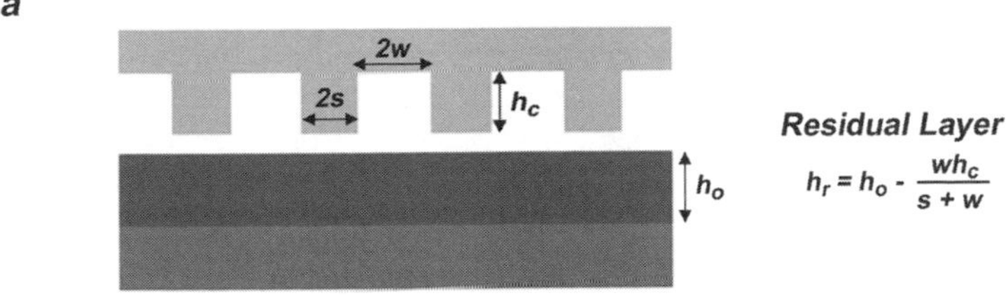

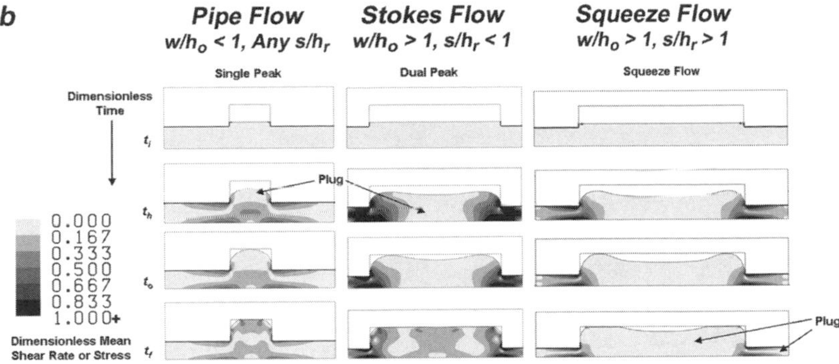

Fig. 5.6 Local cavity filling depends on geometry. (a) Cartoon labeling geometry of single cavity filling. (b) Distribution of shear stress, dependent on geometry, distinguishes different types of flow. Pipe flow occurs when $w/h_o < 1$. Stokes flow occurs when $w/h_o > 1$ and $s/h_r < 1$. Squeeze flow occurs when $w/h_o > 1$ and $s/h_r > 1$. Reprinted from [63] with permission from IOP Publishing

The directional flow ratio defines the location of the geometric constriction governing flow: the cavity width or space between indenters in single peak deformation, or the polymer film between the indenter and substrate in dual peak deformation. Another dimensionless geometric ratio, the polymer supply ratio s/h_r, can predict the distribution of shear stress between the indenter and substrate during dual peak deformation, where s is half the indenter width and h_r is the residual film thickness after full replication. The residual film thickness can be defined analytically based on mold geometry and conservation of volume [64]. As a mold is pressed further into a film to achieve full replication, the amount of polymer beneath the indenter decreases while the distance the polymer must flow laterally to fill the cavity increases. When $s/h_r < 1$, polymer deformation occurs over short distances and shear flow dominates deformation. When $s/h_r >> 1$, polymer deformation occurs over long distances and squeeze flow dominates deformation. The polymer supply ratio indicates the relative importance of shear or squeeze flow, defining a metric for the amount of polymer that must be displaced from beneath the indenter during molding.

The flow modes of local cavity filling defined by mold geometry allow for predictions of characteristic velocities and fill times to optimize molding processes based on viscous flow theory [63]. Single peak flow, occurring when $w/h_o < 1$ regardless the value of s/h_r, resembles the classic fluid mechanics flow situation of steady laminar flow over a given distance between infinite plates or within a pipe [65]. In micromolding applications with single peak flow, the distance of flow is simply the cavity height, h_c. Single peak flow can occur for various micromolding geometries such as closely spaced lines, cylindrical cavities, square cavities, and rectangular cavities. For a given effective fluid pressure P_{eff}, a characteristic velocity V for generic single peak filling assuming no slip boundary conditions can be defined as

$$V = P_{\mathrm{eff}} \times \frac{D_{\mathrm{h}}^2}{C\eta h_c} \tag{5.1}$$

where D_{h} is the hydraulic diameter dependent on geometry [65], C a constant dependent on geometry, and η the viscosity. P_{eff} can be estimated as the fluid pressure initially in contact with the mold (i.e. pressure applied to the mold times ratio of total mold area divided by mold area initially in contact with polymer) or conservatively as simply the pressure applied to the mold during molding. P_{eff} is also reduced due to head pressure loss [65] for micromolding applications when values of w/h_o near 1 [63]. For single peak flow in closely spaced lines as defined in Fig. 5.6a, where P_{eff} is the fluid pressure initially in contact with the mold, the characteristic velocity and corresponding fill time t can be estimated as

$$V = \left(P\frac{s+w}{s}\right) \times \frac{(2w)^2}{12\eta h_c} \tag{5.2}$$

$$t = \frac{12\eta h_c^2 s}{P(2w)^2(s+w)}. \tag{5.3}$$

Dual peak flow fill times in micromolding applications can be predicted by either Stokes or squeeze flow [63]. Most micromolding applications in dual peak flow are governed by squeeze flow due to the desire to reduce the residual layer thickness prior to demolding. Stokes flow is not commonly encountered in micromolding processes. Dual peak squeeze flow, occurring when $w/h_o > 1$ and $s/h_r > 1$, resembles a lubrication problem where a thin fluid is squeezed between two parallel plates [66, 67] from an initial thickness h_o to a final residual thickness h_r. The situation of cavity filling during molding differs from true squeeze flow as the contact area between mold and polymer increases laterally during processing [61, 63, 68]. A characteristic velocity for generic dual peak squeeze flow filling assuming no slip boundary conditions can be defined as

$$V = \frac{CP_{eff}}{\eta s_{eff}^2} \frac{h_o^2 h_r^2}{h_o + h_r} \tag{5.4}$$

where s_{eff} is an arbitrary feature width and P_{eff} the pressure, with both P_{eff} and s_{eff} dependent on effective mold-polymer contact area. To account for the slowing of imprint as squeeze flow filling progresses while still maintaining reasonable estimates of simulated fill times, analytical predictions of fill times can use a weighted average of contact area to estimate P_{eff} and s_{eff} [63]. A characteristic squeeze flow filling velocity and time for periodic lines as defined in Fig. 5.6a can be estimated as the time required to squeeze a fluid by infinite plates of width twice the tool width weighted by the ratio of indenter width to tool width:

$$V = \frac{2P}{\eta(s+w)^2} \frac{h_o{}^2 h_r{}^2}{h_o + h_r} \times \left(\frac{s+w}{s}\right) \tag{5.5}$$

$$t = \frac{\eta(s+w)^2}{2P} \left(\frac{1}{h_r{}^2} - \frac{1}{h_o{}^2}\right) \times \left(\frac{s}{s+w}\right). \tag{5.6}$$

The ability to estimate flow timescales based on mold geometry allows predictions prior to processing of the governing physical forces that will drive polymer flow. The Capillary number, $Ca = \eta V/\sigma$ where σ is the surface tension of the polymer fluid, provides a relative measure of viscous forces to capillary forces. When $Ca >> 1$, molding processes are driven by viscous forces while when $Ca << 1$, molding processes are driven by capillary forces. Figure 5.7 shows a flow driving mechanism regime map [63] and examples of dual peak flow profiles driven by viscous/viscoelastic [55] and capillary forces [60]. The flow driving mechanism regime map defines Ca by estimating the geometric-dependent characteristic velocity V and correctly predicts the flow profile and governing physics of all reported experiments and simulations of micromolding components of size 100 nm–10 mm [63].

Predicting polymer flow in the capillary regime is more complex than predicting flow in the viscous regime, as capillary forces can speed up or slow down the fill time during capillary flows due to surface tension and the chemical wetting interaction between mold and polymer [69, 70]. Surface chemistry is of fundamental importance in capillary flows. For a wetting fluid having a small contact angle with the mold, the polymer will wet the master surface and climb the indenter sidewalls in a dual peak mode, independent of w/h_o. This wetting behavior has been observed in many squeeze flow geometries with film thickness near 100 nm [60, 61, 71] and will generally result in faster fill times than analytical predictions based on viscous flow theory. However, some interfacial configurations will be inhibited due to surface tension forces, requiring higher pressures or longer fill times for full replication than analytical predictions based on viscous flow theory [70]. The deformation behavior for a non-wetting fluid is more complex than for a wetting fluid, and can depend on surface tension, contact angle, viscosity, pressure, film thickness, and cavity spacing. For a non-wetting fluid with large contact angle, polymer flow will be governed by the ratio of pressure and surface tension and will generally require longer fill times than analytical predictions based on viscous flow theory.

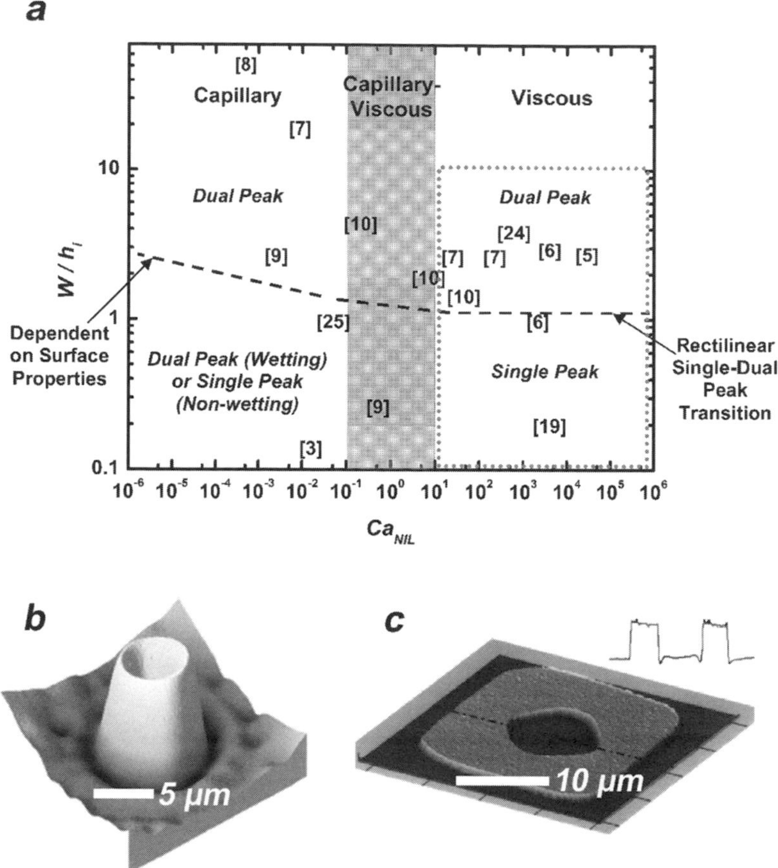

Fig. 5.7 (a) Predictions of viscous and capillary flow based on geometric-dependent *Ca*. Reprinted from [63] with permission from IOP Publishing. (b) Dual peak deformation with smooth profile from viscous flow. Reprinted with permission from [55] © 2003 IEEE. (c) Dual peak deformation with vertical sidewalls from capillary flow. Reprinted from [60] with permission from Elsevier © 2000

5.4.2 Nonuniform Long Range Polymer Transport

Practical micro- and nanomolding applications replicate multiple feature sizes and shapes in a single molding operation, resulting in correlation and crosstalk of stress and flow fields [72]. Many studies of large-scale flow field effects with nonuniform molding tools have noted difficulties in simultaneously replicating patterns with large and small features in close proximity, requiring much longer molding times for full replication compared to molding times of uniform patterns [61, 73, 74]. The nonuniform distribution of viscous polymer flow over large areas limits full replication and residual layer uniformity due to incomplete filling and tool warping. A few alternatives to compression micro- and nanomolding have been proposed to overcome the limitations of nonuniform molding. UV-NIL or SFIL, laser-assisted

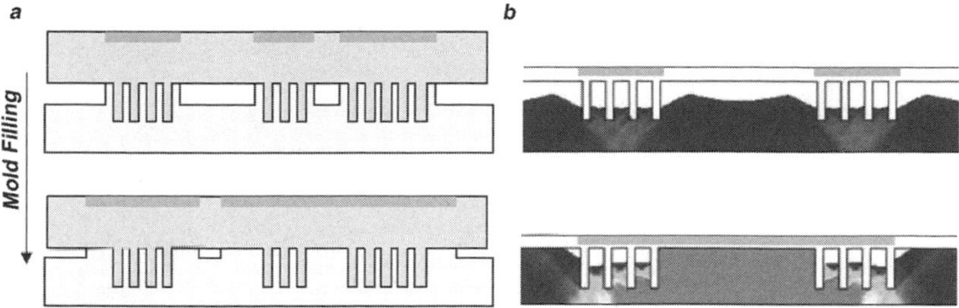

Fig. 5.8 Densely packed regions in nonuniform mold filling approximated as solid regions of a mold, indicated by the solid line above the mold features. (a) After completion of pipe flow, the filled region is modeled as a solid area [53,61]. Reprinted from [61] with permission from Elsevier © 2001, with adaptation from [53]. (b) Simulations show the slow filling pipe flow regions can be modeled as a solid area prior to completion of local pipe flow. Reprinted from [77] with permission from Elsevier © 2005

NIL, and one-step imprint-photolithography displace a smaller volume of material and/or imprint at lower viscosities than compression molding based on thermal cycling [4, 75, 76]. These techniques, however, require additional expensive tooling and limit material selection for micro- and nanomolded applications.

An understanding of local cavity filling enables improved understanding of nonuniform long range polymer transport. In single step molding of large and small features, regions of densely packed features will fill independently of surrounding areas with local cavity fill times predicted by viscous flow theory [72]. The order of filling is determined by local geometry and effective indenter widths, with global mold filling time governed by the maximum effective indenter width [72]. Figure 5.8a shows predictions of nonuniform mold filling where small width cavities quickly fill before full mold replication [53, 61]. The filled small cavities increase the area of mold-polymer contact, increasing the effective maximum indenter width governing the remaining fill time. The solid line above the mold features in Fig. 5.8 indicates the effective maximum indenter width. Areas of slow-filling constrained single peak flow can also be approximated as a filled area representing a maximum indenter width, since ample polymer supply is available to locally fill the cavities and excess polymer must be transported to other regions of the mold. Figure 5.8b shows simulations where small cavities fill after large cavities [77]. The simulations show an evenly distributed stress field beneath the small width cavities prior to filling, indicating the validity of a squeeze flow approximation for filling of the surrounding areas. Approximating densely packed small width cavity regions as a solid indenter characterizing nonuniform molding agrees well with reports that long range residual layer uniformity resulting from polymer flow between sections of different pattern density is independent of pattern size [74].

5.5 Design Rules for New Molding Processes

Previous micro- and nanomolding processes and process parameters were learnt empirically, relying on ad-hoc or recipe-based approaches due to a lack of comprehensive understanding of polymer flow at length scales from 100 nm to 1 mm.

With detailed knowledge of local cavity flow and long range polymer transport in nonuniform filling, it is possible to predict polymer flow for full mold replication prior to mold fabrication or experimental validation. This predictive power enables better design of molds and more appropriate selection of materials and molding process parameters with less investment of time and capital than required of recipe-based approaches to process optimization. Optimized mold layouts can improve uniformity of components by spatially balancing stress distribution in mold and polymer during molding. Optimized process parameters can improve throughput time and yield of processes. This section introduces practical design rules for molds and molding processes to guide high yield manufacturing of micro- and nanomolded components.

The design space of molding processes encompasses four main areas: mold geometry, material selection, film thickness, and process parameters. An optimization routine for a specific molding process analyzes local cavity flow and fill times and the direction of nonuniform long range polymer transport for a given mold geometry, material, film thickness, and embossing pressure and temperature. As processing permits, the mold can then be redesigned, a new material or film thickness can be selected, and/or the molding pressure and temperature can be modified to optimize the physical driving mechanism of flow, mold fill time, and distribution of stress in polymer and mold. In general, mold fill times are optimized by molding with low M_W polymers, large film thickness, and high molding pressure and temperature. However, depending on the molding application, one or more areas of the molding design space may be fixed. For example, lithographic molding applications may minimize the residual film thickness, the molded application may require use of a specific high M_W polymer, cured polymer molds or tooling may limit the maximum allowable molding pressure, or sensitive components beneath the molded layer may limit the molding temperature.

A case study best illustrates the process of designing molding processes. Figure 5.9 shows a 2 cm × 2 cm repeat unit of a wafer-scale mold for studying cell growth and alignment on microfeatures. The mold is composed of eight 500 μm × 500 μm regions each filled with 4 μm tall lines of width and spacing indicated in the figure. The mold creates 4 μm deep trenches in 15 μm thick low M_W polymer after molding with 20 MPa load at temperature $T_g + 10°C$.

The first step in optimizing molding processes is to identify and analyze areas of different local cavity filling. The mold shown in Fig. 5.9 has eight distinct areas of local cavity flow. For each area of local cavity flow, the geometry of a unit cell is identified and a local residual layer is calculated following conservation of volume. Next, the local cavity flow mode is identified for each area according to the directional flow ratio w/h_o and the polymer supply ratio s/h_r [63]. Independent fill times and characteristic velocities according to pipe, Stokes, or squeeze flow are then predicted for each area according to local geometry and appropriate polymer viscosity. The viscosity of low M_W polymers is easily determined by modeling the polymer as a Newtonian fluid with Williams-Landel-Ferry (WLF) temperature dependence of viscosity [78]. The effective viscosity of high M_W polymers during molding operations is often less than predictions of the zero shear Newtonian fluid

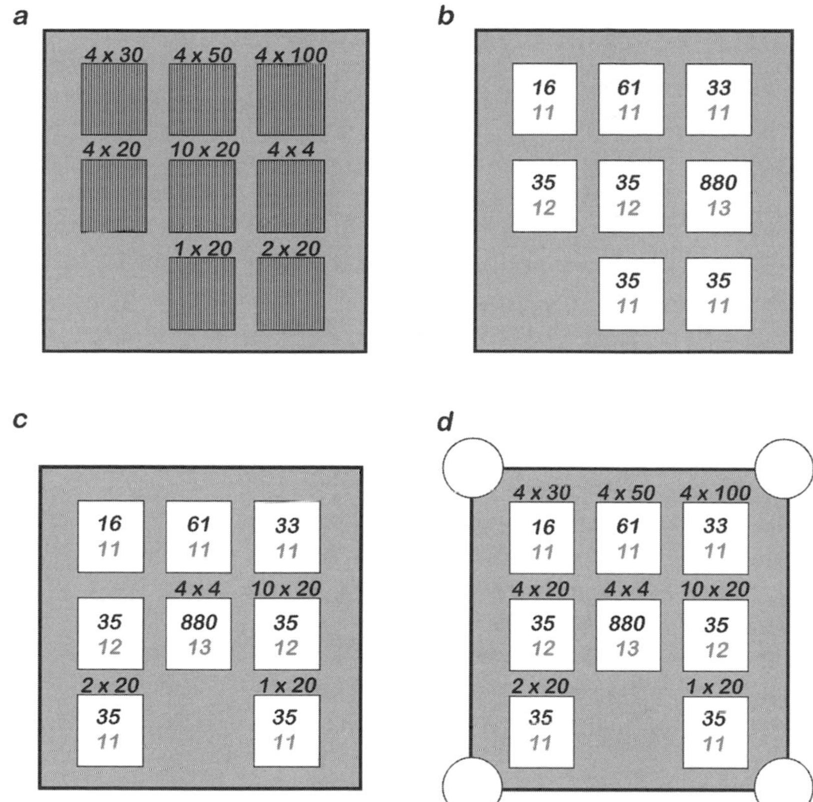

Fig. 5.9 Case study optimizing mold layout of square repeat unit 2 cm × 2 cm for molding of 15 μm film with Newtonian viscosity 10^9 Pa-sec under 20 MPa applied load. (a) Original mold design for studying cell alignment and the effects of feature density. Each 500 μm × 500 μm field is composed of periodic lines with height 4 μm and feature width and spacing indicated. The upper left field corresponds to indenter width $2s = 4$ μm and cavity width $2w = 30$ μm. (b) Local cavity analysis for the original mold layout following [63] with local fill time in seconds (top) and local residual film thickness in μm (bottom) indicated. (c) An improved mold layout to direct spatially balanced long range polymer transport. (d) Optimal mold design integrated with solid support structures at the corners of the repeat unit. The support structures prevent mold bending between repeat units of a wafer-scale mold. The 400 μm diameter support structures are designed for squeeze flow filling of the repeat unit to match the pipe flow fill time of the central region of the mold

viscosity due to shear thinning [68, 79]. A simple way to account for shear thinning is to use an effective Newtonian viscosity at an appropriate shear rate to estimate the shear-rate dependent viscosity of shear thinning power law fluids. Molding measurements from T_g + 10 °C to T_g + 70 °C suggest effective shear rates ranging from 10^0 s^{-1} at low temperatures to 10^2s^{-1}at high temperatures [68]. Figure 5.9b lists the predicted fill time and local residual layer for each area of the mold using a Newtonian viscosity for the low M_W polymer of 10^9 Pa-sec.

Investigations of nonuniform long range polymer transport directly build upon the analysis of local cavity filling. Figure 5.9b shows the initial mold layout with the patterned area of 4 μm × 4 μm feature width and spacing requiring dramatically longer times to fill and having much thicker local residual layer than surrounding areas. This patterned area of high density features with large local residual layer will require large amounts of polymer to be transported to surrounding areas. The high density features can be modeled as a solid indenter in squeeze flow to accurately predict polymer flow to surrounding areas. The unsymmetric mold layout shown in Fig. 5.9b results in unbalanced stresses in mold and polymer that can cause long range mold bending and incomplete replication [43]. Redesign of the mold layout can reduce mold bending by balancing long range polymer flow. Figure 5.9c shows an improved mold layout, where polymer flow is directed from the central region of the mold to surrounding areas. Support posts added on the periphery of the repeat unit will also reduce mold bending and improve uniformity of long range polymer flow [41]. Figure 5.9d shows a fully optimized mold layout with solid support posts of 400 μm diameter at the four corners of the repeat unit. The support post diameter is designed to experience local squeeze flow on the timescale to match pipe flow filling of the central region of 4 μm × 4 μm patterns.

The last analytical step with the original mold features and process parameters is determining the physical driving mechanism governing flow and optimizing fill time. The area of the mold with slowest local cavity filling will dominate the filling of the mold as a whole, where slowest local filling will generally occur for the smallest cavity width in pipe flow or the largest indenter width in squeeze flow. Calculation of the capillary number $Ca = \eta V/\sigma$ based on the smallest characteristic velocity determines the governing physics of flow, driven by either capillary or viscous/viscoelastic forces. The micromolding process of Fig. 5.9d is dominated by viscous flow with $Ca >> 1$. For capillary driven flows of $Ca << 1$, special care must be taken to account for surface chemistry and interaction between mold and polymer. Wetting fluids in capillary flow can result in enhanced mold-polymer adhesion while non-wetting fluids can require longer fill times or higher pressures for full replication. If the mold setup allows, capillary flows can be avoided by increasing the molding pressure and/or the initial film thickness, resulting in faster viscous flow processes dominating deformation. Increasing the molding temperature will also result in faster fill times, though cycle times increase for heating and cooling. Any material or film thickness changes require reiteration of the optimization procedure to properly predict flow timescales in the local geometries.

For the viscous pipe flow limited filling of the case study depicted in Fig. 5.9, changes in film thickness will not reduce fill times since pipe flow filling is independent of film thickness. Increasing the molding temperature 10–20 °C would allow for controlled, optimized molding of the microfeatures. Molding at T_g + 30 °C of a master tool as designed in Fig. 5.9d would allow for full mold replication in less than 60 sec. The process design routine analyzing the fill times of local cavity filling, the direction of long range nonuniform polymer flow, and the physical driving mechanism of flow allows for optimized micro- and nanomolding processes.

5.6 Summary

This chapter introduces micro- and nanomanufacturing of polymers via molding, where the manner of material flow, heat transfer, and subsequent solidification distinguishes various types of molding processes. Most micro- and nanomolded applications are manufactured by compression molding with thermal cycling (NIL) or ultraviolet post-processing (UV-NIL). Both NIL and UV-NIL processes have fabricated a variety of applications with feature sizes from 1 nm to 1 mm. Micro- and nanomolding have fabricated functional applications in microfluidics, optics, and other areas. Micro- and nanomolding have also been used for high resolution lithographic patterning for applications in high density data storage and next generation ICs.

High yield replication during micro- and nanomolding requires simultaneous polymer flow over a range of disparate length scales, from single features of size 1 nm–1 mm in local cavity flow to full replication of 100–300 mm diameter wafer-scale molds. By analyzing polymer flow during local cavity filling and nonuniform long range polymer flow, it is possible to predict the physical driving mechanism governing flow and develop guidelines for optimized processing via micro- and nanomolding.

Micro- and nanomolding processes offer a low cost, scalable alternative to silicon based microfabrication that capitalize on the high resolution, ease of processing, and wide range of mechanical, optical, or chemical properties of polymers. With efforts to reduce defects and improve throughput via roll-to-roll manufacturing, micro- and nanomolding can potentially imitate the success of traditional plastic molding that enabled widespread fabrication of plastic components for common, everyday applications. Successful high resolution, high yield micro- and nanomolding processes can enable widespread fabrication of nanotechnology-related products.

References

1. Heckele, M. and W.K. Schomburg, *Review on micro molding of thermoplastic polymers.* Journal of Micromechanics and Microengineering, 2004. **14**: pp. R1–R14.
2. Chou, S.Y., P.R. Krauss, and P.J. Renstrom, *Imprint of sub-25 nm vias and trenches in polymers.* Applied Physics Letters, 1995. **67**(21): pp. 3114–3116.
3. Chou, S.Y., P.R. Krauss, and P.J. Renstrom, *Imprint lithography with 25-nanometer resolution.* Science, 1996. **272**(5258): pp. 85–87.
4. Colburn, M., et al., *Step and flash imprint lithography: a new approach to high-resolution patterning.* in *SPIE Emerging Lithographic Technologies III.* 1999: SPIE.
5. Haverkorn von Rijsewijk, H., P. Legierse, and G. Thomas, *Manufacture of laser vision video disks by a photopolymerization process.* Philips Technical Review, 1982. **40**: pp. 287–297.
6. Xia, Y. and G.M. Whitesides, *Soft lithography.* Annual Review of Materials Science, 1998. **28**: pp. 153–184.
7. Xing, R., Z. Wang, and Y. Han, *Embossing of polymers using a thermosetting polymer mold made by soft lithography.* Journal of Vacuum Science and Technology B, 2003. **21**(4): pp. 1318–1322.

8. Schulz, H., et al., *Master replication into thermosetting polymers for nanoimprinting.* Journal of Vacuum Science and Technology B, 2000. **18**(6): pp. 3582–3585.
9. Hirai, Y., et al., *Imprint lithography for curved cross-sectional structure using replicated Ni mold.* Journal of Vacuum Science and Technology B, 2002. **20**(6): pp. 2867–2871.
10. Mela, P., et al., *The zeta potential of cyclo-olefin polymer microchannels and its effects on insulative (electrodeless) dielectrophoresis particle trapping devices.* Electrophoresis, 2005. **26**: pp. 1792–1799.
11. Jaszewski, R.W., et al., *The deposition of anti-adhesive ultra-thin teflon-like films and their interaction with polymers during hot embossing.* 1999. **143**(1–4): pp. 301–308.
12. Schift, H., et al., *Controlled co-evaporation of silanes for nanoimprint stamps.* 2005. **16**(5): pp. S171–S175.
13. Jung, G.Y., et al., *Vapor-phase self-assembled monolayer for improved mold release in nanoimprint lithography.* 2005. **21**(4): pp. 1158–1161.
14. Houle, F.A., et al., *Template-resist surface adhesion studies in uv-nanoimprint lithography.* In *Nanoimprint and Nanoprint Technology*. 2006. San Francisco, CA.
15. Hirai, Y., S. Yoshida, and N. Takagi, *Defect analysis in thermal nanoimprint lithography.* Journal of Vacuum Science and Technology B, 2003. **21**(6): pp. 2765–2770.
16. Mendels, D.A. *The build-up and relaxation of internal stresses during cool-down in a single nano-imprint lithography cell.* In *Nanoimprint and Nanoprint Technology*. 2006. San Francisco, CA.
17. Guo, L.J., *Recent progress in nanoimprint technology and its applications.* Journal of Physics D: Applied Physics, 2004. **37**: pp. R123–R141.
18. McGraw, G.J., et al., *Polymeric microfluidic devices for the monitoring and separation of water-borne pathogens utilizing insulative dielectrophoresis.* Proceedings of SPIE, 2005. **5715**: pp. 59–68.
19. Simmons, B.A., et al., *The development of polymeric devices as dielectrophoretic separators and concentrators.* MRS Bulletin, 2006. **31**: pp. 120–124.
20. Johansson, F., et al., *Axonal outgrowth on nano-imprinted patterns.* Biomaterials, 2006. **27**: pp. 1251–1258.
21. Guo, L. and C. Chao, *Polymer microring resonators fabricated by nanoimprint technique.* Journal of Vacuum Science and Technology B, 2002. **20**: pp. 2862–2866.
22. Vettiger, P., et al., *The "Millipede"-nanotechnology entering data storage.* IEEE Transactions on Nanotechnology, 2002. **1**(1): pp. 39–55.
23. Huang, L.R., et al., *Continuous particle separation through deterministic lateral displacement.* Science, 2004. **304**: pp. 987–990.
24. Lee, G., et al., *Microfabricated plastic chips by hot embossing methods and their applications for DNA separation and detection.* Sensors and Actuators B, 2001. **75**: pp. 142–148.
25. Charest, J., et al., *Hot embossing for micro patterned cell substrates.* Biomaterials, 2004. **25**: pp. 4767–4775.
26. Charest, J.L., et al., *Polymer cell culture substrates with combined nanotopographical patterns and micropatterned chemical domains.* Journal of Vacuum Science and Technology B, 2005. **23**(6): pp. 3011–3014.
27. Gadegaard, N., et al., *Applications of nano-patterning to tissue engineering.* Microelectronic Engineering, 2006. **83**: pp. 1577–1581.
28. Sun, Y. and S.R. Forrest, *Organic light emitting devices with enhanced outcoupling via microlenses fabricated by imprint lithography.* Journal of Applied Physics, 2006. **100**: pp. 073106.
29. Nilsson, D., S. Balslev, and A. Kristensen, *A microfluidic dye laser fabricated by nanoimprint lithography in a highly transparent and chemically resistant cyclo-olefin copolymer (COC).* Journal of Micromechanics and Microengineering, 2005. **15**: pp. 296–300.
30. Liang, R.C., et al., *Microcup displays: Electronic paper by roll-to-roll manufacturing processes.* Journal of the Society for Information Display, 2003. **11**(4): pp. 621–628.
31. Guo, L., P. Krauss, and S. Chou, *Nanoscale silicon field effect transistors fabricated using imprint lithography.* Applied Physics Letters, 1997. **71**: pp. 1881–1883.

32. McAlpine, M.C., R.S. Friedman, and C.M. Lieber, *Nanoimprint lithography for hybrid plastic electronics.* Nano Letters, 2003. **3**(4): pp. 443–445.
33. Austin, M.D. and S.Y. Chou, *Fabrication of 70 nm channel length polymer organic thin-film transistors using nanoimprint lithography.* Applied Physics Letters, 2002. **81**(23): pp. 4431–4433.
34. Cardinale, G.F., et al., *Fabrication of a surface acoustic wave-based correlator using step-and-flash imprint lithography.* Journal of Vacuum Science and Technology B, 2004. **22**(6): pp. 3265–3270.
35. Luo, G., et al., *Nanoimprint lithography for the fabrication of interdigitated cantilever arrays.* Nanotechnology, 2006. **17**: pp. 1906–1910.
36. Li, M., et al., *Fabrication of circular optical structures with a 20 nm minimum feature size using nanoimprint lithography.* Applied Physics Letters, 2000. **76**(6): pp. 673–675.
37. Wang, J.J., et al., *Free-space nano-optical devices and integration: Design, fabrication, and manufacturing.* Bell Labs Technical Journal, 2005. **10**(3): pp. 107–127.
38. Chou, S. and P. Krauss, *Imprint lithography with sub-10 nm feature size and high throughput.* Microelectronic Engineering, 1997. **35**: pp. 237–240.
39. Khang, D. and H. Lee, *Wafer-scale sub-micron lithography.* Applied Physics Letters, 1999. **75**: pp. 2599–2601.
40. Li, N., W. Wu, and S.Y. Chou, *Sub-20-nm alignment in nanoimprint lithography using moire fringe.* Nano Letters, 2006. **6**(11): pp. 2626–2629.
41. Jung, G.-Y., et al., *Circuit fabrication at 17 nm half-pitch by nanoimprint lithography.* Nano Letters, 2006. **6**(3): pp. 351–354.
42. McClelland, G.M., et al., *Nanoscale patterning of magnetic islands by imprint lithography using a flexible mold.* Applied Physics Letters, 2002. **81**(8): pp. 1483–1485.
43. McClelland, G.M., et al., *Contact mechanics of a flexible imprinter for photocured nanoimprint lithography.* Tribology Letters, 2005. **19**(1): pp. 59–63.
44. Austin, M.D., et al., *Fabrication of 5 nm linewidth and 14 nm pitch features by nanoimprint lithography.* Applied Physics Letters, 2004. **84**(26): pp. 5299–5301.
45. Hua, F., et al., *Polymer imprint lithography with molecular-scale resolution.* Nano Letters, 2004. **4**(12): pp. 2467–2471.
46. Hua, F., et al., *Processing dependent behavior of soft imprint lithography on the 1-10-nm scale.* IEEE Transactions on Nanotechnology, 2006. **5**(3): pp. 301–308.
47. Xu, Q., et al., *Approaching zero: Using fractured crystals in metrology for replica molding.* Journal of American Chemical Society, 2005. **127**: pp. 854–855.
48. Keddie, J.L., R.A.L. Jones, and R.A. Cory, *Size-dependent depression of the glass-transition temperature in polymer-films.* Europhysics letters, 1994. **27**(1): pp. 59–64.
49. Dalnoki-Veress, K., et al., *Molecular weight dependence of reductions in the glass transition temperature of thin, freely standing polymer films.* Physical Review E, 2001. **63**: pp. 031801.
50. De Gennes, P.G., *Glass transitions in thin polymer films.* The European Physical Journal E, 2000. **2**: pp. 201–205.
51. Ngai, K.L., *Mobility in thin polymer films ranging from local segmental motion, Rouse modes to whole chain motion: A coupling consideration.* The European Physical Journal E, 2002. **8**: pp. 225–235.
52. Alcoutlabi, M. and G.B. McKenna, *Effects of confinement on material behaviour at the nanometre size scale.* Journal of Physics: Condensed Matter, 2005. **17**: pp. R461–R524.
53. Cross, G.L.W., *The production of nanostructures by mechanical forming.* Journal of Physics D: Applied Physics, 2006. **39**: pp. R363–R386.
54. Hirai, Y., et al., *Simulation and experimental study of polymer deformation in nanoimprint lithography.* Journal of Vacuum Science and Technology B, 2004. **22**(6): pp. 3288–3293.
55. Cross, G.L.W., et al., *Mechanical aspects of nanoimprint patterning.* IEEE-Nano (San Francisco), 2003. **2**: pp. 494–497.
56. Cross, G.L.W., et al., *The mechanics of nanoimprint forming.* Materials Research Society Symposium Proceedings, 2004. **841**: pp. R1.6.1 to R1.6.12.

57. Cross, G.L.W., B.S. O'Connell, and J.B. Pethica, *Influence of elastic strains on the mask ratio in glassy polymer nanoimprint.* Applied Physics Letters, 2005. **86**(8): pp. 081902-1 to 081902-3.
58. Rowland, H.D. and W.P. King, *Polymer deformation and filling modes during microembossing.* Journal of Micromechanics and Microengineering, 2004. **14**: pp. 1625–1632.
59. Juang, Y.-J., L.J. Lee, and K.W. Koelling, *Hot embossing in microfabrication. Part I: Experimental.* Polymer Engineering and Science, 2002. **42**(3): pp. 539–550.
60. Heyderman, L.J., et al., *Flow behaviour of thin polymer films used for hot embossing lithography.* Microelectronic Engineering, 2000. **54**: pp. 229–245.
61. Scheer, H.C. and H. Schulz, *A contribution to the flow behaviour of thin polymer films during hot embossing lithography.* Microelectronic Engineering, 2001. **56**: pp. 311–332.
62. Hirai, Y., et al., *Study of the resist deformation in nanoimprint lithography.* Journal of Vacuum Science and Technology B, 2001. **19**(6): pp. 2811–2815.
63. Rowland, H.D., et al., *Impact of polymer film thickness and cavity size on polymer flow during embossing: toward process design rules for nanoimprint lithography.* Journal of Micromechanics and Microengineering, 2005. **15**: pp. 2414–2425.
64. Landis, S., et al., *Stamp design effect on 100 nm feature size for 8 inch nanoimprint lithography.* Nanotechnology, 2006. **17**: pp. 2701–2709.
65. Fox, R.W. and A.T. McDonald, *Introduction to fluid mechanics.* 5th ed. 1998, New York: John Wiley & Sons, Inc. 762.
66. Macosko, C.M., *Rheology: Principles, measurements, and applications.* 1994, New York: Wiley-VCH. 550.
67. Engmann, J., C. Servais, and A.S. Burbidge, *Squeeze flow theory and applications to rheometry: A review.* Journal of Non-Newtonian Fluid Mechanics, 2005. **132**: pp. 1–27.
68. Schulz, H., et al., *Impact of molecular weight of polymers and shear rate effects for nanoimprint lithography.* Microelectronic Engineering, 2006. **83**: pp. 259–280.
69. Colburn, M., et al., *Ramifications of lubrication theory on imprint lithography.* Microelectronic Engineering, 2004. **75**: pp. 321–329.
70. Reddy, S., P.R. Schunk, and R.T. Bonnecaze, *Dynamics of low capillary number interfaces moving through sharp features.* Physics of Fluids, 2005. **17**: pp. 122104.
71. Jeong, J.-H., et al., *Flow behavior at the embossing stage of nanoimprint lithography.* Fibers and Polymers, 2002. **3**(3): pp. 113–119.
72. Rowland, H.D., et al., *Simulations of nonuniform embossing: The effect of asymmetric neighbor cavities on polymer flow during nanoimprint lithography.* Journal of Vacuum Science and Technology B, 2005. **23**(6): pp. 2958–2962.
73. Gourgon, C., et al., *Influence of pattern density in nanoimprint lithography.* Journal of Vacuum Science and Technology B, 2003. **21**(1): pp. 98–105.
74. Schulz, H., M. Wissen, and H.C. Scheer, *Local mass transport and its effect on global pattern replication during hot embossing.* Microelectronic Engineering, 2003. **67–68**: pp. 657–663.
75. Cheng, X. and L.J. Guo, *One-step lithography for various size patterns with a hybrid mask-mold.* Microelectronic Engineering, 2004. **71**: pp. 288–293.
76. Xia, Q., et al., *Ultrafast patterning of nanostructures in polymers using laser assisted nanoimprint lithography.* Applied Physics Letters, 2003. **83**(21): pp. 4417–4419.
77. Young, W.-B., *Analysis of the nanoimprint lithography with a viscous model.* Microelectronic Engineering, 2005. **77**: pp. 405–411.
78. Williams, M.L., R.F. Landel, and J.D. Ferry, *Temperature Dependence of Relaxation Mechanisms.* Journal of American Chemical Society, 1955. **77**: pp. 3701–3707.
79. Schulz, H., et al., *Choice of the molecular weight of an imprint polymer for hot embossing lithography.* Microelectronic Engineering, 2005. **78–79**: pp. 625–632.
80. Shen, X.J., L.-W. Pan, and L. Lin, *Microplastic embossing process: experimental and theoretical characterizations.* Sensors and Actuators A, 2002. **97–98**: pp. 428–433.

Abbreviations

MEMS	microelectromechanical systems
IC	integrated circuit
T_g	glass transition temperature
M_W	molecular weight
NIL	nanoimprint lithography
UV-NIL	ultraviolet nanoimprint lithography
SFIL	step and flash nanoimprint lithography
PDMS	polydimethylsiloxane
NEMS	nanoelectromechanical systems
FETs	field effect transistors
w	cavity half width
H_o	initial film thickness
s	indenter half width
h_r	residual film thickness
h_c	cavity height
P_{eff}	effective fluid pressure
V	characteristic velocity
D_h	hydraulic diameter
C	arbitrary constant
η	polymer viscosity
t	fill time
s_{eff}	effective feature half width
Ca	Capillary number
σ	surface tension
WLF	Williams-Landel-Ferry

Chapter 6
Temperature Measurement of Microdevices using Thermoreflectance and Raman Thermometry

Thomas Beechem and Samuel Graham

Abstract Device temperature is often a primary factor in the proper operation, reliability, and lifetime of both MEMS and microelectronics. Thus, the measurement and verification of operational temperature is often an integral aspect the design and improvement of microdevices for commercial applications. Raman thermometry and thermoreflectance are two techniques commonly employed in the measurement of temperature at small length scales since they are noncontact in nature and their spatial and temporal resolution is on par with the needs of current device architectures. This work provides a summary in the physical basis, experimental methodology, and application of each of these techniques with respect to the analysis of microdevices.

Keywords: Thermoreflectance · Raman Thermometry · MEMS · Microelectronics · Thermal Management

6.1 Introduction

The continued development of micro- and nanoscale systems (MEMS/NEMS) over the last several decades has put new demands on critical metrology techniques necessary to measure temperature in these devices[1–3]. Accurate temperature measurement with spatial and temporal resolution capable of resolving critical phenomena in device architectures is necessary to both validate device performance, device reliability, and to implement changes in device design. While a plethora of methods exist to measure temperature with a spatial resolution greater than 10 μm, it has become more difficult to perform such tasks with a resolution which is less than 1 μm, a regime which is of importance to current day MEMS and NEMS.

S. Graham
The George W. Woodruff School of Mechanical Engineering, Georgia Institute of Technology, 771 Ferst Drive, Atlanta, GA 30332-0405
e-mail: sgraham@gatech.edu

P. J. Hesketh (ed.), *BioNanoFluidic MEMS.*

At its most fundamental level, the lattice temperature of a material is a measure of average atomic energy which is stored in vibrational motion called phonons. Direct monitoring of this motion is intractable and hence the overwhelming majority of temperature measurements instead are indirect in nature. These indirect methods monitor not the temperature itself, but rather, a separate temperature dependent phenomenon. The dependent phenomena, in turn, arise as the major energy carriers, photons, electrons, and phonons; themselves have a statistical population dependent upon the local temperature. Thus it is not surprising that oftentimes the nature of interaction of these energy carriers is temperature dependent as well. As a consequence, mensuration of temperature most often centers on the monitoring of the interaction between these energy carriers.

In semiconductor devices, these interactions are observed most often in one of three distinct manners, namely, through direct contact of a probe with the device, or in the measurement of the electrical, or optical response of the device during operation [4]. Regardless of the manner employed, the ideal temperature measurement has a spatial resolution capable of identifying even the smallest temperature gradients while simultaneously having a temporal resolution great enough to capture the entirety of transient behavior. In addition, the measurement itself should have little or no effect on device function while being easily incorporated to a diverse range of packaging orientations. As there is no technique capable of meeting each of these criterions, semiconductor characterization is then a balance of amenities and liabilities as is shown below.

Direct contact measurements range in complexity from use of a standard thermocouple to the incorporation of advanced AFM tips capable of sensing temperature with nanometer resolution [5]. With respect to mciroscale devices, contact methods which are used often consist of scanning thermal microscopy in which a small scanning probes capable of sensing temperature is rastered across a device surface. Such a technique is limited spatially by the tip size and potential interactions of water condensation at the contact point, and is ideal for mapping operations. However, its temporal resolution is constrained by the heat transport dynamics between the tip, surrounding medium, and surface. In addition, the technique is applicable only when the layer of interest is accessible which may not be the case for some packaging configurations. An excellent review of this type of metrology was given by Majumdar in 1999 [6].

While direct contact methods rely on heat transfer to occur until two surfaces are isothermal, electrical methods instead rely on the changes in the transport of electrons to probe the temperature of the material. Examples for which the electrical response of a system has been incorporated to measure temperature in microdevices includes the use of forward voltage at a p–n junction, overall resistance change of a metal resistor, and incorporation of a transducer mounted directly onto the device [4, 7, 8]. Due to the high transport speeds of the electrons themselves, high temporal resolution is easily obtainable; however, spatial resolution is limited as the acquired data are averages between electrical contacts. However, as only circuit components are needed to obtain a measurement, this genre of methods are well suited for applications where packaging makes the active surfaces of the device inaccessible.

Optical methods monitor the temperature dependence of photons either emitted (spontaneous or stimulated) or reflected from the region of interest. This thermal dependence arises as several parameters affecting this radiation, i.e. the emissivity, reflectivity, as well as the electronic and lattice band structure, are themselves temperature dependent. This has led to a host of different measurement techniques including pyrometry, interferometry, thermoreflectance, and Raman spectroscopy to name a scant portion of those incorporated. An excellent review of each of these techniques is provided by Zhang [9]. As the wave nature of the radiation serves as the probe in these methods, most far field applications have spatial resolutions limited only by the wavelength of the monitored photons ranging from 1 μm for visible light to nearly 10 μm for infrared imaging techniques [10, 11]. Theoretically, the temporal resolution of these techniques is limited only by the interaction time between the device and the photons, ∼1 fs, however, in practice the resolution is limited by the experimental equipment employed with the best reports being on the order of a picosecond [12, 13]. Additionally, these techniques also rely on an optically viable surface for measurement, a stipulation which is not always fulfilled in multilayered and packaged devices.

As stated previously, none of the above methodologies serve as the ideal for characterization of temperature in microdevices and as such it is necessary to weigh each of the techniques advantages before implementation. This review, however, is limited in scope to only two optical measurement techniques, namely thermoreflectance and Raman spectroscopy, methods which have garnered great interest as each has the ability to both map devices while simultaneously working with nanosecond resolution [14–17]. Subsequent sections will then focus on each of these methods beginning with a short explanation of the physics behind the underlying temperature dependent phenomenon. Calibration techniques and experimental procedures will then be discussed followed by a summary of the technique's implementation into the characterization of microdevices. Following these descriptions, a summary will be proffered in which future temperature measurement trends will be ruminated upon.

6.2 Temperature Measurement of Microdevices by Means of Thermoreflectance

6.2.1 Physical Basis of the Thermoreflectance Technique

Active layers of semiconductor devices are composed of materials in specific crystallographic arrangements. These arrangements, which by definition are periodic, give rise to specific band structures defining the allowable states of the lattice vibrations (phonons) and electrons. The manner and extent to which the states of these bands are filled is directly dependent upon the statistical distribution of the species involved as phonons and electrons follow Bose-Einstein and Fermi-Dirac distributions, respectively [18]. These distributions and the shape of the band structures themselves, in turn, are each functions of temperature [19]. Thus any entity interacting with these distributions will "feel" the effect of this thermally dependent distribution thus providing a manner to estimate temperature.

In most thermoreflectance measurements, the monitored device is bombarded with incident radiation normal to the surface in question. At impingement with the surface, this radiation will be reflected, transmitted, or absorbed into the material. The extent to which each of these possibilities occurs is dependent upon the manner in which the incident photons interact with the distributed electrons and phonons in the crystal lattice. The interaction between the photons and the interacting species is dependent upon several parameters as is shown in functional form for the reflectivity below:

$$\rho = f(\lambda, T, \theta, \psi) \tag{6.1}$$

where ρ is the spectral directional reflectivity, λ is the wavelength of the reflected light, T is the temperature, while θ and ψ are the zenith and azimuthal angles, respectively [9]. In practical measurement, this multivariate dependency is easily removed with the use of a monochromatic radiation source fixed at a constant angle with respect to the surface thus making the reflectivity a function of temperature alone, $\rho = f(T)$, and hence a useful probe.

Capitalizing on this singular dependency, changes in the reflectivity can be used to measure temperature by monitoring the intensity ratio between incident and reflected electromagnetic fields upon exposure of a surface to radiation. This ratio is termed, the reflectance, and is mathematically described according to the relation shown below:

$$R(T) = \frac{E^*_{\text{ref}} E_{\text{ref}}}{E^*_{\text{inc}} E_{\text{inc}}} = \rho^2(T) \tag{6.2}$$

where $R(T)$ is the reflectance , and E_{ref} and E_{inc} are measurements of the reflected and incident field with the * corresponding to their complex conjugates [20]. As seen from Equation (6.2), with a change in temperature both the reflectivity and the ratio of the intensities (reflectance) will be affected. Consequently, if the temperature change is moderate, an estimation of the variation in reflectance can be estimated using a first order Taylor expansion where T_0 is a known reference temperature [21]:

$$R(T) = R(T_0) + \frac{\partial R}{\partial T}(T - T_0) \tag{6.3}$$

Thus the temperature may be estimated based on the change in the reflectance as is shown below:

$$\Delta T = \frac{1}{\beta} \frac{\Delta R}{R(T_0)} \tag{6.4}$$

where $\Delta T = T - T_0$, $\Delta R = R(T) - R(T_0)$, and $\beta = (\partial R/\partial T)[1/R(T_0)]$ is termed the thermoreflectance coefficient [9]. Assuming that the thermoreflectance coefficient has been found through proper calibration (a topic which is discussed in the subsequent subsection), Equation (6.4) then allows for the measurement of temperature through an observation in the change in reflectance. The thermoreflectance coefficient β is typically quite small, on the order of 10^{-5}–10^{-3}/°C for most common materials, thus making its calibration and the accurate measurement of the reflectance vital for true estimations of temperature [9, 22].

6.2.2 Experimental Methodology

A typical experimental setup for thermoreflectance metrology of electrically driven microdevices is shown in Fig. 6.1. Monochromatic laser light initially travels through a beam splitter where upon a portion of the beam continues on toward the sample while the remainder immediately travels to the photodetector to be used as a reference. This reference allows measurements to be independent of variations in the laser power which can cause great problems due to the relatively small magnitude of the thermoreflectance coefficient. That portion of the light traveling to the surface is then polarized by a quarter wave plate before being focused onto the sample using a standard microscope objective. The scattered light is then collected by the same objective and returns whence it came upon which it passes through the quarter wave plate once again thus insuring that only truly reflected and not emitted radiation is able to pass on into the photodetector to be measured. Notice that since the intensity of the light is measured, a consistent focus point is needed for reliable results. This can at times be challenging both due to mechanical drift of the stage but more often as a consequence of problems resulting from thermal expansion.

In order to measure the actual device performance, oftentimes the system is coupled with the electrical driver via a lock-in amplifier [15]. By coupling the photodetector and pulse generator through the lock-in amplifier, it is possible to synchronize when the thermal measurements are taken relative to the electrical loading. In such a manner, it is possible to obtain transient thermal measurements of devices during

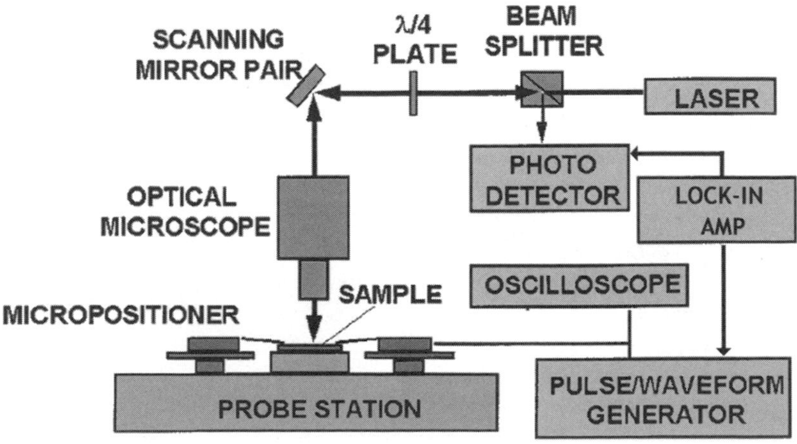

Fig. 6.1 Experimental set up for measurement of microdevices using thermoreflectance metrology Adapted from Cahill et al. [23]

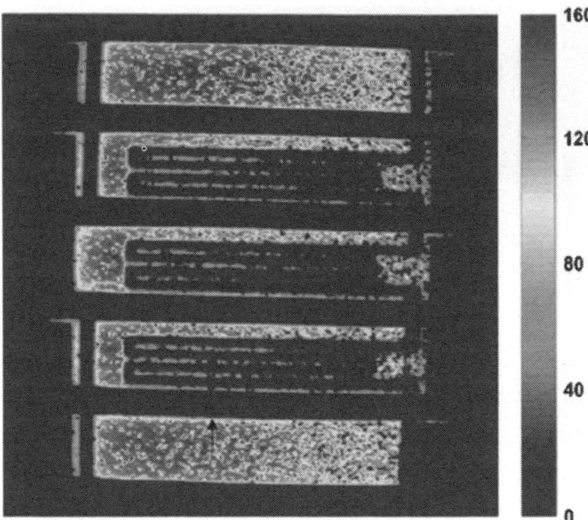

Fig. 6.2 Thermal map of polysilicon resistors obtained using thermoreflectance metrology as reported by Tessier et al. [21]

operation. Additionally, these single point measurements can be extended to entire maps through acquisition of a matrix of points along the surface of the device. These multiple measurements are acquired by moving the device relative to the focus point via a mechanical stage or through translation of the actual focus point via mirrors. In such a manner the entire 2D thermal response of the device may be acquired as shown in Fig. 6.2.

6.2.3 Calibration of the Thermoreflectance Coefficient

As most materials thermoreflectance coefficients are extremely small, usually less than 1 part in 1000 for every degree change, accurate calibration is both vital and challenging. Due to this difficulty, it is not uncommon for researchers to rely on previously published thermoreflectance standards either directly measured or obtained from first principles calculations. In microdevices, however, standard material properties are oftentimes invalid as residual stress resulting from the processing scheme or divergence from bulk behavior due to the thinness of layer may change the optical properties. Consequently, direct calibration of the measured device is quite often necessary.

The most straightforward methodology incorporates a heated chuck to raise the entire device temperature where upon a thermoreflectance measurement is taken and correlated to the temperature standard. The temperature standard is found through use of a thermocouple either embedded in the chuck or directly mounted to the backside of the device. After a series of thermoreflectance measurements taken across the entire temperature range, the coefficient is deduced from the resulting reflectance-temperature curve [24]. While straightforward in its application, problems arise as thermal expansion effects can induce great uncertainty into the calibration due to problems with consistent focus of the same point [25].

This problem can be circumvented if the heated region is in some manner reduced in size as expansion is directly proportional to volume. While alleviating the problem of expansion effects, this approach distinctly complicates the calibration process as standard heating via macroscopic conduction and temperature sensing using thermocouples is no longer applicable. Rather heating and temperature sensing of the standard must now take place in the micro level regime. Electrical solutions have most often been employed in response to these challenges. For example, in measuring metal interconnect structures, Ju and Goodson performed their calibration in a two step process by first correlating temperature and resistance using standard heating methods. Knowing resistivity as a function of temperature, electrical pulses were then sent through the device to locally heat the material as simultaneous reflectance and resistivity measurements allowed for accurate calibration of the reflectance coefficient [16, 25]. If the material in question is not an electrically conductive, a similar procedure can be employed, however, it becomes necessary to embed an electrical heater which can serve as both as a local heater and temperature sensor. This integrated sensor approach has been employed recently with success by Tessier et al. to calibrate the thermoreflectance coefficient for a polysilicon device using a CCD collector. The approach, however, requires additional micromachining of the device which may effect operation [21].

6.2.4 Applications of Thermoreflectance in Temperature Measurement of Microdevices

As stated previously, temperature is a key factor in determining device failure, reliability, and overall lifetime. It is, therefore, imperative that devices can then be thermally characterized in a manner that is quick, accurate, and unobtrusive. While this most often takes the form of directly measuring temperature in some manner, a complete characterization necessitates understanding those parameters which affect the temperature distribution, namely, thermal conductivity and diffusivity. In response to these dualistic requirements, incorporation of the thermoreflectance technique has occurred in two distinct complementary fashions, one of which directly measures temperature and a second which evaluates the thermal conductivity using the laser as both energy source and probe. The following is a brief, and by no means exhaustive, exposition describing the implementation of each in the thermal evaluation of microdevices and the materials composing these devices.

Direct measurement of temperature in semiconductor devices using thermoreflectance techniques began in earnest in the early 1990s with publication of several studies each focusing on the steady state distribution of temperature in a range of different semiconductor devices. For example, lasers were examined independently by Epperlein and Mansanares et al. whereby hot spots were located in the active region of the device and correlated to device degradation [26, 27]. Meanwhile, hot spots were also examined in the operation of integrated circuits through work headed by Claeys et al. In these studies, the first to maps of actual devices were acquired through use of a scanning procedure with simultaneous accounting for the effects of the passivation layers [28–30]. Building from these studies, Ju and Goodson then examined the transient thermal behavior through investigations of metallic interconnects and silicon based electronics with 10 ns resolution [15, 16, 25].

Recent investigations have extended the capabilities of the technique even further through the use of ever more capable equipment and novel methods. For example, Tessier et al. reports through the use of charge coupled devices (CCD) in lieu of photodetectors, an improved spatial resolution of 360 nm, nearly a 60% that of the incident radiation which was utilized [21]. In addition, Christofferson and Shakouri have reported a novel method whereby an infrared laser is used to tunnel through the substrate of a thermoelectric device as is shown in Fig. 6.3 thus providing a measurement of underside temperature, an application which could be of great use in flip-chip applications [31]. Finally, entire transient maps of MOSFETS have been obtained which may be of great use in identifying and observing the failure mechanisms present in microdevices [32].

Beyond measuring temperature, reflectance measurements have been used to deduce a host of other parameters as well including thermal: conductivity, diffusivity, and boundary conductance, a measure of interfacial resistance to heat flow. To deduce these material dependent properties, a modification of the thermoreflectance technique is made whereby instead of external electrical energy heating the sample, it is now the incident radiation's role to both heat and monitor the sample. More explicitly, this approach, typically termed "pump and probe," uses a brief high intensity exposure of radiation to heat the sample and then measures the temperature of the surface by measuring the reflectance change of the surface relative to a second much less intense beam. By changing the path length of the second "probe" beam, through use of a dovetail mirror as shown in Fig. 6.4, the temporal behavior of

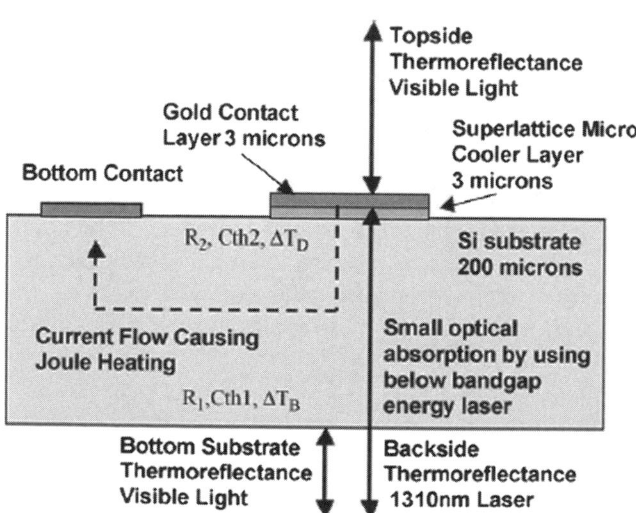

Fig. 6.3 Schematic showing "underside" temperature measurement of microdevice for use in flip chip characterization as reported by Christofferson and Shakouri [31]

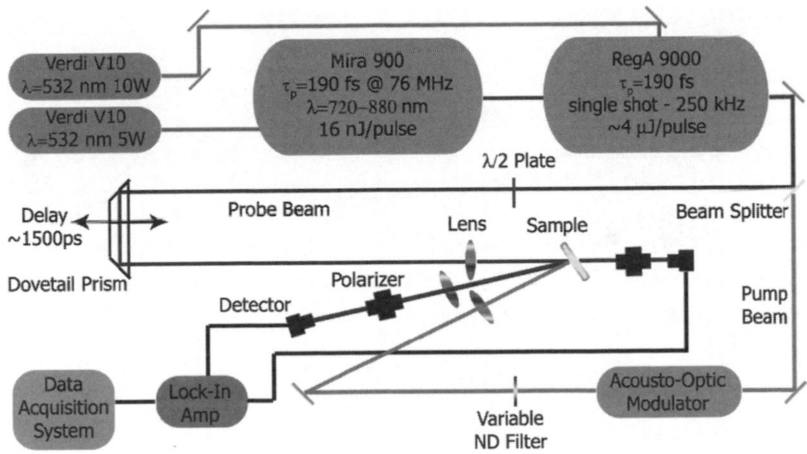

Fig. 6.4 Experimental set up of typical "pump and probe" technique as reported by Hopkins et al. [34]. The moveable mirror allows for picosecond resolution

the temperature at the surface can be enumerated to picosecond resolution. As the exciting pump beam is much larger in size than the probe beam, one dimensional conduction dominates and knowledge of this temperature evolution provides the necessary conditions for solution of the parabolic two step heat conduction equation whereby the thermal diffusivity (conductivity) is found through fitting of the data [33].

The modern incarnation of the technique began when Paddock and Easley measured the thermal diffusivity of a variety of thin films in the 1980s [35]. Subsequent research then expanded the technique to measure a variety of metals and semi-conductor materials in order to examine changes in thermal properties with reductions in length scale and dimension [12, 36]. In addition to standard thermal properties, pump and probe measurements have also been used on multi-layered systems allowing for measurement of the thermal boundary conductance [37, 38]. An excellent review of pump and probe techniques is supplied by Norris et al. [39].

6.3 Temperature Measurement of Microdevices by Means of Raman Spectroscopy

6.3.1 Physical Basis of the Raman Technique

Similar in approach to a thermoreflectance measurement, the typical Raman investigation occurs through bombardment of a surface with radiation while concomitantly observing the photonic energy reaching a detector. In a thermoreflectance investigation, temperature is estimated by that quantity of radiation untainted by the examined surface, as that which is measured is perfectly reflected. Raman spectroscopy, in contrast, examines "tainted" radiation scattered by the volume. Despite a seemingly definitive difference, the two techniques are similar, however, with respect to the fact that the degree to which this interaction, or lack of interaction, occurs distinctly depends upon the distribution of phonons and electrons in the material lattice. As, again, these distributions are temperature dependent so too will the interactions and hence the resulting reflection, and Raman scattered radiation. Thus the key to examining the temperature dependence of the Raman scattered radiation is to understand the interaction between the incident radiation and the material lattice.

Upon impingement of incident radiation, a photon with energy ε_i, may either be reflected, absorbed, or transmitted. In Raman spectroscopy, our concern rests solely upon an interaction in which this photon is absorbed by either an electron or in a dramatically more unlikely case a phonon [40]. As part of this event, the absorbing species (i.e. the electron or phonon) is promoted from its equilibrium state of energy ε_0 to an excited state of energy ε_L resulting in a non-equilibrium distribution of the excited entity. Seemingly simultaneously, thermalization will occur whereby this excited entity will "relax" back to its equilibrium state and distribution. Most often this occurs directly thereby inducing the "re-emission" of a photon of energy equal to that which was incident ($\varepsilon_L-\varepsilon_0=\varepsilon_i$) in a process known as Rayleigh scattering.

In a small statistical subset of these occurrences, however, an intermediate event takes place by which the excited species either absorbs or emits an *additional* energy carrier (phonon or electron) in the process of thermalizing as is shown in Fig. 6.5. As a consequence of this intermediate event, the types of which are summarized in Table 6.1, the excited species moves to a secondary non-equilibrium state of energy ε_m which is unequal to either its original ground (ε_0) or excited level (ε_L). Upon

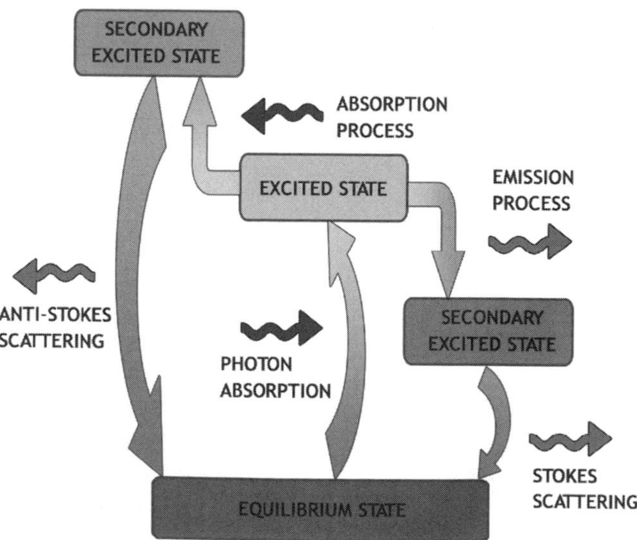

Fig. 6.5 Process flow diagram of Raman scattering where upon 3 separate events give rise to inelastic scattering

Table 6.1 Scattering cascade giving rise to the raman effect. Although events numbered 3 through 6 are possible it is generally assumed that events 1 and 2 significantly dominate [40]

	Event 1	Event 1 $\Delta\varepsilon$	Event 2	Event 2 $\Delta\varepsilon$	Event 3	Event 3 $\Delta\varepsilon$	Scattering type
1	Photon (ε_i) absorption by electron	$\varepsilon_L - \varepsilon_0 = \varepsilon_i$	Phonon (ε_D) absorption by electron	$\varepsilon_L + \varepsilon_D = \varepsilon_m$	Electron relaxation and photon emission	$\varepsilon_m - \varepsilon_0 = \varepsilon_f$	Anti-Stokes
2	Photon (ε_i) absorption by electron	$\varepsilon_L - \varepsilon_0 = \varepsilon_i$	Phonon (ε_D) emission by electron	$\varepsilon_L - \varepsilon_D = \varepsilon_m$	Electron relaxation and photon emission	$\varepsilon_m - \varepsilon_0 = \varepsilon_f$	Stokes
3	Photon (ε_i) absorption by phonon	$\varepsilon_L - \varepsilon_0 = \varepsilon_i$	Phonon (ε_D) absorption by phonon	$\varepsilon_L + \varepsilon_D = \varepsilon_m$	Phonon relaxation and photon emission	$\varepsilon_m - \varepsilon_0 = \varepsilon_f$	Anti-Stokes
4	Photon (ε_i) absorption by phonon	$\varepsilon_L - \varepsilon_0 = \varepsilon_i$	Phonon (ε_D) emission by phonon	$\varepsilon_L - \varepsilon_D = \varepsilon_m$	Phonon relaxation and photon emission	$\varepsilon_m - \varepsilon_0 = \varepsilon_f$	Stokes
5	Photon (ε_i) absorption by phonon	$\varepsilon_L - \varepsilon_0 = \varepsilon_i$	Electron (ε_D) absorption by phonon	$\varepsilon_L + \varepsilon_D = \varepsilon_m$	Phonon relaxation and photon emission	$\varepsilon_m - \varepsilon_0 = \varepsilon_f$	Anti-Stokes
6	Photon (ε_i) absorption by phonon	$\varepsilon_L - \varepsilon_0 = \varepsilon_i$	Electron (ε_D) emission by phonon	$\varepsilon_L - \varepsilon_D = \varepsilon_m$	Phonon relaxation and photon emission	$\varepsilon_m - \varepsilon_0 = \varepsilon_f$	Stokes

relaxation of the excited entity from energy ε_m to its original equilibrium energy ε_0, a "new" photon of energy $\varepsilon_m - \varepsilon_0 = \varepsilon_f$ will be emitted. Due to the intermediate reaction, the emitted photon will have an energy unequal to that of the incident radiation ($\varepsilon_f \neq \varepsilon_i$) leading to what is known as inelastic scattering and the so called Raman effect. By monitoring the difference between ε_f and ε_i through the shift in incident and exiting photon frequencies, an examination of the intermediate reaction can take place thus providing insight into the temperature dependent characteristics of the crystal lattice.

The nature of this temperature dependence may be illustrated through incorporation of classical theory to describe the interaction between the incident radiation and the crystal lattice. We begin this analysis by defining two properties determined by the electronic distribution within a crystal lattice in equilibrium, namely, the dipole moment, P, and electric polarizability, α. These properties will change with deviation in equilibrium interatomic spacing (i.e. stressing or thermal expansion) of the lattice or due to the presence time-dependent vibrational motions (i.e. phonons) around these equilibrium positions. The electric polarizability is a second order tensor response function which represents the volume and shape of the charge distribution in the lattice. When a photon with electric field, E, is incident on the lattice, the induced dipole moment is given by,

$$P = \varepsilon_0 \alpha E \tag{6.5}$$

where ε_0 is the free-space permittivity. At any finite temperature, the presence of phonons causes the electric polarizability tensor to constantly change with time. These changes may be described through a Taylor series expansion about the equilibrium position of the lattice atoms

$$\alpha = \alpha_0 + \frac{\mathrm{d}\alpha}{\mathrm{d}q} q + \ldots \tag{6.6}$$

where α_0 is the polarizability at the equilibrium lattice spacing and $q = q_0 \cos(\omega_0 t)$ is the time-dependent change in the lattice spacing due to phonon vibrations with amplitude q_0 and frequency ω_0. Realizing that the incoming electric field, $E = E_0 \cos(\omega)$, can be written as an oscillatory function of amplitude E_0 and frequency ω, using Equation (6.6), we may now write Equation (6.5) in an expanded form

$$P = \varepsilon_0 \alpha_0 E_0 \cos(\omega t) + \frac{\mathrm{d}\alpha}{\mathrm{d}q} \varepsilon_0 q_0 \cos(\omega_0 t) E_0 \cos(\omega t) \tag{6.7}$$

where ω_0 is the frequency of the phonon vibration and ω is the vibrational frequency of the incident photon. Applying a trigonometric identity to (6.7) leads to the following relation,

$$p = \varepsilon_0 \alpha_0 E_0 \cos(\omega t) + \frac{d\alpha}{dq}\frac{\varepsilon_0 q_0 E_0}{2}\cos((\omega_0 - \omega)t) + \frac{d\alpha}{dq}\frac{\varepsilon_0 q_0 E_0}{2}\cos((\omega_0 + \omega)t) \tag{6.8}$$

The first term on the right hand side of Equation (6.8) accounts for Rayleigh scattering of photons. The second and third terms result in Stokes and anti-Stokes Raman scattering, respectively, where the photons are shifted away from their incident frequency ω by an amount equal to the optical phonon frequency ω_0 [41]. From Equation (6.8), it is readily seen that the resulting Raman shift is directly dependent upon this phonon vibrating at ω_0. The temperature dependence of the Raman signal is then an exercise in examining the thermal behavior of this phonon. The following subsections describe the mechanisms by which this behavior may be captured by the Raman signal.

6.3.2 *Stokes/Anti-Stokes Intensity Ratio as a Measurement of Temperature*

At a given frequency shift ($|\omega - \omega_0|$), the intensity of a Raman signal will be proportional to the number of phonons at frequency ω_0 present to take part the scattering processes [40]. In an anti-stokes process, the number of phonons which may be absorbed at a given temperature can be calculated from the Bose-Einstein distribution function as shown below,

$$N_0 = \frac{1}{\exp(\hbar\omega_0 / k_B T) - 1} \tag{6.9}$$

where $\hbar$ is modified Planck's constant, and k_B the Boltzmann constant [18]. Similarly for a Stokes process, the total number of phonons will be this equilibrium distribution, N_0, plus the emitted phonon for a cumulative total of N_0+1. Thus the ratio in intensity of these two signals provides a measurement of temperature as described below,

$$\frac{I_{AS}}{I_S} \cong \frac{N_0}{N_0 + 1} \cong C \exp\left(\frac{-\hbar\omega_0}{k_B T}\right) \tag{6.10}$$

where C is a calibration factor, while I_{AS} and I_S are the Anti-Stokes and Stokes intensities, respectively [42–44].

6.3.3 *Stokes Shift as a Measurement of Temperature*

Quantitative (extensive) measurements, such as those which compare intensities, may be difficult to implement as the results will be dependent upon a host of experimental variables which determine signal strength. Consequently, it is often desirable to analyze instead a qualitative (intensive) aspect of the acquired signal. In Raman spectroscopy, this is accomplished by examining the manner in which the vibratory aspects of the analyzed phonon change with temperature.

The vibratory nature of a given phonon mode can be modeled by considering a spring-oscillator system assuming the force laws between the atoms or molecules in the lattice are known. The solution for this type of classical oscillator reveals that the resulting lattice vibrational frequencies vary with the interatomic forces [20]. As the lattice is heated, cooled, or stressed, the equilibrium positions of the atoms are displaced, resulting in an overall volumetric expansion or contraction of the lattice and a change in interatomic forces due to anharmonic effects [45]. These changes in the interatomic forces subsequently modify the phonon vibrational frequencies (ω_0) resulting in a change of the measured Raman frequency shift.

In addition to this volumetric contribution, interactions between the phonons themselves augment the frequency shift as well [45]. This occurs as the mere presence of a phonon will alter the equilibrium spacing of the atoms in a lattice. With a change in equilibrium spacing, an associated altering of the interatomic forces will occur. Due to this change in the interatomic forces, the frequency of oscillation of both this and other phonons will be modified thus affecting the shift of the Stokes and Anti-Stokes scattering. As the phonons responsible for this modification are governed by the Bose-Einstein distribution of thermal occupation, the resulting Raman shift varies due to this contribution in a completely temperature dependent manner wholly free of stress effects [14]. This contribution due to the presence of phonon interaction is typically termed the explicit contribution.

The total frequency shift in the Raman signal is then some superposition of the explicit and volumetric contributions. The two effects may be superposed in order to measure temperature according to the following empirical relation,

$$\Delta\omega(T) = A(T - T_0) \tag{6.11}$$

where A is a separately obtained calibration constant, T the temperature being measured, T_0 a known reference temperature, and $\Delta\omega(T)$ the stokes shift at a given temperature. Finally, the Stokes shift is typically used in lieu of the Anti-Stokes, as at room temperature its scattering cross section is much greater thus allowing for smaller integration times.

6.3.4 Stokes Linewidth as a Measurement of Temperature

The volumetric contribution to the Stokes peak shift is affected by both thermal expansion and mechanical stressing. In the presence of thermomechanical loading, the final shift will be a convolution of each effect causing significant errors in the measurement of either temperature or stress [14]. In response to this difficulty, the linewidth (FWHM) of the Stokes peak which is solely temperature dependent is an integral tool in the investigation of microdevices where thermal stress evolution may reach high levels [46, 47]

The origin of Stokes linewidth arises according to the Heisenberg uncertainty principle. According to this principle a measured species, in this case the phonon, may only be measured within a certain energy band ($\Delta\varepsilon$) if its availability to be measured (i.e. lifetime) is finite. This is described mathematically according to the energy-time uncertainty relation:

$$\Gamma \approx \Delta\varepsilon = \frac{\hbar}{\tau} \tag{6.12}$$

where Γ is the width of the Raman line, $\hbar = 5.3 \times 10^{-12}\,\mathrm{cm^{-1}s}$ is modified Planck's constant, and τ is the scattering time for a phonon [48]. From Equation (6.12), one can see that the measured linewidth of a Raman peak will then vary with scattering time of the phonon mode. The scattering time of this phonon mode is dependent upon a variety of factors including microstructural defects, material boundaries, and most importantly other phonons. It is this dominant phonon–phonon scattering which gives rise to the temperature dependence of the linewidth as the number of phonons available for scattering is once again dependent upon the temperature deferent Bose-Einstein population distribution. Most simply, as the temperature increases so too does the number of phonons present thereby increasing the likelihood of a scattering event. This increased likelihood reduces the phonon lifetime thus increasing the linewidth. The temperature can then be measured with appropriate calibration via a relation like that shown below:

$$\Gamma = B(T - T_0)^2 + C(T - T_0) + \Gamma_0 \tag{6.13}$$

where B and C are calibration constants and Γ_0 is the linewidth at a reference temperature.

6.3.5 Calibration and Experimental Procedure

A typical experimental set up for a micro-Raman backscattering experiment is shown in Fig. 6.6. Monochromatic laser light, typically 488 or 514 nm insured through use of a rejection filter, is spatially collimated where upon it is focused using standard microscopic optics resulting in a beam spot of as little as 1 μm [49]. Upon impingement with the surface, the scattered radiation is collected by the microscope objective and the light is transferred through holographic filters thus negating the overwhelming Rayleigh signal. After filtering, the scattered radiation then passes through a slit which aligns the radiation with the dovetail filter. The dovetail filter in turn disperses the light into its constituent components via Bragg diffraction thus allowing measurement of the light across the spectrum via a multi-channel CCD collector. Finally, in the analysis of microdevices, the sample is oftentimes mounted in a moveable x–y stage allowing for a mapping procedure to be performed with up to 1 μm spatial resolution.

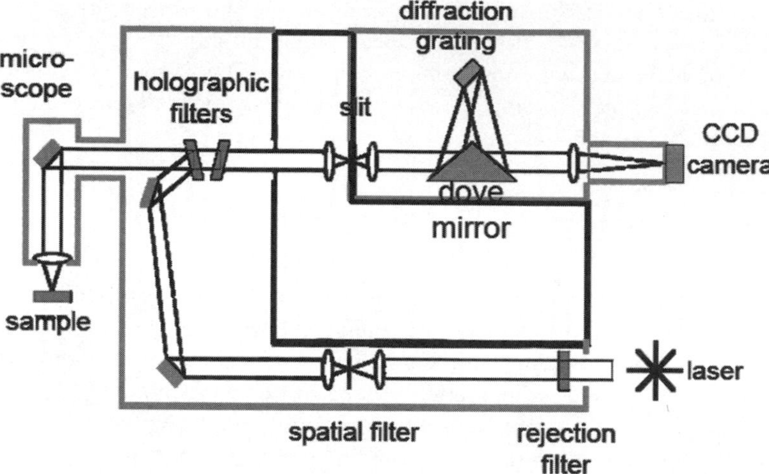

Fig. 6.6 Typical arrangement of micro-Raman backscattering experiment in which light is both scattered and collected via a microscope and subsequently measured utilizing a dispersive grating and CCD device

Unlike that which was seen with regards to thermoreflectance, the calibration of the Raman signal is relatively straightforward most often requiring only a thermal chuck and thermocouple. The calibration takes place by heating the sample to a proscribed temperature where upon several Raman spectra are acquired with the associated peak shifts and linewidths found using data fitting software. Repeating this analysis across the entire range of interest, one finds that the peak red-shifts with an increase temperature while the linewidth broadens as shown in Fig. 6.7. By tracking this broadening and shifting behavior, curves of the form of Equations (6.12) and (6.13) can be fit to the data as shown in Fig. 6.8 thus providing

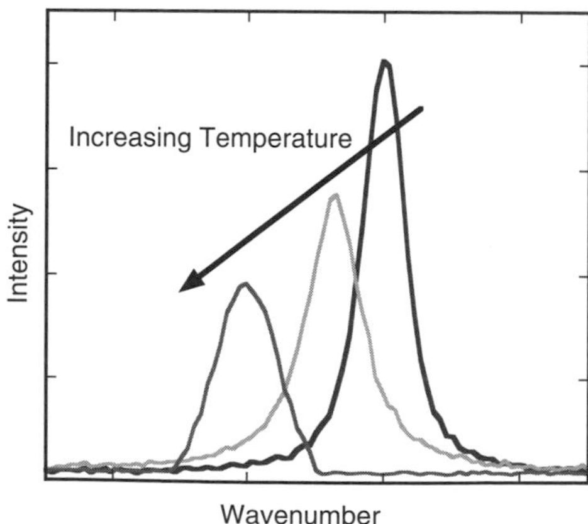

Fig. 6.7 Representative Raman response with an increase in temperature. Notice that the peak shifts to lower wavenumber and the linewidth broadens

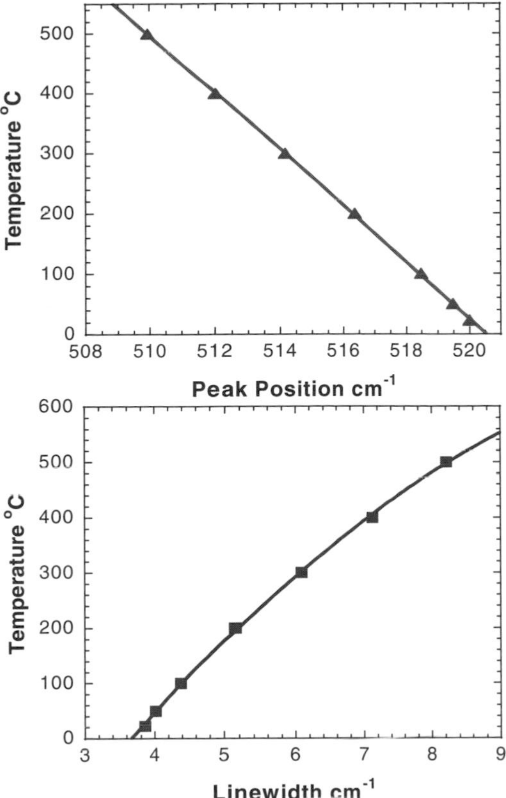

Fig. 6.8 Calibration curves of Temperature versus Peak Position (top) and Linewidth (bottom) for a silicon sample. The curve allows direct estimation of temperature from the Raman Spectra

a direct correlation between the temperature and Raman spectra. A similar procedure is undertaken for calibrations of the temperature versus the Stokes/Anti-Stokes intensity ratio. However, as this measurement is intrinsically extensive in nature, high correlation between the acquired data and the expected response is most difficult to acquire [42–44]. In addition, the ratio changes extremely slowly at higher temperatures making accurate measurement even more difficult in this thermal realm [50]. As a consequence of these difficulties, as well as the longer experimental times necessary to acquire both the Stokes and Anti-Stokes signal, most current investigations implement the Stokes shift or linewidth for the acquisition of temperature.

6.3.6 Applications of Raman Spectroscopy in Temperature Measurement of Microdevices

The implementation of Raman spectroscopy to the characterization of devices has scaled directly with the availability of relatively inexpensive laser sources and optics. Hence there is little work prior to 1990 concerning the use of Raman as a non-contact temperature probe. Much like the development of thermoreflectance, those studies which were performed focused on the thermal behavior of laser diodes [51, 52]. In these initial investigations, the ratio of Stokes to Anti-Stokes intensity was incorporated to measure the temperature distribution extending away from the active region of the device. Using this approach, degradation was found to be directly correlated to facet heating thus illustrating the technique's usefulness while providing a framework for much of the future work that continues to this day.

Building on this approach of degradation analysis through thermal mapping, Raman spectroscopy was then extended to the analysis of field effect transistors (FET). It has been in the analysis of these transistors where the technique has firmly found its niche even to this day. This trend began in the analysis of silicon based FET's with the acquisition of temperature distributions of 1 μm resolution across the source to drain channel as is shown in Fig. 6.9 [49, 53]. Through these distributions, temperature was seen to peak near the pinch off region where the electric field is at its maximum thus indicating direct electro-thermal coupling. In addition, due to the small temperature increases in the devices measured and thus the corresponding difficulty in using the Stokes/Anti-Stokes ratio, these studies mark the first use of Stokes peak shift in the estimation of temperature. This has proven to be a harbinger as nearly all subsequent studies have used the peak shift method due to its increased accuracy and reduced acquisition times [50].

In addition to these Si-based FET's, many studies have incorporated Raman thermography into the investigation of wide bandgap devices and in particular gallium nitride (GaN) based transistors. Gallium nitride devices have garnered much interest in the optoelectronic and communication industries due to their ability to operate at high frequency and power, conditions which lead to large thermal loads which in turn affect reliability [55–57]. These studies follow much the same structure as those performed upon Si-based FET's where upon the peak shift of the Raman signal is used to deduce a temperature profile (both 1D and 2D) with micron resolution along the source to drain channel to within ~10–20°C [58–61]. Kuball et al. report maximum temperature rises, confirmed by finite element models, of nearly 200°C

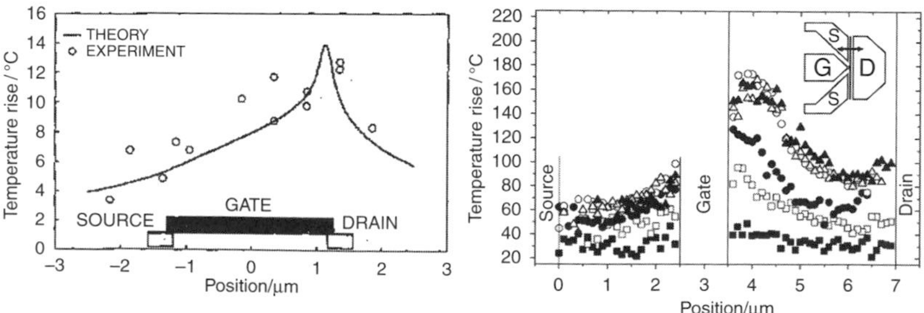

Fig. 6.9 Temperature profiles along source gate drain channel for a Si based FET (left) and GaN based FET (right). The maximum temperature occurs near the gate on the drain side in the location of the maximum electric field
Figures are courtesy of Ostermeir et al. and Rajasingam et al. [53, 54]

in GaN based FET's while seeing similar profiles (Fig. 6.9) as those reported in silicon based transistors [62]. These profiles in turn can then be used to estimate the electric field strength independent of electrical characterization [54]. In addition, the same group has also correlated the location of hot spots to defects in the underlying substrate highlighting the importance of total device thermal management [63]. As only device lifetime is affected by these hot spots and not necessarily electrical performance, Raman thermography can then be used in the identification of devices which may fail prematurely [64]. Finally, analysis on these structures has recently been extended to the transient domain with temporal resolution of 200 ns using a pulsed laser synchronized with electrical loading of the device [17].

Similar to microelectronics, MEMS devices rely on efficient thermal management for proper operation and reliability and as a consequence Raman thermometry has been incorporated into their characterization. For example Kearney et al. utilizes the peak shift of the Raman signal to obtain temperature distributions in a silicon based electro-thermo actuator for design optimization [65]. Heated atomic force cantilevers have been characterized by Raman spectroscopy as well in order to accurately determine their temperature versus power characteristics, a parameter vital for their proper operation in sensing and data storage applications [5].

In the great majority of applications of Raman thermometry, the peak shift has been utilized as the predominant estimation of temperature. However, in the presence of thermal stress which is present in many MEMS and microelectronic devices, peak shift estimations of the temperature will be in error as the signal will be a convolution due both to the mechanical and thermal loading as is seen in Fig. 6.10 [46, 66, 67]. This effect may be circumvented through use of linewidth of the Stokes peak which is stress insensitive to the first order and has recently been implemented as an accurate measurement of temperature in devices [47, 68]. Using this additional parameter, simultaneous maps of temperature and stress have been made on constrained polysilicon electro-thermal beams thus portending the use of Raman spectroscopy as a tool for the visualization of degradation mechanisms resulting from thermal loading.

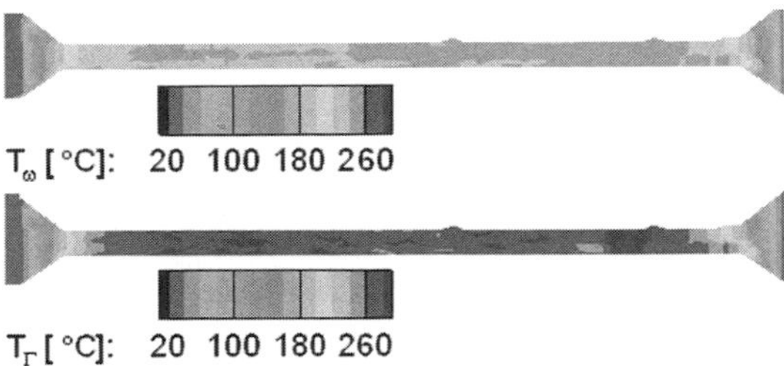

Fig. 6.10 Temperature maps of polysilicon beam heaters using the peak shift (top) and the linewidth (bottom). The peak shift records an erroneous lower temperature due to the presence of compressive thermal stress

6.4 Summary and Conclusions

Commensurate with the greater dissipation required with ever decreasing length scales and power densities, thermal management is a key factor in the determination of the reliability and lifetime of many microelectronics and MEMS. As a consequence, the accurate measurement of temperature at these small length scales is vitally important for continued market adoption and improvement of these devices. Optical methods have proved quite adept at the acquisition of these temperatures due to their ability to probe materials with high spatial and temporal resolution in a manner which does not disrupt operation.

The nature of this ability arises as the optical properties of a material are dependent upon the interplay of incident radiation and temperature dependent populations of both the electronic and lattice energy carriers. Hence, the interaction itself is temperature dependent as well. In a thermoreflectance measurement, much as its name implies, this interaction is measured through a change in reflected intensity which in turn can be correlated to temperature. Meanwhile, in Raman spectroscopy the degree to which light is inelastically scattered provides a measure of the temperature dependent characteristics of the lattice and hence an estimation of temperature itself. In either case, the technique is limited only by the availability of an optical surface, the wavelength of the incident light, and the time of interaction between the incident radiation and the material.

Yet in practice as devices become ever smaller and device features march onward to sizes comparable to the wavelength of extreme ultraviolet light and beyond, no longer will standard far field optical techniques remain viable. It is then necessary for characterization tools which may capture the salient aspects of heat transfer in these reduced domains to be developed at a rate comparable to that of the device themselves. As the fundamental temperature dependent interactions will not change at a quantum level, it is not the nature of the techniques which must necessarily change but rather the manner in which the interplay between incident radiation and material is both induced and detected.

References

1. Y.-T. Cheng and L. Lin, In *Springer Handbook of Nanotechnology*, edited by B. Bhushan (Springer: Berlin, New York, 2004), pp. 1111–1132.
2. S. Singhal, T. Li, A. Chaudhari, A. W. Hanson, R. Therrien, J. W. Johnson, W. Nagy, J. Marquart, P. Rajagopal, J. C. Roberts, E. L. Piner, I. C. Kizilyalli, and K. J. Linthicum, Microelectronics Reliability **46**, 1247–1253 (2006).
3. W. Zhu, J. Zhu, S. Nishino, and G. Pezzotti, Applied Surface Science **252**, 2346–2354 (2006).
4. D. L. Blackburn, Semiconductor Thermal Measurement and Management Symposium, 2004. Twentieth Annual IEEE, 70–80 (2004).
5. J. Lee, T. Beechem, T. L. Wright, B. A. Nelson, S. Graham, and W. P. King, Journal of Microelectromechanical Systems **15**, 1644–1655 (2006).
6. A. Majumdar, Annual Review of Materials Science **29**, 505–585 (1999).
7. J. Altet, S. Dilhaire, S. Volz, J. M. Rampnoux, A. Rubio, S. Grauby, L. D. Patino Lopez, W. Claeys, and J. B. Saulnier, Microelectronics Journal **33**, 689–696 (2002).

8. J. Altet, A. Rubio, E. Schaub, S. Dilhaire, and W. Claeys, Solid-State Circuits, IEEE Journal **36**, 81–91 (2001).
9. Z. Zhang, Annual Review of Heat Transfer **11**, 351–411 (2000).
10. A. N. Smith and P. M. Norris, In *Handbook of Heat Transfer*, edited by A. Bejan and A. Krauss (John Wiley & Sons, Inc., 2003), pp. 1309–1358.
11. P. W. Webb, Circuits, Devices and Systems, IEE Proceedings G **138**, 390 (1991).
12. N. Taketoshi, T. Baba, and A. Ono, Bulletin of the National Research Laboratory of Metrology(Japan) **49**, 111–114 (2000).
13. K. T. Tsen, J. G. Kiang, D. K. Ferry, and H. Morkoc, Applied Physics Letters **89**, 112111–112111 (2006).
14. T. Beechem, S. Graham, S. Kearney, L. Pinney, and J. Serrano, Simultaneous Mapping of Temperature and Stress in Microdevices Utilizing Micro-Raman Spectroscopy, Review of Scientific Instruments, **78**, 061301-1-061301-9(2007).
15. S. Ju, O. W. Kading, Y. K. Leung, S. S. Wong, and K. E. Goodson, Electron Device Letters, IEEE **18**, 169–171 (1997).
16. Y. S. Ju and K. E. Goodson, Electron Device Letters, IEEE **18**, 512–514 (1997).
17. M. Kuball, G. J. Riedel, J. W. Pomeroy, A. Sarua, M. J. Uren, T. Martin, K. P. Hilton, J. O. Maclean, and D. J. Wallis, Electron Device Letters, IEEE **28**, 86–89 (2007).
18. G. Chen, *Nanoscale Energy Transport and Conversion* (Oxford University Press, New York, NY, 2005).
19. R. Rosei and D. W. Lynch, Physical Review B **5**, 3883–3894 (1972).
20. C. Kittel, *Introduction to Solid State Physics*, 8th ed. (Wiley, Hoboken, NJ, 2005).
21. G. Tessier, M. L. Polignano, S. Pavageau, C. Filloy, D. Fournier, F. Cerutti, and I. Mica, Journal of Physics D: Applied Physics **39**, 4159–4166 (2006).
22. W. Claeys, S. Dilhaire, S. Jorez, and L. D. Patino-Lopez, Microelectronics Journal **32**, 891–898 (2001).
23. D. Cahill, K. Goodson, and A. Majumdar, Journal of Heat Transfer **124**, 223–241 (2002).
24. T. Q. Qiu, C. P. Grigoropoulos, and C. L. Tien, Experimental Heat Transfer **6**, 231–241 (1993).
25. Y. S. Ju and K. E. Goodson, Journal of Heat Transfer **120**, 306–313 (1998).
26. P. W. Epperlein, Japanese Journal of Applied Physics **32**, 5514 5522 (1993).
27. A. M. Mansanares, J. P. Roger, D. Fournier, and A. C. Boccara, Applied Physics Letters **64**, 4 (1994).
28. W. Claeys, S. Dilhaire, and V. Quintard, Microelectronic Engineering **24**, 411 (1994).
29. W. Claeys, S. Dilhaire, V. Quintard, J. P. Dom, and Y. Danto, Quality and Reliability Engineering International **9**, 303–308 (1993).
30. V. Quintard, G. Deboy, S. Dilhaire, T. Phan, D. Lewis, and W. Claeys, Microelectronic Engineering **31**, 291–298 (1996).
31. J. Christofferson and A. Shakouri, Microelectron. Journal **35**, 791–796 (2004).
32. P. L. Komarov, M. G. Burzo, G. Kaytaz, and P. Raad, Proceedings to the Pacific Rim/ASME International Electronic Packaging Technical Conference and Exhibition on Integration and Packaging of MEMS, NEMS and Electronic Systems (InterPACK), 17–22.
33. T. Q. Qiu and C. L. Tien, International Journal of Heat Mass Transfer **37**, 2789–2797 (1994).
34. P. Hopkins and P. Norris, Applied Physics Letters **89**, 131909 (2006).
35. C. Paddock and G. L. Eesley, Journal of Applied Physics **60**, 285–290 (1986).
36. J. L. Hostetler, Microscale Thermophysical Engineering **1**, 237–244 (1997).
37. A. N. Smith, Microscale Thermophysical Engineering **4**, 51–60 (2000).
38. R. J. Stoner and H. J. Maris, Physical Review B **48**, 16373–16387 (1993).
39. P. M. Norris, A. P. Caffrey, R. J. Stevens, J. M. Klopf, J. T. McLeskey, and A. N. Smith, Review of Science Instrumentation **74**, 400–406 (2003).
40. R. Loudon, Advances in Physics **13**, 423 (1964).
41. D. A. Long, *Raman Spectroscopy* (McGraw-Hill, 1977).
42. M. Balkanski, R. F. Wallis, and E. Haro, Physical Review B **28**, 1928 (1983).
43. T. R. Hart, R. L. Aggarwal, and B. Lax, Physical Review B **1**, 638 (1970).

44. R. Tsu and J. G. Hernandez, Applied Physics Letters **41**, 1016–1018 (1982).
45. G. Lucazcau, Journal of Raman Spectroscopy **34**, 478–496 (2003).
46. M. Abel and S. Graham, ASME InterPack2005 **IPACK2005-73088**, 1–8 (2005).
47. M. Abel, S. Graham, J. Serrano, S. Kearney, and L. Phinney, Journal of Heat Transfer, In Press (2006).
48. B. Di Bartolo, *Optical Interactions in Solids* (John Wiley & Sons, New York, 1968).
49. G. Abstreiter, Applied Surface Science **50**, 73–78 (1991).
50. J. B. Cui, J. Ristein, and L. Ley, Physical Review Letters **81**, 429–432 (1998).
51. H. Brugger and P. W. Epperlein, Applied Physics Letters **56**, 1049 (1990).
52. S. Todoroki, M. Sawai, and K. Aiki, Journal of Applied Physics **58**, 1124 (1985).
53. R. Ostermeir, K. Brunner, G. Abstreiter, and W. Weber, Electron Devices, IEEE Transactions **39**, 858–863 (1992).
54. S. Rajasingam, J. W. Pomeroy, M. Kuball, M. J. Uren, T. Martin, D. C. Herbert, K. P. Hilton, and R. S. Balmer, Electron Device Letters, IEEE **25**, 456–458 (2004).
55. H. Kim, V. Tilak, B. M. Green, J. A. Smart, W. J. Schaff, J. R. Shealy, and L. F. Eastman, physica status solidi(a) **188**, 203–206 (2001).
56. M. Meneghini, L. Trevisanello, G. Meneghesso, E. Zanoni, F. Rossi, M. Pavesi, U. Zehnder, and U. Strauss, Superlattices and Microstructures **40**, 405–411 (2006).
57. M. Kuball, M. J. Uren, and T. Martin, In *Reliability Optimization for Wide Bandgap Devices: Recent Developments in High-Spatial Resolution Thermal Imaging of GaN Devices* (Bethesda, Maryland, USA, 2005), p. 246.
58. M. Kuball, S. Rajasingam, A. Sarua, M. J. Uren, T. Martin, B. T. Hughes, K. P. Hilton, and R. S. Balmer, Applied Physics Letters **82**, 124 (2002).
59. I. Ahmad, V. Kasisomayajula, D. Y. Song, L. Tian, J. M. Berg, and M. Holtz, Journal of Applied Physics **100**, 113718 (2006).
60. R. Aubry, C. Dua, J. C. Jacquet, F. Lemaire, P. Galtier, B. Dessertenne, Y. Cordier, M. A. DiForte-Poisson, and S. L. Delage, European Physical Journal of Applied Physics **30**, 77–82 (2005).
61. Y. Ohno, M. Akito, S. Kishimotoa, K. Maezawa, and T. Mizutani, Phys. Stat. Sol. C **0**, 57–60 (2002).
62. M. Kuball, J. M. Hayes, M. J. Uren, T. Martin, J. C. H. Birbeck, R. S. Balmer, and B. T. Hughes, IEEE Electron Device Letters **23**, 7–9 (2002).
63. J. Pomeroy, M. Kuball, D. Wallis, A. Keir, P. Hilton, R. Balmer, M. Uren, and T. Martin, Applied Physics Letters **87**, 103508 (2005).
64. M. Kuball, J. W. Pomeroy, S. Rajasingam, A. Sarua, M. J. Uren, T. Martin, A. Lell, and V. Harle, Physica Status Solidi(a) **202**, 824–831 (2005).
65. S. P. Kearney, L. M. Phinney, and M. S. Baker, Journal of Microelectromechanical Systems, **15**, 314–321 (2006).
66. M. Abel, Thesis, Georgia Institute of Technology, (2005).
67. J. Kim, J. A. Freitas Jr, P. B. Klein, S. Jang, F. Ren, and S. J. Pearton, Electrochemical and Solid-State Letters **8**, G345 (2005).
68. J. R. Serrano, L. M. Phinney, and S. P. Kearney, Journal of Micromechanics and Microengineering **16**, 1128–1134 (2006).

Chapter 7
Stereolithography and Rapid Prototyping

David W. Rosen

Abstract Stereolithography is a "rapid prototyping" or additive manufacturing process that has been used to fabricate 3D and high aspect ratio microstructures for customized packaging of microfluidic devices and microsensors. Micro-stereolithography denotes a collection of processes and technologies that can fabricate devices and packages with improved resolution, with features down to 1 μm in size. We show the feasibility of the integration of stereolithography with micromachined devices, using alignment, cleaning and dicing tests as these are applied to chemical sensors, interdigitated electrodes, and an AFM cantilever fluid cell package. Advantages of stereolithography and micro-stereolithography are based on their capability to fabricate complex 3D shapes. Packages fabricated with these processes provide designed small volumes that eliminate the dead volume that can occur in reaction chambers. The fabrication of MEMS packages on a wafer level scale decreases the manufacturing and assembly time in addition to being cost efficient.

Keywords: Stereolithography · Rapid prototyping · MEMS packages · Sensor packages · Micro-fluidics

7.1 Rapid Prototyping

Rapid prototyping (RP) technologies comprise a class of manufacturing technologies that enable parts and devices of virtually any shape to be fabricated directly from 3D Computer-Aided Design (CAD) models. The common characteristic of these technologies is that they build parts layer-by-layer in an additive manner. Originally, the field began as a means to produce prototype mechanical parts rapidly and, hence, was called "rapid prototyping." The first commercially available RP

D. W. Rosen
G. W. Woodruff School of Mechanical Engineering, Georgia Institute of Technology, Atlanta, GA 30332 404-894-9668 404-894-9342
e-mail: david.rosen@me.gatech.edu

P. J. Hesketh (ed.), *BioNanoFluidic MEMS.*

machines became available in the late 1980s. Stereolithography (SL) was arguably the first technology to be commercialized. In this technology, an ultra-violet laser traces part cross-sections on the top surface of a vat of liquid photopolymer, causing it to cure in the shape of the cross-section. By repeatedly curing cross-sections, parts get built layer by layer.

In 2006, it has been reported that more than 25 companies are marketing RP machines worldwide. The total market for machines and materials is approximately $530 million. More than 11,000 RP machines have been installed worldwide. As a result, the industry is significant, but relatively small compared to the machine tool, molding, and electronics manufacturing industries. Good resources are available for monitoring this industry, including the Castle Island web site [8] and the annual Wohlers Report [37]. Also, good references are available on RP technologies, machines, vendors, and applications, including the books edited by Paul Jacobs [22, 23].

The first applications of RP technologies were as prototypes of mechanical parts, essentially "prints" of CAD models that people could look at, hold in their hands, and convey design ideas to others. Other applications emerged, some quickly and others over many years. RP parts have been used as marketing samples, parts for soliciting production manufacturing quotes, patterns for casting processes, and even as tools and molds. Some companies have used RP parts to test assembly lines, ensuring that the production parts can be assembled properly. Other companies have installed RP parts in products, in cases where production parts were not yet available. Field upgrades were performed to swap the RP parts for production parts.

More recently, several production manufacturing applications have emerged, igniting interest in the idea of "rapid manufacturing," actually using RP machines to perform production manufacturing. Because these technologies can be utilized for much more than just making prototypes, it is reasonable to choose a name that better conveys their nature; hence, I use the term "Additive Manufacturing" or AM to highlight the additive nature of the processes and their potential application in production manufacturing.

In this chapter, the focus will be on two variants of the stereolithography technology, the commercial laser-scanning SL and mask-projection SL commonly used for micro-scale fabrication. Examples will be provided in the area of micro-fluidics and micro-sensors. Additionally, the usage of SL parts as molds for packages and devices made from PDMS is presented, along with some example devices.

7.2 Stereolithography

7.2.1 Technology Description

The basic idea of additive manufacturing is to add material and process it to form a patterned deposit. Most AM processes deposit material in layers (mostly planar layers), but differ on the method of forming the layer. For our purposes here, the stereolithography technology will be used to illustrate the part building process.

A schematic of a typical commercial SL machine is shown in Fig. 7.1. Parts are manufactured by fabricating cross sectional contours, or slices, one on top of another. In commercial SL machines, these slices are created by tracing 2D contours of a CAD model in a vat of photopolymer resin with a laser. The optics system includes a laser, focusing and adjustment optics, and two galvanometers that change the laser beam's position in the vat. The part to be built rests on a platform that is dipped into the vat of resin. After each slice is created, the platform is lowered, the surface of the vat is recoated, then the laser starts to trace the next slice of the CAD model, building the prototype from the bottom up. During the part preparation phase, the SL machine user has the opportunity to specify many process variables, including layer thickness, resin parameters, and the amount of inter-layer bonding. A more complete description of the stereolithography process may be found in [22, 23].

In general, the process of depositing material, processing the material with an energy source, and repeating is typical for many AM technologies. Other technologies only deposit material (usually molten, forming layers upon solidification) or deposit and process material simultaneously.

The process developed to fabricate micro-fluidics packages is shown in Fig. 7.2. The part or package to be fabricated in SL is modeled in CAD, then built on a SL machine. It is cleaned in an ultrasonic bath, typically with an alcohol or TPM solvent, until all liquid resin is removed from the part surfaces. The part is assembled

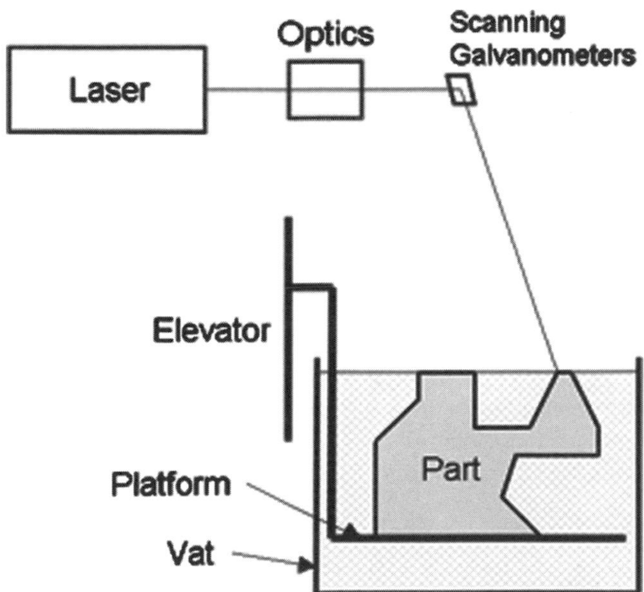

Fig. 7.1 Schematic of SL Machines

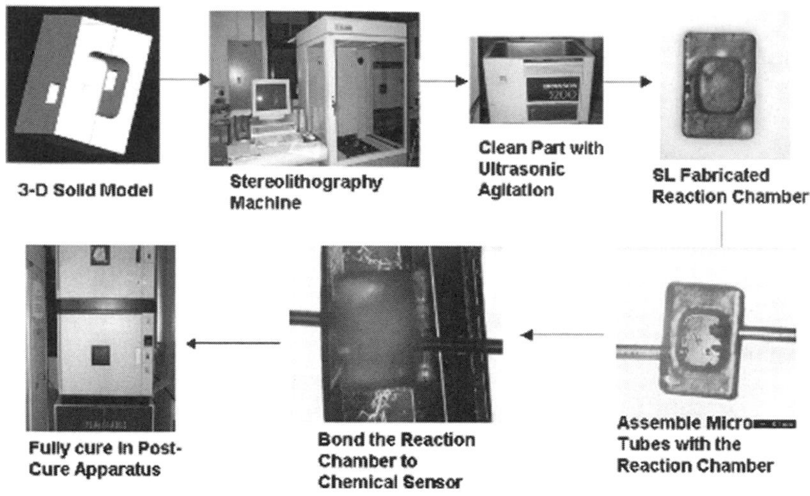

Fig. 7.2 Illustration of micro-fluidics system fabrication

with a die or other device components, often using SL resin as a bonding agent. After assembly, the package is placed in the Post-Cure Apparatus (PCA) for complete curing using blanket UV radiation.

7.2.2 Materials

The stereolithography process makes use of liquid, ultraviolet (UV) curable photopolymers as their primary materials. Frequently, these materials are called simply resins. Upon irradiation by a laser beam, these materials undergo a chemical reaction to become solid. This reaction is called photopolymerization, and is typically complex, involving many participating species. SL resins are similar to resists used in microelectronics, such as SU-8, in that both types of materials are photopolymers.

Photopolymers were developed in the late 1960s and soon became widely applied in several commercial areas, most notably the coating and printing industry. Many of the glossy coatings on paper and cardboard, for example, are photopolymers. Additionally, UV curable photopolymers are used in dentistry, such as for sealing the top surfaces of teeth in order to fill in deep grooves and prevent cavities.

Various types of radiation may be used to cure commercial photopolymers, including gamma rays, x-rays, electron beams, UV, and visible light, although UV and electron beam are the most prevalent. In SL systems from 3D Systems, UV radiation is used exclusively although, in principle, other types could be used. In the SLA-250, a helium–cadmium (HeCd) laser is used with a wavelength of 325 nm. In contrast, the solid-state lasers used in the other SL models are Nd-YVO_4, which are frequency-tripled to a wavelength of 354 nm.

Thermoplastic polymers that are typically injection molded have a linear or branched molecular structure that allows them to melt, solidify, melt, etc. In contrast, SLA photopolymers are cross-linked thermosets and, as a result, do not melt and exhibit much less creep and stress relaxation than do thermoplastics.

Free-radical photopolymerization was the first type that was commercially developed and was the first type developed for SL using acrylate chemistry. Acrylates form long polymer chains once the photoinitiator becomes "reactive," building the molecule linearly by adding monomer segments. Cross-linking typically happens after the polymer chains grow enough so that they become close to one another. Acrylate photopolymers exhibit high photospeed (react quickly when exposed to UV radiation), but have a number of disadvantages including significant shrinkage and a tendency to warp and curl. As a result, they are rarely used in isolation now.

The most common cationic photopolymers are epoxies, although vinylethers are also commercially available. Epoxy monomers have rings, as shown in Fig. 7.3. When reacted, these rings open, resulting in sites for other chemical bonds. Ring-opening is known to impart minimal volume change on reaction, because the number and types of chemical bonds are essentially identical before and after reaction [23]. As a result, epoxy SLA resins typically have much smaller shrinkages and much less tendency to warp and curl. Almost all commercially available SLA resins are combinations of epoxies and acrylates.

Polymerization is the process of linking small molecules (monomers) into larger molecules (polymers) comprised of many monomer units [22]. Polymerization of SLA monomers is an exothermic reaction, with heats of reaction around 85 kJ/mole

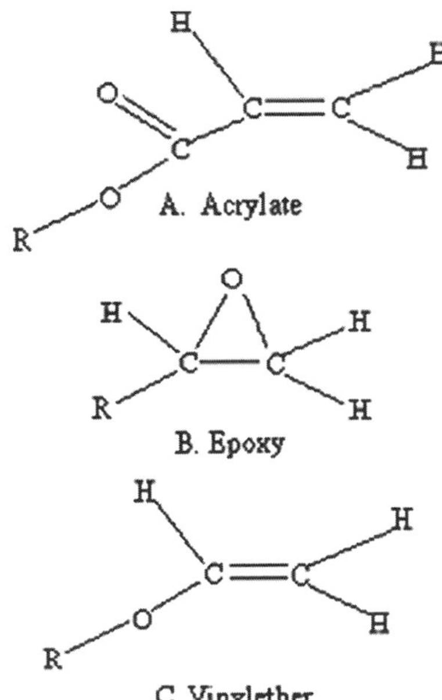

Fig. 7.3 Molecular Structure of SL Monomers

for an example acrylate monomer. Despite high heats of reaction, a catalyst is necessary to initiate the reaction. As described earlier, a photoinitiator acts as the catalyst.

Schematically, the free radical-initiated polymerization process can be illustrated as shown in Fig. 7.4 [22]. On average, for every two photons (from the laser), one radical will be produced. That radical can easily lead to the polymerization of over 1000 monomers, as shown in the intermediate steps of the process, called propagation. Polymerization terminates from one of three causes, recombination, disproportionation, or occlusion. In general, longer polymer molecules are preferred, yielding higher molecular weights. This indicates a more complete reaction. In Fig. 7.4, the P–I term indicates a photoinitiator, the $-I\bullet$ symbol is a free radical, and M in a monomer.

Cationic photopolymerization shares the same broad structure as free radical polymerization, where a photoinitiator generates a cation as a result of laser energy, the cation reacts with a monomer, propagation occurs to generate a polymer, and a termination process completes the reaction. A typical catalyst for a cationic polymerization is a Lewis Acid, such as BF_3 [36]. Cationic photopolymerization received little attention early on, but that changed during the 1990s as a result of the interest in SL technology.

Basic raw materials such as polyols, epoxides, (meth) acrylic acids and their esters, diisocyanates etc. are used to produce the monomers and oligomers used for radiation curing. Most of the monomers are multifunctional monomers or polyol polyacrylates which give a crosslinking polymerization. The main chemical families of oligomers are polyester acrylate, epoxy acrylates, urethane acrylates, amino acrylates (used as a photoaccelerator in the photoinitiator system) and cycloaliphatic epoxies [13].

Resin suppliers create ready-to-use formulations by mixing the oligomers and monomers with a photoinitiator, as well as other materials to affect reaction rates and part properties. In practice, photosensitizers are often used in combination with the photoinitiator to shift the absorption towards longer wavelengths. In addition, supporting materials may be mixed with the initiator to achieve improved solubility in the formulation. Furthermore, mixtures of different types of photoinitiators may also be employed for a given application. Thus, photoinitiating systems are, in practice, often highly elaborate mixtures of various compounds which provide optimum performance for specific applications [11].

Other additives facilitate the application process and achieve products of good properties. A reactive diluent, for example, is usually added to adjust the viscosity

$P\text{-}I \rightarrow \text{-}I\bullet$ (free radical formation)

$I\bullet + M \rightarrow I\text{-}M\bullet$ (initiation)

$I\text{-}M\bullet \rightarrow \rightarrow I\text{-}M\text{-}M\text{-}M\text{-}M\ldots\text{-}M\bullet$ (propagation)

$\rightarrow I\text{-}M\text{-}M\text{-}M\text{-}M\ldots\text{-}M\text{-}I$ (termination)

Fig. 7.4 Free Radical Polymerization Process

of the mixtures to an acceptable level for application [15]; it also participates in the polymerization reaction.

7.2.3 Modeling

As a laser beam is scanned across the resin surface, it cures a line of resin to a depth that depends on many factors. It is also important to consider the width of the cured line as well as its profile. The shape of the cured line depends on resin characteristics, laser energy characteristics, and the scan speed. We will investigate the relationships among all of these factors in this subsection.

The first concept of interest here is irradiance, the radiant power of the laser per unit area, $H(x,y,z)$. As the laser scans a line, the radiant power is distributed over a finite area. Consider a laser scanning a line along the x-axis at a speed V_s, where the z-axis is oriented perpendicular to the resin surface and is positive into the resin [22]. Assume that the coordinate origin is located such that the point of interest, $p\prime$, has an x coordinate of 0. The irradiance at any point x,y,z in the resin is related to the irradiance at the surface, assuming that the resin absorbs radiation according to the Beer-Lambert Law. Finally, assume that the lasers used in SL machines are Gaussian lasers. Then, the general form of the irradiance equation is given as:

$$H(x, y, z) = H(x, y, 0)\mathrm{e}^{-z/D_p} \tag{7.1}$$

where the exponential term models the resin's attenuation of laser energy. The irradiance at a point (x,y) on the resin surface is given by:

$$H(x, y, 0) = \frac{2P_L}{\pi W_0^2}\mathrm{e}^{-2[(x^2+y^2)/W_0^2]} \tag{7.2}$$

where: P_L=output power of laser [W], V_s=scan speed of laser [m/s], W_0=radius of laser beam focused on the resin surface [cm], D_p=depth of penetration of laser into a resin until a reduction in irradiance of $1/e$ is reached (key resin characteristic) [cm].

However, it is exposure, not irradiance, that is of interest since the level of **exposure** at a point determines whether or not the resin at that point gels (solidifies). Exposure is the energy per unit area and can be determined at point p on the resin surface by appropriately integrating Eqn. 7.2 along an entire scan line, as shown in Eqn. 7.3. The scan starts at a time t_s and ends at time t_e.

$$E(y, 0) = \int_{t_s}^{t_e} \frac{2P_L}{\pi W_0^2}\mathrm{e}^{-2[(x^2(t)+y^2(t))/W_0^2]}\mathrm{d}t \tag{7.3}$$

To integrate Eqn. 7.3, it is convenient to assume that the scan is along the x-axis from point (x_s,y) to point (x_e,y). After performing an appropriate change of variables, the exposure received at point $(0,y)$ is:

$$E(y,0) = \frac{P_L}{\sqrt{2\pi} W_0 V_s} e^{-2y^2/W_0^2} [\mathrm{erf}(b) - \mathrm{erf}(a)] \tag{7.4}$$

where $a = \sqrt{2}/W_0 x_s$ and $b = \sqrt{2}/W_0 x_e$. Exposure at an arbitrary point in the vat can be modeled by adding the exponential attenuation term from Eqn. 7.1:

$$E(y,z) = \frac{P_L}{\sqrt{2\pi} W_0 V_s} e^{-2y^2/W_0^2} e^{-z/D_p} [\mathrm{erf}(b) - \mathrm{erf}(a)] \tag{7.5}$$

It turns out that if a scan vector is longer than several times the laser beam diameter, that the scan is effectively infinitely long. This is due to the fast attenuation of Gaussian laser beams. Changing the limits of integration in Eqn. 7.3 from t_s to t_e to $-\infty$ and $+\infty$, respectively, and changing variables from time to distance, x, gives an equation for exposure distribution on the resin surface:

$$E(y,0) = \sqrt{\frac{2}{\pi}} \frac{P_L}{W_0 V_s} e^{-2y^2/W_0^2} \tag{7.6}$$

Combining this with the resin attentuation term yields the fundamental general exposure equation:

$$E(x,y,z) = \sqrt{\frac{2}{\pi}} \frac{P_L}{W_0 V_s} e^{-2y^2/W_0^2} e^{-z/D_p} \tag{7.7}$$

The profile of a cured scan line can be determined readily. Starting with Eqn. 7.7, the locus of points in the resin that is just at its gel point, where $E = E_c$, is denoted by $y*$ and $z*$. Eqn. 7.7 can be rearranged, with $y*$, $z*$, and E_c substituted to give Eqn. 7.8.

$$e^{2y^{*2}/W_0^2 + 2z^{*2}/D_p} = \sqrt{\frac{2}{\pi}} \frac{P_L}{W_0 V_s E_c} \tag{7.8}$$

Taking natural logarithms of both sides yields

$$2\frac{y^{*2}}{W_0^2} + \frac{z^{*2}}{D_p} = \ln\left[\sqrt{\frac{2}{\pi}} \frac{P_L}{W_0 V_s E_c}\right] \tag{7.9}$$

which is the equation of a parabolic cylinder in $y*$ and $z*$, as is clear from the following form.

$$ay^{*2} + bz^{*} = c$$

where a, b, and c are constants, immediately derivable from Eqn. 7.9.

As is probably intuitive, the width of a cured line of resin is the maximum at the resin surface; i.e., y_{max} occurs at z=0. To determine line width, we start with the line shape function derived earlier, Eqn. 7.9. Setting z=0 yields Eqn. 7.10:

$$L_w = 2y^2 = W_0^2 \ln\left(\sqrt{\frac{2}{\pi}}\frac{P_L}{W_0 V_s E_c}\right) \quad (7.10)$$

Line width is denoted L_w and is 2* y_{max}. With a lot of substitution and algebra, it is possible to show that the line width can be computed by Eqn. 7.11.

$$L_w = W_0 \left(\frac{2C_d}{D_p}\right)^{1/2}$$

where C_d=cure depth=depth of resin cure as a result of laser irradiation [cm].

As a result, two important aspects become clear. First, line width is proportional to the beam spot size. Second, if a greater cure depth is desired, line width must increase, all else remaining the same. This becomes very important when performing process planning, particularly if fine features need to be fabricated.

7.3 Micro-Fluidics and Micro-Sensor Examples

For small volume manufacturing of packaged sensors or microfluidics, the production rate can be increased if the packages are integrated with the sensors on the wafer level. Using stereolithography technology to integrate the packages directly on top of the sensors on a silicon wafer can eliminate the assembly time for each individual cell. In addition the process provides a seal between the package and the sensors. Some similar integration of microfluidics on top of some micromachined silicon dies has been done before (Ikuta et al. 1994, 1998, 1999). However, the details of how a practical system uses this technology needs to be addressed. These issues were: (1) the cleaning method and process of the sensing areas which have been contaminated with resin or solvent; (2) the dicing method for the packaged sensors on a wafer; (3) the functionality of the sensors after the packaging and cleaning processes; (4) the alignment between the location of the laser scanning and the sensor location. In order to investigate the feasibility of integrating and manufacturing the high aspect ratio stereolithography structures on top of the micromachined devices, many processes details need to be investigated.

7.3.1 Experiments

For the examples presented in this section, the stereolithography machine, SLA 3500 (3D Systems, Rock Hill, SC), was used to fabricate 0.3 cm^3 measurement cells

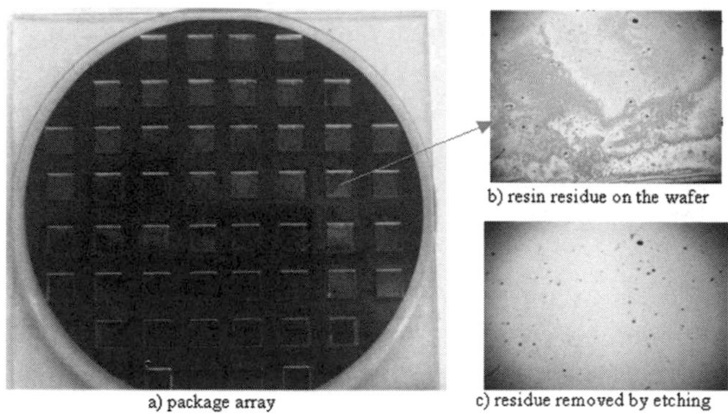

Fig. 7.5 Array of sensor housings built on a wafer [34]

on a 3-inch silicon wafer as shown in Fig. 7.5a. The resin, SL 7510 (Huntsman), was used as the stereolithography material.

7.3.1.1 Remnant Removal

After building the stereolithography packages around the micromachined sensors, some resin residue remained on the surface of the sensing areas as in Fig. 7.5b [34]. The surface residue may affect sensor functionality, so the residue must be removed without causing any damage to the devices. Since the SL resin is a cross-linked photopolymer, it cannot be removed easily with another solvent. Thus, oxygen plasma etching (LFE Barrel Etcher) was used to remove the thin layer of resin on the silicon surface. After oxygen plasma etching for about 1 hour, the whole packaged wafer was almost free of resin residue. A picture of the same measurement cell after the cleaning is shown in Fig. 7.5c.

7.3.1.2 Dicing

After sensor arrays on a silicon wafer have been fabricated and packaged with SL, each individual measurement cell needs to be separated. Dicing of plain silicon is well understood, however, dicing of silicon with stereolithography polymer is not normally done. A diamond dicing saw was used to dice the packaged silicon wafer to individual cells. The diced measurement cells were diced successfully [33].

7.3.1.3 Alignment of Wafers and Devices to SL Packages

In order to build packages on wafers, it is necessary to align the wafer and the SL machine. SL machines do not have position feedback capability, so alignment must be planned into the fabrication process. Two broad approaches to alignment have been developed in our work. Low precision alignment can be achieved by simply

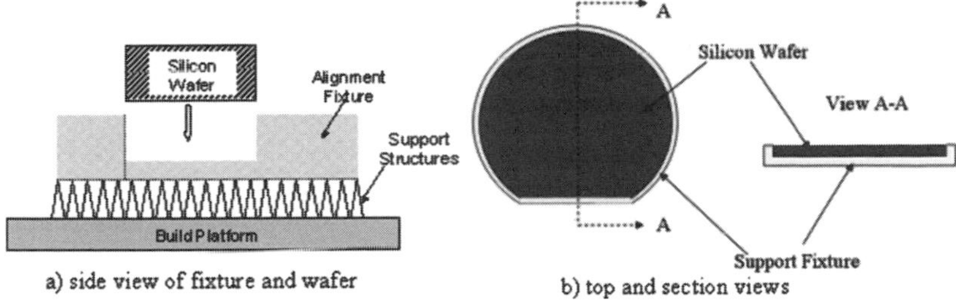

Fig. 7.6 Low precision alignment approach to aligning wafers to SL builds

building a fixture for the silicon wafer, as shown in Fig. 7.6. The fixture is fabricated with a cavity into which the wafer will be placed. When the SL machine builds the fixture to the level of the top of the wafer, the SL machine is paused and the wafer is inserted into the fixture cavity in the SL machine. Then, the SL machine build process can be resumed and additional devices or parts of the fixture can be built.

The second approach achieves much higher precision and utilizes a mask aligner, a typical machine in standard photolithography practice [10]. The first step in this approach is to fabricate a fixture, as in the low precision approach. The fixture should be removed from the SL machine and the wafer placed into it. Subsequent steps include:

- Fabricate an array of packages or devices using the SL machine that serves additionally as a mask. Holes should be designed into the mask/array that facilitate alignment.
- Align the SL mask/array to the wafer using a mask aligner.
- Bond mask/array to wafer by coating the bottom surface of the mask/array with SL resin and exposing the assembly to UV radiation.

An example of the alignment process is shown in Fig. 7.7. First, the array of SL packages is fabricated, which also serves as the alignment mask. The mask assembly is placed on top of the wafer with devices already fabricated. The mask aligner is used to manually align mask and wafer.

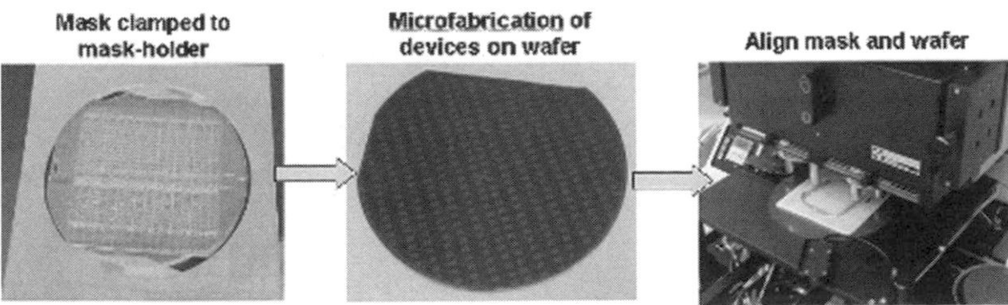

Fig. 7.7 High precision alignment of wafer and SL package array

7.3.2 *Results*

7.3.2.1 Typical Device Fabrication Process

As described, an array of measurement cells was fabricated with stereolithography on top of a 3-inch silicon wafer, as shown in Fig. 7.5a. Oxygen plasma etching was used to clean the silicon surface. In Fig. 7.8, an array of measurement cell lids is shown with inlet and outlet nozzles of 1 mm diameter, which can be connected to standard size tubing. The lid was bonded on top of the silicon wafer package to enclose the measurement cells. A diamond dicing saw was used to dice the modified silicon wafer into individual cells. This is a demonstration of packaging micromachined sensors on a wafer level scale.

In order to ensure the functionality of the packaged devices, the packaging techniques described above were applied to some real micromachined devices, such as arrays of chemical sensors and interdigital electrodes, and an atomic force microscope probe.

7.3.2.2 Chemical Sensor

An array of devices very similar to that shown in Figs. 7.5 and 7.8 was fabricated. Each measurement cell has an internal volume of about 1 mm^3. After assembly, impedance measurements were taken by a 1260 Impedance/Gain Phase Analyzer (Solatron, Hampshire, UK). Under the room temperature of 16°C, driven frequency of 1 KHz, voltage of 10 mV, the modified impedance measurement of the sensor with purging 1000 ppm Ammonia gas in Argon was shown in Fig. 7.9 [34]. This result showed that the integration of the stereolithography reaction chamber did not affect the functionality of the sensor. In addition, the small dead volume of the reaction chamber decreased the response time of the sensor.

Fig. 7.8 Lid array with I/O nozzles [34]

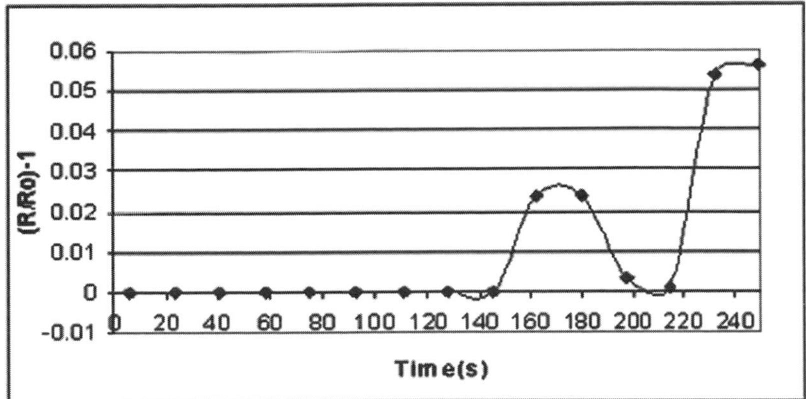

Fig. 7.9 Impedance measurement of packaged chemical sensor [34]

7.3.2.3 Biochemical Sensor

A cyclic voltammetry measurement of a stereolithography fabricated measurement cell around an interdigitated electrodes array was taken and shown in Fig. 7.10 [34]. The peaks of the graph showed that the platinum electrodes and contact pads were still active after integration of stereolithography packages and cleaning processes.

7.3.2.4 Build on AFM Cantilever Beam

A stereolithography fabricated fluid measurement cell was built around an atomic force microscope cantilever beam with a nano probe. The measurement cell has a volume of about 4 μL with a wall thickness of about 400 μm. The fluid measurement cells with atomic force microscope (AFM) probe can be used for tests such as DNA test, bio-cell test and so on, which must be carried out under buffer liquid.

Similar to the examples above, the fluid measurement cell with AFM probe was tested after fabrication and cleaning with oxygen plasma etcher. The cantilever beam

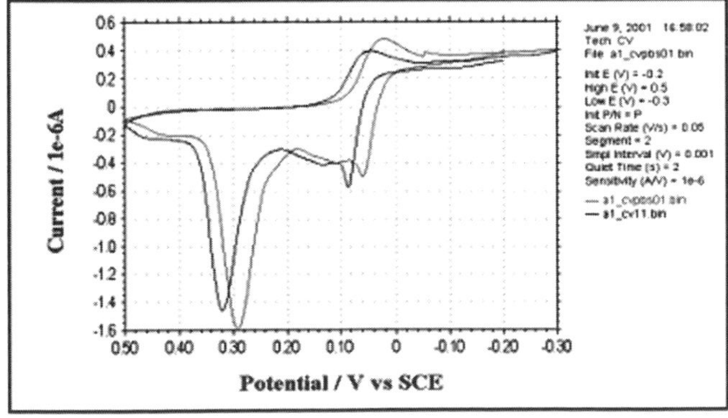

Fig. 7.10 Cyclic voltammetry measurement of a SL packaged interdigitated electrode device [34]

with integrated measurement cell was mounted on an atomic force microscope and a frequency test performed. The device performed successfully [33].

7.4 Micro-Stereolithography

7.4.1 Introduction

A considerable amount of research has been performed to extend the general SL technology into the micro-scale. Three broad methods have been investigated:

- laser scanning, similar to macro-scale SL,
- mask-projection SL, where dynamic masks are used to irradiate entire part cross-sections at one time, and
- two-photon polymerization, that is similar to laser scanning SL, except that the photoinitiators require two photons, instead of just one, to become active, which enables very fine resolution in the SL process.

Laser scanning technologies for the micro-scale typically have scanned the vat in X, Y, and Z directions, rather than scanning the laser beam, since the focal length must be so short in order to achieve small beam spot sizes. The Integrated Hardening method of Ikuta and Hirowatari [17] was one of the first developments in this area. They used a laser spot focused to a 5 μm diameter and the resin vat is scanned underneath it to cure a layer. Examples of devices built with this method include tubes, manifolds, and springs and flexible microactuators [30] and fluid channels on silicon [18]. Takagi and Nakajima [31] have demonstrated the use of this technology for connecting MEMS gears together on a substrate. The artifact fabricated using micro-SL can be used as a mold for subsequent electroplating followed by removal of the resin [21]. Indeed, this current research has been able to achieve sub-1 μm minimum feature size. The capability of building around inserted components has also been proposed for components such as ultrafiltration membranes and electrical conductors. Applications include fluid chips for protein synthesis [19] and bioanalysis [20]. The bioanalysis system was constructed with integrated valves and pumps that include a stacked modular design, $13 \times 13\,\mathrm{mm}^2$ and 3 mm thick, each of which has different fluid function. However, the full extent of integrated processing on silicon has not yet been demonstrated. The benefits of greater design flexibility and lower cost of fabrication will be realized.

Mask-projection SL (MPSL) was also developed during the 1990s. Several groups in Japan and Europe pursued this technology. The basic idea is to project an image displayed on a dynamic mask onto the resin surface in the vat in order to cure a part cross-section. The main advantage of this method is speed: since an entire part cross-section can be cured at one time, it can be faster than scanning a laser beam. Dynamic masks can be realized by LCD screens, by spatial light modulators, or, more recently, by Digital Micromirror Devices (DMD), such as the Digital Light Processing (DLP™) chips manufactured by TI.

A schematic and photograph of our MPSL system is shown in Fig. 7.11. The MPSL process starts with the CAD model of the object to be built. The object is sliced at various heights and the cross-sections of the slices are stored as bitmaps. These bitmaps are displayed on a dynamic pattern generator and are imaged onto the resin surface in order to cure a layer. The layer is built on a platform which is lowered into a vat of resin to coat the cured layer with a fresh layer of resin and the next layer, corresponding to the next cross section is cured on top on it. Likewise, by curing layers one over the other, the entire micro part is built.

MPSL systems have been realized by several groups around the world. Some of the earlier systems utilized LCD displays as their dynamic mask [5, 26], while another early system used a spatial light modulator [9, 14]. The remaining systems all used DMD's as their dynamic masks [1, 2, 4, 16, 24, 28]. These latest systems all use UV lamps as their radiation source, while other have used lamps in the visible range [1, 4] or lasers in either the UV [9, 14].

A wide range of applications have been pursued by these researchers. Mechanical objects were common, although some micro-fluidics packaging was reported. Independently, Bertsch et al. [3], and Sun et al. [27], used ceramic particle-filled resins to fabricate green ceramic parts which were subsequently fired to produce fully dense ceramic parts with feature sizes less than 10 μm in size. A good overview of micro-SL technology, systems, and applications is the book by Varadan et al. [35].

Most of the research presented in these papers is experimental. In [24], we presented the MPSL system that we developed. The system comprises of broadband UV lamp as the light source, a Digital Micromirror Device (DMD) from Texas Instruments as a dynamic mask and an automated XYZ stage from ASI imaging. We cure parts out of the DSM SOMOS 10120 resin with our system. We modeled the lateral dimensions of a layer cured using our MPSL system in terms of the process parameters. The irradiation of the resin surface has been modeled using the ray tracing approach. The curing characteristics of the resin have been empirically modeled by plotting its working curve. These models were used to formulate a

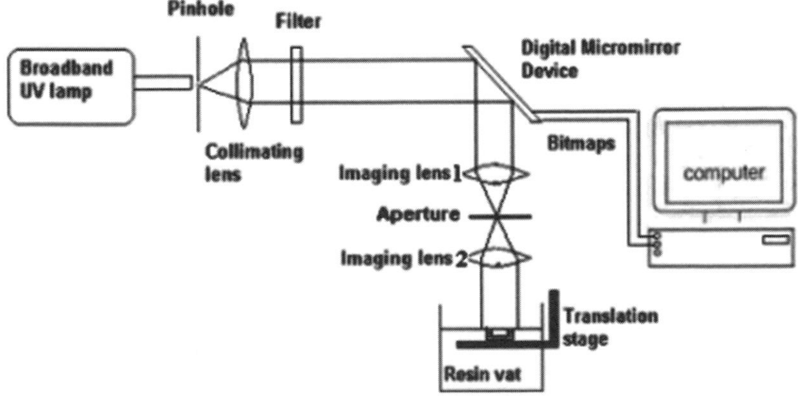

Fig. 7.11 Schematic of the MPSL optical system [23]

process-planning method to cure a layer with the required lateral dimensions. This method was used to generate the bitmap to be displayed on the DMD and compute the time for which it should be imaged onto the resin surface to cure the desired layer. Using this process planning method it is possible to cure layers within 3% error in their lateral dimensions.

The vertical dimension of a MPSL part built by curing dimensionally accurate layers over one another is not equal to the algebraic summation of the individual layer thicknesses and involves some errors. These errors are a result of unwanted cure due to print through errors. In [25], we proposed a method called the "Compensation Zone approach" to compensate for this unwanted curing. This method entails subtracting a tailored volume (Compensation Zone) from underneath the CAD model in order to compensate for the increase in the Z dimension that would occur due to print-through. By controlling the process parameters, including the thickness of the Compensation Zone, it is possible, in theory, to eliminate the print-through errors completely.

7.4.2 Compensation Zone Modeling

The primary process variables under user control are:

- Thickness of the Compensation Zone, given by the function $Z_c(x,y)$
- Thickness of every layer given by the function $LT_k(x,y)$, where $LT_k(x,y)$ is the thickness of the kth layer from bottom
- Exposure supplied to cure every layer, given by function $E_k(x,y)$

In Fig. 7.12, it can be seen that the thickness of every layer is not constant. For example, the bottom layer of the boss protrusion is cylindrical and has a non-zero compensation zone to compensate for extra exposure leaking through from layers above. Hence, we denote the layer thicknesses by functions of lateral coordinates as $LT_k(x,y)$. The exposure distribution for the bottom layer of the protrusion must correspond to the desired thickness profile. The central region does not have a compensation zone since there are no layers directly above it. No overcure is necessary

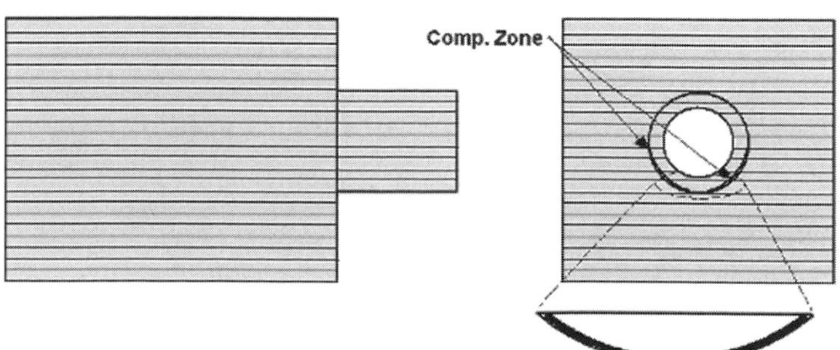

Fig. 7.12 Sliced part showing compensation zone

since no layers are directly below it either. In order to cure Layer 4, an exposure as shown in Fig. 7.4 will have to be supplied. The overcure (OC) required to bind the layer to the layer underneath it will be required only at the middle portion of the layer. At its edges, OC will be zero. In order to cure this layer, the general exposure equation for the kth pixel (projected micro-mirror) is given by Eqn. 7.12.

$$E_k(x,y) = E_c e^{(LT_k(x,y)+OC)/D_p} \tag{7.12}$$

The exposure received along the bottom surface of the part should be equal to E_c. Summing up all of the exposures from all layers at the kth pixel gives:

$$\sum_{k=1}^{n} E_c e^{(LT_k(x,y)+OC)/D_p} e^{-(\sum_{m=1}^{k} LT_m(x,y)+Z_c(x,y))/D_p} = E_c \tag{7.13}$$

Canceling the term E_c, we get the relation between the layer thickness $LT_k(x,y)$ and Compensation Zone $Z_c(x,y)$:

$$\sum_{k=1}^{n} e^{(LT_k(x,y)+OC)/D_p} e^{-(\sum_{m=1}^{k} LT_m(x,y)+Z_c(x,y))/D_p} = 1 \tag{7.14}$$

The height of the part is the summation of the thickness of every layer and the thickness of the Compensation Zone. The height of the part will be given by Eqn. 7.15.

$$h(x,y) = \sum_{k=1}^{n} LT_k(x,y) + Z_c(x,y) \tag{7.15}$$

In order to build a part with accurate vertical dimensions, process variables of layer thickness and compensation zone should be selected so that Eqns. 7.12, 7.14, and 7.15 are satisfied. As such, the Compensation Zone approach can be represented as a problem of solving simultaneous nonlinear equations.

These equations are solved independently at every discrete point on the part's bottom surface. For every discrete point (x_1,y_1), the problem has $n+1$ independent variables, corresponding to the thicknesses of n layers $(LT_k(x_1,y_1))$ and the thickness of the Compensation Zone $(Z_c(x_1,y_1))$. $E_k(x_1,y_1)$ is a dependent variable, dependent on $LT_k(x_1,y_1)$ as given by Eqn. 7.12. Only two constraints, given by Eqns. 7.14 and 7.15, are to be satisfied by the independent variables. So, for a part with the number of layers $n>1$, the problem is under-constrained. Multiple combinations of layer thicknesses and Compensation Zone thickness can be used to avoid print-through. Additional constraints would have to be imposed on the layer thicknesses and the Compensation Zone thickness so as to have a problem with unique solution. The solution can then be computed using some root finding algorithm.

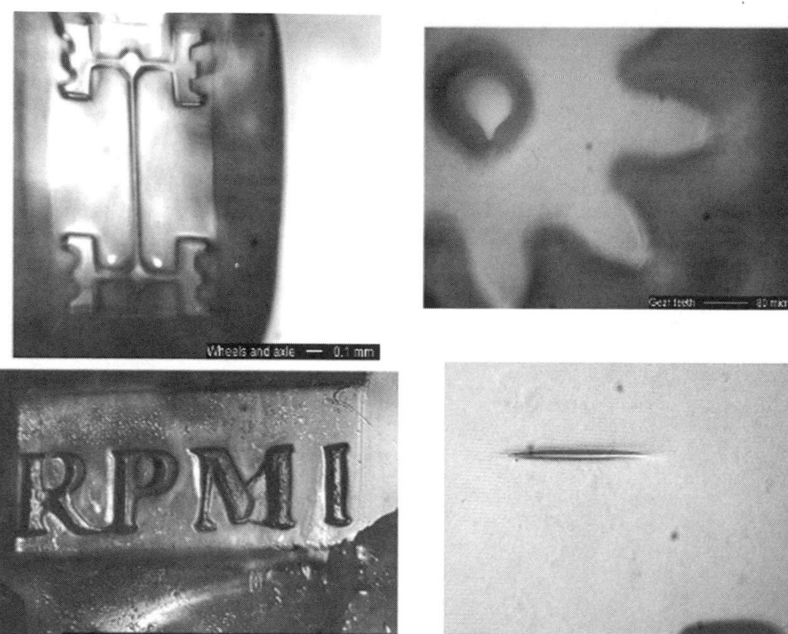

Fig. 7.13 Example parts built using MPSL system [24]

7.4.3 Examples

The pictures taken under an optical microscope of some of the micro parts cured using our system are presented in Fig. 7.13. In Fig. 7.13a, the four wheels and axle of a micro-SUV is shown. This is a nine-layer part. The axle is 57 μm in width and is overhanging. Fig. 7.13b is the close-up of the teeth of a micro spur gear. The thickness of the teeth at the pitch circle diameter of the gear is measured to be 40 μm. Fig. 7.13c is a single layer part that shows the logo of our laboratory. Fig. 7.13d shows a two layered, three pixel wide rib which was cured to validate our layer cure model for very small features. The experimentally measured width of the line is 6 μm while the designed width was 6.2 μm, helping to validate our process model.

7.5 PDMS Molding with SL Molds

Poly Di-Methyl Siloxane (PDMS) is an elastomer compound that can be defined on glass or silicon wafers using procedures like spin coating, dip coating etc. The surface of the elastomer can be activated by exposing it to an oxygen plasma environment. The activated PDMS coated substrate can then be bonded to glass substrate or SU8 coated on glass [12] (on which microchannel structures have been previously defined using photolithography). This bonding process can be an efficient and rapid technique for fabricating microchannels.

PDMS packages and flow channels can be fabricated easily by using molds produced in SL. The typical procedure for PDMS molding in SL molds is as follows.

The channels or package geometry is modeled in CAD, then built using a SL machine. In our work, the SL mold was filled with PDMS solution (10 parts Sylgard 184 silicone elastomer base (Dow Corning, Midland, MI) to 1 part curing agent). After degassing in a vacuum desiccator overnight, the PDMS was cured at 60°C for 1 h. The PDMS was easily removed from the mold, and leak-free contact was made with silicon wafers or dies using finger pressure contact.

7.5.1 Fluid Flow Manifold for a Microvalve

The electromagnetic microvalve fabricated for an integrated direct methanol micro-fuel cell [6, 7] is shown in Fig. 7.14. Figure 7.15 shows the parts that have been built with stereolithography for microvalve testing in a methanol water mixture. The figure shows a 12 mm × 12 mm chip, which has 12 microvalves. The sealing in the microvalve structure was accomplished using PDMS. The PDMS was molded in the SL part to define the fluidic channels and housing for testing the microvalve [29]. Testing demonstrated that the assembled structure did not leak until a pressure of 57.4 kPa was reached.

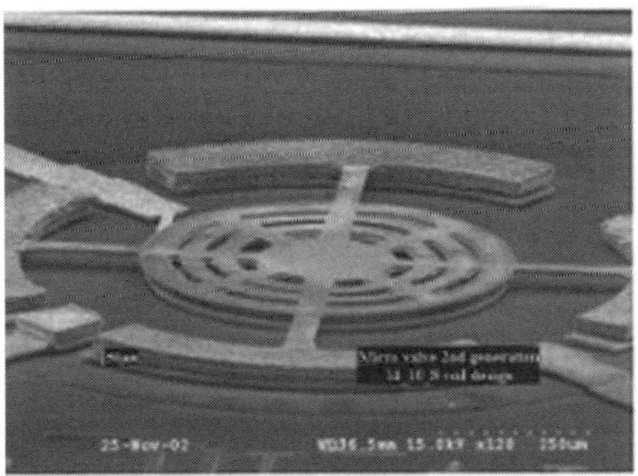

Fig. 7.14 SEM micrograph of completed microvalve [7]

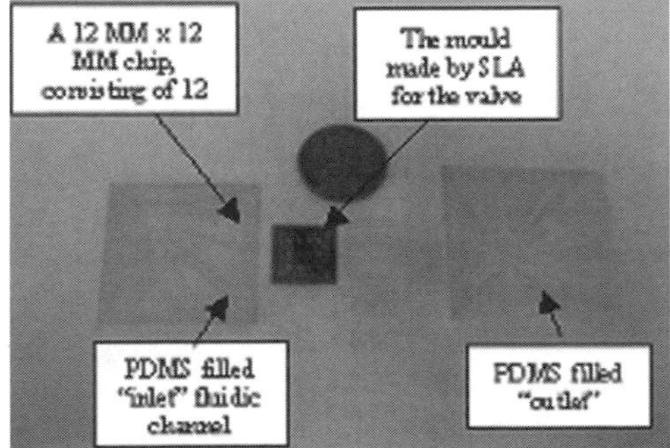

Fig. 7.15 The parts for the whole microvalve system [7]

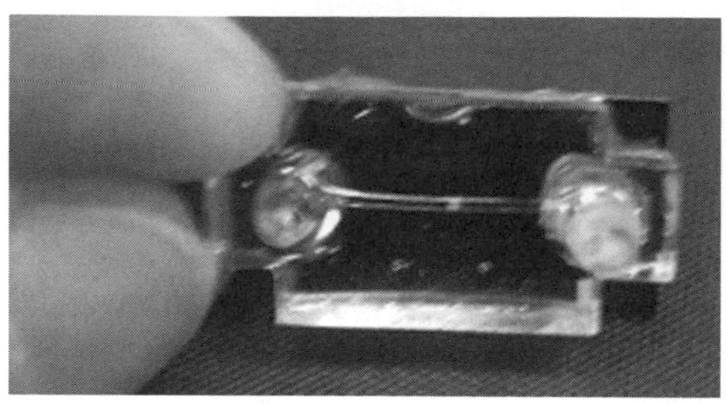

Fig. 7.16 Photo of bioassay on-a-chip device [32]

7.5.2 *Bioassay on a Chip*

A biofluidics system was developed for detecting the bacteriophage MS2, a simulant for biothreat viruses, such as smallpox using the PDMS molding method [32]. The system is shown in Fig. 7.16. A PDMS channel was fabricated and positioned over a platinum trench interdigitated array (IDA) electrode, Ag/AgCl reference electrode, and platinum auxiliary electrode on a silicon dioxide chip. The PDMS channel was 300 μm deep, 500 μm wide, and 12 mm long. The detection approach utilized a microbead-based electrochemical immunoassay.

Experimental results demonstrated that no leakage was observed for flow rates of 0.01∼10 mL/min. This cell achieved a sensitivity of 90 ng/mL MS2 with a rapid response time.

Acknowledgments The author gratefully acknowledges the assistance of many people: SL microfluidics (Drs. James Gole, Peter Hesketh, Lenward Seals, L. Angela Tse), MPSL (Mr. Ameya Limaye), PDMS molding (Mr. J.S. Bintoro, Dr. Sangkyum Kim), as well as funding from many sources: School of Mechanical Engineering at Georgia Tech; Rapid Prototyping & Manufacturing Institute member companies (3D Systems, Baxter Healthcare, DSM Somos, Ford, Huntsman, Pratt & Whitney, Siemens), the Manufacturing Research Center at Georgia Tech, the Microelectronics Research Center at Georgia Tech, DARPA, Motorola, the National Science Foundation through grants DMI-9618039 and IIS-0120663, and the US Department of Education GAANN fellowship program.

References

1. Beluze L, Bertsch A, Renaud P (1999) "Microstereolithography: a new process to build complex 3D objects," SPIE Symposium on Design, Test and Microfabrication of MEMS/MOEMS, Vol. 3680, pp. 808–817.
2. Bertsch A, Zissi S, Jezequel J, Corbel S, Andre J (1997) "Microstereolithography using liquid crystal display as dynamic mask-generator," *Microsystems Technologies*, pp. 42–47.
3. Bertsch A., Lorenz H., Renaud P. (1999) "3D Microfabrication by combining microstereolithography and thick resist UV lithography," *Sensors and Actuators*, Vol. 73, pp. 14–23.
4. Bertsch A, Bernhard P, Vogt C, Renaud P (2000) "Rapid prototyping of small size objects," *Rapid Prototyping Journal*, Vol. 6, Number 4, pp. 259–266.
5. Bertsch A, Jiguet S, Renaud P (2004) "Microfabrication of ceramic components by microstereolithography," *Journal of Micromechanics and Microengineering*, Vol. 14, pp. 197–203.

6. Bintoro JS, Hesketh PJ (2005) "An electromagnetic actuated on/off microvalve fabricated on top of a single wafer," *Journal of Micromechanics and Microengineering*, Vol. 15, Number 6, pp. 1157–1173.
7. Bintoro JS, Luharuka R, Hesketh PJ (2003) "A structure of bistable electromagnetic actuated microvalve fabricated on a single wafer, implementing the SLA and PDMS technique," ASME International Mechanical Engineering Congress and RD & D Expo, paper IMECE2003-43857, Washington, DC, November 15–21.
8. Castle Island, http://home.att.net/~castleisland/.
9. Chatwin C, Farsari M, Huang S, Heywood M, Birch P, Young R, Richardson J (1998) "UV microstereolithography system that uses spatial light modulator technology," *Applied Optics*, Vol. 37, pp. 7514–7522.
10. Choudhury A (2003) "Process development for a silicon carbide micro four-point probe," Masters Thesis, Georgia Institute of Technology.
11. Crivello JV, Dietliker K (1998) "Photoinitiators for Free Radical, Cationic & Anionic Photopoly-merisation," 2nd Edition, Vol. III In *Chemistry & Technology of UV & EB Formulation for Coatings, Inks & Paints*, Edited by Bradley G., John Wiley & Sons, Inc., Chichester & New York.
12. Cui L, Morgan H (2000) "Design and fabrication of traveling wave dielectrophoresis structures," *Journal of Micromechanics and Microengineering*, Vol. 10, pp. 72–79.
13. Dufour P (1993) "State-of-the-art and Trends in Radiation Curing", In Radiation Curing in Polymer Science and Technology–Vol I: Fundamentals and Methods, Edited by Fouassier J. P. and Rabek J. F., Elsevier Applied Science, London & New York.
14. Farsari M, Huang S, Birch P, Claret-Tournier F, Young R, Budgett D, Bradfield C, Chatwin C (1999) "Microfabrication by use of spatial light modulator in the ultraviolet: experimental results", *Optics Letters*, Vol. 24, No. 8, pp. 549–550.
15. Fouassier JP (1993) "An Introduction to the Basic Principles in UV Curing", In Radiation Curing in Polymer Science and Technology–Vol I: Fundamentals and Methods, Edited by Fouassier J. P. and Rabek J. F., Elsevier Applied Science, London & New York.
16. Hadipoespito G, Yang Y, Choi H, Ning G, Li X (2003) "Digital Micromirror device based microstereolithography for micro structures of transparent photopolymer and nanocomposites", Proceedings of the 14th Solid Freeform Fabrication Symposium, Austin TX, pp. 13–24.
17. Ikuta K, Hirowatari K (1993) "Real three dimensional microfabrication using stereolithography and metal molding," *Proceedings of the IEEE MEMS*, pp. 42–47.
18. Ikuta K, Hirowatari K, Ogata T (1994) "Three dimensional micro integrated fluid systems fabricated by micro stereo lithography," *Proceedings of the IEEE MEMS*, pp. 1–6.
19. Ikuta K, Ogata T, Tsubio M, Kojima S (1996) "Development of mass productive microsterolithography," *Proceedings of the MEMS*, pp. 301–305.
20. Ikuta K, Maruo S, Fujisawa T, Fukaya Y (1998) "Chemical IC chip for dynamical control of protein synthesis," *Proceedings of the Interior Symposis Micromechatronics and Human Science, IEEE*, pp. 249–254.
21. Ikuta K, Maruo S, Fujisawa T, Yamada A (1999) "Micro concentrator with opto-sense micro reactor for biochemical IC chip family," *Proceedings of the IEEE MEMS*, pp. 376–380.
22. Jacobs PF (1992) *Rapid Prototyping & Manufacturing, Fundamentals of Stereolithography*, Society of Manufacturing Engineers, Dearborn, MI.
23. Jacobs PF (1996) *Stereolithography and other RP&M Technologies*, Society of Manufacturing Engineers, Dearborn, MI.
24. Limaye A, Rosen DW (2004) "Quantifying dimensional accuracy of a mask projection micro stereolithography system," Proceedings of the Solid Freeform Fabrication Symposium, Austin, TX, August 2–4.
25. Limaye A, Rosen DW (2006) "Compensation zone approach to avoid *Z* errors in mask projection stereolithography builds," *Rapid Prototyping Journal*, Vol.12, No. 5, pp. 283–291.

26. Monneret S, Loubere V, Corbel S (1999) "Microstereolithography using dynamic mask generator and a non-coherent visible light source", Proceedings of SPIE, Vol. 3680, pp. 553–561.
27. Sun C, Fang N, Zhang X (2002) "Experimental and numerical investigations on micro-stereolithography of ceramics," *Journal of Applied Physics*, Vol. 92, Number 8, pp. 4796–4802.
28. Sun C, Fang N, Wu D, Zhang X (2005) "Projection micro-stereolithography using digital micro-mirror dynamic mask", *Sensors and Actuators A*, Vol. 121, pp. 113–120.
29. Sutanto, J, Hesketch, PJ, Berthelot, YH (2006) "Design, microfabrication and testing of a CMOS compatible bistable electromagnetic microvalve with latching/unlatching mechanism on a single wafer," *Journal of Micromechanics and Microengineering*, Vol. 16, Number 2, pp. 266–275.
30. Suzumori K, Koga A, Haneda R (1994) "Microfabrication of integrated FMA's using stereo lithography," *Proceedings of the IEEE*, pp. 136–141.
31. Takagi T, Nakajima N (1994) "Architecture combination by microphotoforming process," *Proceedings of the IEEE MEMS*, pp. 211–216.
32. Thomas JH, Kim SK, Hesketh PJ, Halsall HB, Heineman WR (2004) "Bead-based electrochemical immunoassay for bacteriophage MS2," *Analytical Chemistry*, Vol. 76, Number 10, pp. 2700–2707.
33. Tse AL (2002) "MEMS packaging with stereolithography," Masters Thesis, Georgia Institute of Technology.
34. Tse AL, Hesketh PJ, Gole JL, Rosen DW (2003) "Rapid prototyping of chemical sensor packages with stereolithography," *Microsystems Technology*, Vol. 9, pp. 319–323.
35. Varadan VK, Jiang S, Varadan VV (2001) Microstereolithography and other Fabrication Techniques for 3D MEMS, Wiley.
36. Wilson JE (1974) *Radiation Chemistry of Monomers, Polymers, and Plastics*, Marcel Dekker, New York.
37. Wohlers, T (2006) Wohlers Report, Wohlers Associates, Inc., Fort Collins, CO.

Symbols and Abbreviations

Symbol	Designation
$\mathbf{C_d}$	Cure depth
$\mathbf{D_p}$	Depth of penetration
E	Exposure (energy per unit area)
$\mathbf{E_c}$	Critical exposure
H	Irradiance (radiant power per unit area)
$\mathbf{L_w}$	Cured line width in stereolithography
LT	Layer thickness
OC	Overcure
$\mathbf{P_L}$	Laser power
$\mathbf{V_s}$	Laser scan speed
$\mathbf{W_0}$	Radius of Gaussian laser beam
$\mathbf{Z_c}$	Compensation zone thickness

Chapter 8
Case Studies in Chemical Sensor Development

Gary W. Hunter, Jennifer C. Xu and Darby B. Makel

Abstract The need for chemical sensor technology has increased in recent years generating the need for the development of new, advanced sensor technology. This book chapter provides a series of case studies related to the development and application of chemical sensors centering on microfabricated sensors for aerospace applications. Each case study discusses the development of a sensor or sensor system, including technology challenges, and illustrates a major theme related to chemical sensor development. These major themes suggest possible strategies that can be employed to address technical challenges in the area of sensor development. The chapter closes with a discussion that goes beyond sensor development to application approaches for including sensor technology into operating systems. It is concluded that sensor technology distributed throughout the vehicle with attributes such as ease of use, reliability, and orthogonality can significantly advance intelligent systems

Keywords: Sensor · Microfabricated · MEMS · Fire · Leak · Sensor array · High temperature · Hydrogen · Hydrocarbon · Carbon monoxide · Carbon dioxide · Packaging · Silicon carbide

8.1 Introduction

The need for chemical sensor technology in applications such as industrial processing, aerospace, and security has increased in recent years. For many practical applications, the sensing elements need to be relatively small in size, robust and should not require large sensing sample volume [1–5]. This book chapter provides a series of case studies related to the development of chemical sensors that illustrate a major theme. Each case study will discuss the development of a sensor or sensor system and relate that development to the theme. The case studies presented in this chapter cover a range of topics in chemical sensor development and application centering on microfabricated sensors to meet the needs of aerospace applications.

G. W. Hunter
NASA Glenn Research Center at Lewis Field, Cleveland, OH 44135

P. J. Hesketh (ed.), *BioNanoFluidic MEMS.*

The discussion of the sensor or sensor system development is meant to emphasize a major theme. These themes include:

1. Sensors and Supporting Hardware Need to be Tailored for the Application.
2. Sensor Structure Determines the Technical Challenges Part 1, Importance of Surface Interface Control.
3. Sensor Structure Determines Technical Challenges Part 2, Microfabrication is Not Just Making Something Smaller.
4. One Sensor or Even One Type Of Sensor Often Will Not Solve the Problem, The Need For Sensor Arrays.
5. Supporting Technologies Often Determine Success in a Sensor Application

The technology development described here is predominately based on silicon-based processing techniques, commonly referred to as MicroElectroMechanical Systems (MEMS) technology, and their use in the fabrication of chemical sensing microsystems [6]. Advantages of these silicon-based processing techniques include the ability to produce micro-sized structures in an identical, highly uniform, and geometrically well-defined manner, as well as the ability to produce three-dimensional structures (microfabrication and micromachining). Thus, an array of identical sensors or sensor arrays can be produced on a relatively small substrate, enhancing the reproducibility of the sensor microsystems.

Microfabrication and micromachining technologies are the basic fabrication tools which are used in the following case studies and will be discussed in more detail in Section 8.2.3. These tools can be used to address a range of technical challenges but they do not in and of themselves solve the technical challenges. Rather, the purpose of the case studies and related themes described in this chapter is to suggest, beyond the use of basic microfabrication tools, possible strategies that can be employed to address technical challenges in the area of chemical sensing applications.

8.2 Case Studies

8.2.1 Sensors and Supporting Hardware Need to be Tailored for the Application: Case Study of Silicon Based Hydrogen Sensor Development

The development of MEMS-based hydrogen sensor technology for aerospace applications was generated in response to hydrogen leaks on the Space Shuttle. In 1990, leaks associated with the Space Shuttle while on the launch pad temporarily grounded the fleet until the leak source could be identified. The prevailing method of leak detection at the time was the use of a mass spectrometer connected to an array of sampling tubes placed throughout the region of interest. Although able to detect hydrogen in a variety of ambient environments, the mass spectrometer had a delay time associated with its detection of a leak and pinpointing the exact location of the leak was problematic. In July 1999, the launch of STS-93 was delayed for two days due to an ambiguous signal using the present leak detection system. An alternate leak detection method, or at least an augmentation to the existing leak detection system, was strongly desired. In response to these hydrogen leak problems, NASA endeavored to improve fuel leak detection capabilities during pre-launch operations and flight.

In particular, efforts were made to develop an automated hydrogen (H_2) leak detection system using point-contact hydrogen sensors. These sensors could be placed throughout a region and provide localized leak detection without the need for the sampling tubes of a mass spectrometer. Being able to multiplex the signal from a number of sensors so as to "visualize" the magnitude and location of the hydrogen leak was also desired. If a number of sensors are to be placed in an area, then size, weight, and power consumption for each sensor becomes an issue. There was also a need for monitoring hydrogen concentration in flight during ascent. However, commercially available sensors, which often needed oxygen to operate or depended upon moisture [7], did not meet the needs of this application and thus the development of new types of sensors was necessary [8].

There are a range of operational requirements that a potential hydrogen sensor must meet in order to be relevant to the needs of Shuttle applications. The hydrogen sensor must be able to detect hydrogen from low concentrations through the lower explosive limit (LEL) that is 4% in air. The sensor must be able to survive exposure to 100% hydrogen without damage or change in calibration. Further, the sensor may be exposed to gases emerging from cryogenic sources thus sensor temperature control is necessary. Operation in inert environments is necessary because the sensor may have to operate in areas purged with helium. Overall, the sensing approach used for this application must be tailored to meet these requirements.

The types of sensors that work in these environments are limited and it is difficult to find one single sensor to completely meet the needs of this application. For example, metal films that change resistance upon exposure to hydrogen through absorption of the hydrogen into the bulk of the metal (hydrogen sensitive resistors) have a response proportional to the square root of the partial pressure of hydrogen. This dependence is due to the sensor detection mechanism: migration of hydrogen into the bulk of the metal changing the bulk conductance of the metal [8]. This results in reduced sensitivity at low hydrogen concentrations but a continued response over a wide range of hydrogen concentrations. In contrast, Schottky diodes, composed of a metal in contact with a semiconductor (MS) or a metal in contact with a very thin oxide on a semiconductor (MOS), have a very different detection mechanism. For a palladium–silicon dioxide–silicon (Pd–SiO_2–Si) MOS Schottky diode hydrogen sensor, hydrogen dissociates on the Pd surface and diffuses to the Pd–SiO_2 interface affecting the electronic properties of the diode resulting in an exponential response of the diode current to hydrogen concentration [9]. This exponential response has higher sensitivity at low concentrations and decreasing sensitivity at higher concentrations as the sensor saturates. Thus, by combining both a resistive sensor and a Schottky diode, sensitive detection of hydrogen throughout the range of interest from low to high concentrations can be accomplished.

Temperature control is necessary for both hydrogen sensor types for an accurate reading. Both types of sensors respond to changes in ambient temperature. If the ambient temperature changes, it is strongly suggested that the sensor system be able to compensate for these changes to provide accurate information on the gas concentration. Further, optimum operation of the sensor is temperature dependent and thus maintaining the sensor at a given temperature to maximize performance is necessary.

In order to address these needs the following sensor approach was developed over a number of years using MEMS-based technology to produce a sensor system with minimal size, weight, and power consumption:

- A Pd alloy Schottky diode on a Si substrate. The Schottky diodes were used to measure lower concentrations of hydrogen, e.g., less than 1% down to the ppm level.
- A hydrogen sensitive resistor on the same chip to measure higher hydrogen concentrations, e.g., greater than 1% up to 100%.
- The combined Schottky diode and resistor sensor system can provide measurements that encompass the range of concentrations of interest.
- Temperature control in the form of a temperature detector and heater integrated on the same chip with the sensors to both monitor and allow control of the sensor temperature.

The resulting packaged sensor structure is pictured in Fig. 8.1. The structure includes a Pd-based Schottky diode, a hydrogen sensitive resistor, a temperature detector, and a heater all incorporated in the same chip. The Schottky diode sensor is fabricated using an n-type silicon wafer on which approximately 50 Å of SiO_2 is thermally grown in the sensor region. The heater and temperature detector are platinum covered with SiO_2. Gold leads are applied by thermal compression bonding and the sensor is mounted on a TO5 header or on a ceramic flat package. The surface area of the Schottky diode is 6.1×10^{-3} cm^2 and the sensor dimensions are approximately 2.2 mm on a side. A hydrogen sensitive resistor is included on the same chip to measure higher hydrogen concentrations. This basic sensor design is specifically tailored for hydrogen leak detection applications with features including sensing mechanisms meant to measure a wide range of hydrogen concentrations; temperature control; and a sensor structure meant to minimize size, weight, and power consumption.

However, in order for the sensor to actually work in the targeted application, further tailoring of the sensor design is necessary. For example, the use of pure Pd at near room temperatures as the hydrogen sensitive metal for either the resistor or Schottky diode in this application is problematic for several reasons. The most serious of these issues involves a phase change that occurs at high hydrogen concentrations, which can lead to hysteresis or film damage. Thus, an approach which used a Pd-alloy rather than pure Pd was taken. The first generation of these sensors used palladium silver (PdAg). The use of PdAg in hydrogen sensing applications was pioneered by Hughes [10]. Palladium silver has advantages over Pd and properties that make it more suitable for this application. Palladium silver is more resistant

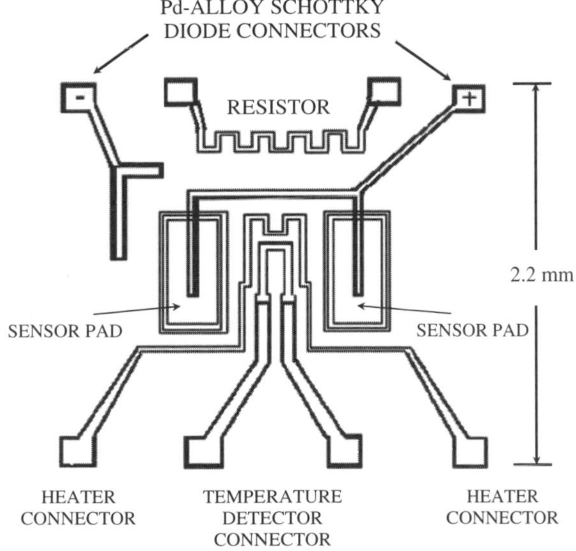

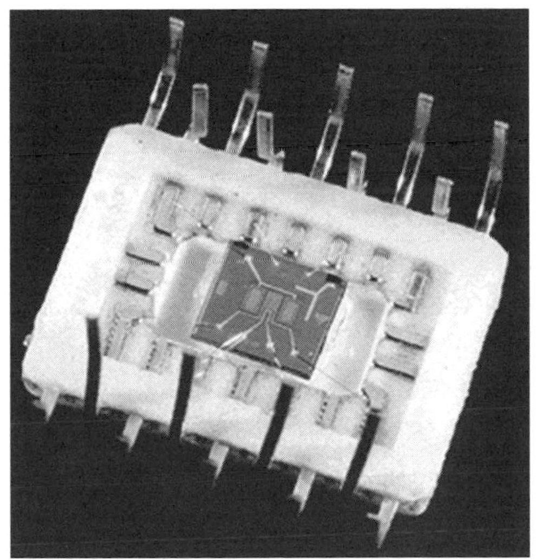

Fig. 8.1 Design and complete package of a Pd alloy hydrogen sensor. The sensor package includes a Schottky diode hydrogen sensor, hydrogen sensitive resistor, temperature detector, and heater. The sensor is designed to be a self contained unit able to measure over a wide range of hydrogen concentrations

to damage from exposure to high hydrogen concentrations than Pd. The sensor responds in an inert environment (no oxygen) to the presence of hydrogen. The presence of oxygen decreases the sensor response but the sensor is still sensitive to low concentrations of hydrogen [11, 12]. Further, a PdAg sensor configuration has been shown to have a sensitivity and response comparable to a mass spectrometer [13].

However, it was found that a first generation Schottky diode sensor using PdAg as the hydrogen sensitive metal did not meet all the needs of aerospace Shuttle applications. At higher temperatures and higher hydrogen concentrations, its calibration changed and the sensor occasionally failed. An example of the degradation of the PdAg sensor response at higher temperatures and higher hydrogen concentrations is shown in Fig. 8.2. The response of a newly fabricated PdAg sensor to 100% hydrogen at 100°C is presented over several cycles. The sensor response decreases with the number of exposures and exhibits a much poorer recovery to the original baseline after each cycle.

One significant reason for this behavior may be seen in Fig. 8.3 [14], which shows an optical micrograph of an as-deposited PdAg Schottky diode sensing pad (Fig. 8.3a) and a sensing pad from the PdAg sensor characterized at 100°C in 100% hydrogen (Fig. 8.3b). The as-deposited PdAg film surface features are uniform and the edges of the film are straight and rectangular. In contrast, the PdAg film after testing shows significant peeling of the PdAg near the edges of the sensor pads' rectangular pattern and a significant amount of surface features. Examination of the region near the edge of the sensor pad in Fig. 8.3b suggests that the PdAg has been removed from the surface leaving only the SiO_2 layer. Overall, a likely reason for the change in the sensor response with heating and exposure to 100% hydrogen is delamination and damage to the PdAg sensing film at higher concentrations and temperatures [14].

Therefore, PdAg as a sensing film did not meet the needs of Shuttle operation

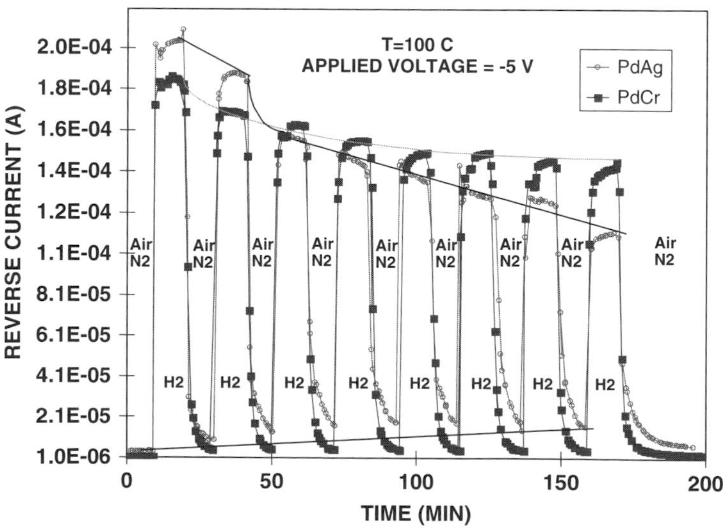

Fig. 8.2 The response of PdAg and PdCr Schottky diode sensors at 100°C to repeated exposures of 9 minutes of air, 1 minute of nitrogen, 10 minutes of 100% hydrogen, and 1 minute of nitrogen. The PdCr diode shows a more repeatable baseline and a more stable response than the PdAg diode [14]

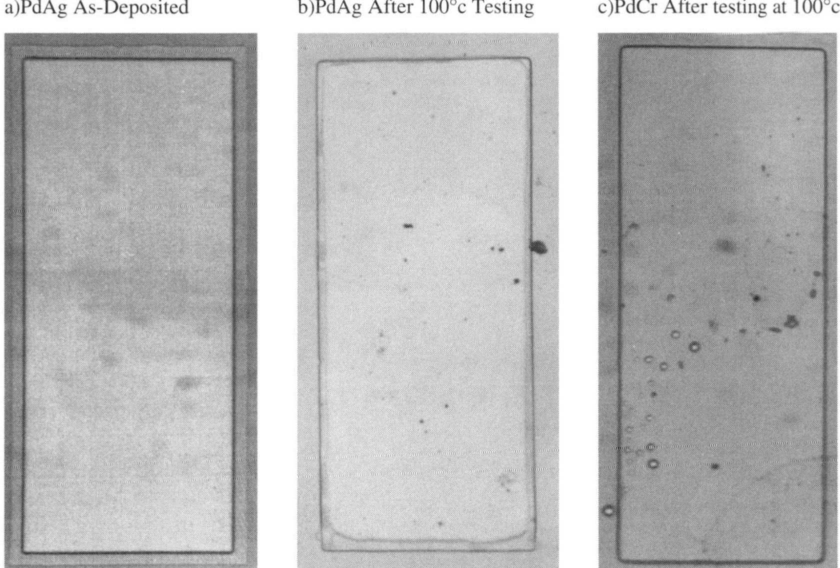

Fig. 8.3 Optical Micrographs (400 ×) of the pad of a) PdAg sensor as deposited; b) PdAg sensor after heating and characterization in 100% H_2 at 100°C; c) PdCr sensor after heating and characterization in 100% H_2 at 100°C. The PdCr sensor does not show delamination of the sensor film near the edges of the sensor pad evident with the PdAg-based sensor [14]

that has the requirement for operation in 100% hydrogen. Nonetheless, PdAg met the needs of a very different application [12]. Ford Motor company required an automated way to determine if there were leaks in the valves and fitting associated with the natural gas powered Ford Crown Victoria while on the assembly line. The approach was to pressurize the tank with nonexplosive concentrations of hydrogen and verify the integrity of the system by looking for hydrogen leaks. The vehicle fuel system was pressurized with 1% hydrogen and 99% nitrogen. The valves and fittings were enclosed with "boots" which included hydrogen sensors and measured gases being emitted from the valves and fittings. The outputs from the various sensors were fed to a central computer-based processing system. The processing system included a visual image of the car and associated valve and fitting system that, if leaks did occur, graphically showed their location and relative magnitude. This complete system received a 1995 R&D 100 Award as one of the 100 most significant inventions of that year. Thus, even though the PdAg based sensor did not meet the needs of the Shuttle application, it did meet the needs of this automotive application.

Further sensor development continued to meet the specific needs of the Shuttle application. This led to the development of a sensor that uses palladium chrome (PdCr) as the hydrogen sensitive alloy. The response of this PdCr Schottky diode to 100% hydrogen at 100°C is also shown in Fig. 8.2. The PdCr sensor is much more stable than the PdAg sensor under these conditions: the PdCr diode response to 100% hydrogen is nearly consistent after the initial exposures with an equally consistent return to a common baseline. Further, the optical micrograph of the corresponding PdCr sensor pad (Fig. 8.3c) shows some degradation of the film surface but the PdCr film does not exhibit the peeling of the sensing film near the edge of the sensing pad that the PdAg film exhibited. A PdCr sensor tested at 75°C also showed good response and adhesion of the film but less resulting surface structure than the PdCr sensor tested at 100°C.

Therefore, PdCr shows significantly improved stability and response over PdAg. The results of these tests and other testing suggest that PdCr is better for applications where the sensor is exposed to higher hydrogen concentrations, while PdAg can be used for lower hydrogen concentration applications.

This has led to the use of PdCr both as a hydrogen sensitive resistor and a Schottky diode material to expand the detection range of the sensor [15]: a Schottky diode provides sensitive detection of low concentrations of hydrogen while the resistor provides sensitivity up to 100% hydrogen (Fig. 8.4). Later versions of the hydrogen leak detection system have included capabilities to process the data and provide relevant data to the user (see Section 8.2.5). As shown in Fig. 8.4, this data processing can convert the raw data and show the users the quantity of interest, e.g., the hydrogen concentration.

The complete hydrogen detection system (two sensors on a chip with supporting electronics) flew on the STS-95 mission of the Space Shuttle (launched October, 1998) and again on STS-96 (launched May, 1999) [16, 17]. Overall, the hydrogen sensor response is seen to generally parallel that of the mass spectrometer on the ground but with a larger signal and quicker response time (perhaps due to the relative location of each measuring device with respect to the hydrogen source). The hydrogen sensor response during the launch phase of flight showed a response near the cut-off of the Shuttle main engine. Near this time, a spike in the hydrogen concentration is observed that decreases with time back to baseline levels. These results are qualitatively consistent with the leakage of very small concentrations of unburnt fuel from the engines into the aft compartment after engine cut-off. Moreover, the advantage of this microsensor approach is that the hydrogen monitoring of the compartment is continuous and, in principle, could be used for real-time health monitoring of the vehicle in flight. Thus, the basic hydrogen sensor structure was demonstrated to be a viable approach to meet the needs of the Shuttle application.

The range of applications for which the basic sensor design has been adapted is shown in Fig. 8.5, and includes the Shuttle mentioned above, the International Space Station, the NASA Helios vehicle, the X-33, and the X-43. It should be noted that the supporting hardware for the various applications is different for each application. In each case, the sensor and hardware, as well as supporting software, were tailored for the application. For example, in the case of the International Space Station (ISS) application, the sensor had to be tailored for a high oxygen and high humidity environment. Since this is a Criticality 1 function which involves protection from potential crew or mission threatening conditions, the complete system of sensors and hardware needed to be designed as triply redundant and associated software

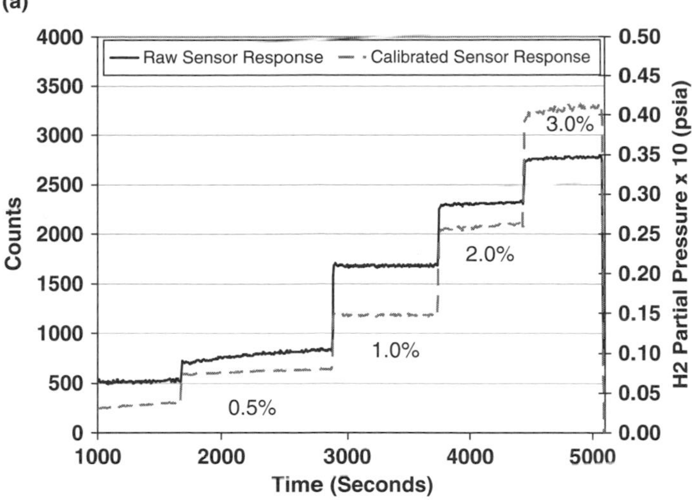

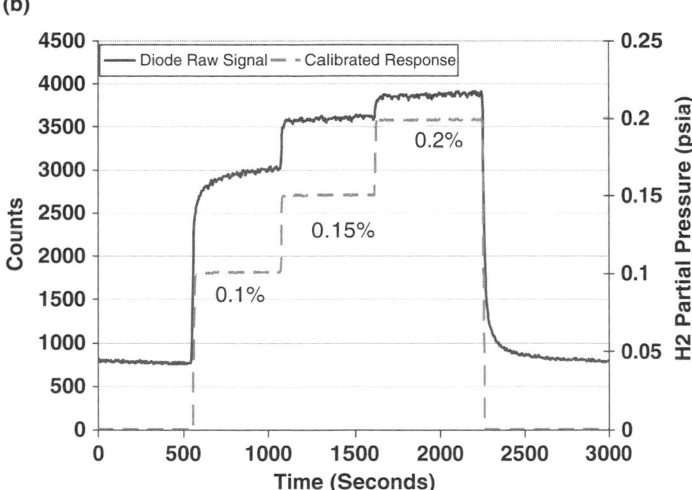

Fig. 8.4 Response to hydrogen of a PdCr alloy based a) Resistor and b) Schottky diode. Both the raw signal and calibrated response are shown. The responses of the two sensor types are complementary [15]

safeguards. This application also required shielding the electronics from the humid operating environment. The ISS sensor system is generally exposed to a pressurized environment with a constant flow of gas. This is contrasted with the Shuttle application which has varying pressure from atmospheric pressure on the launch pad to near vacuum during ascent. The humidity in the Shuttle application can range from high, while on the launch pad exposed to varying Florida weather, to low humidity at

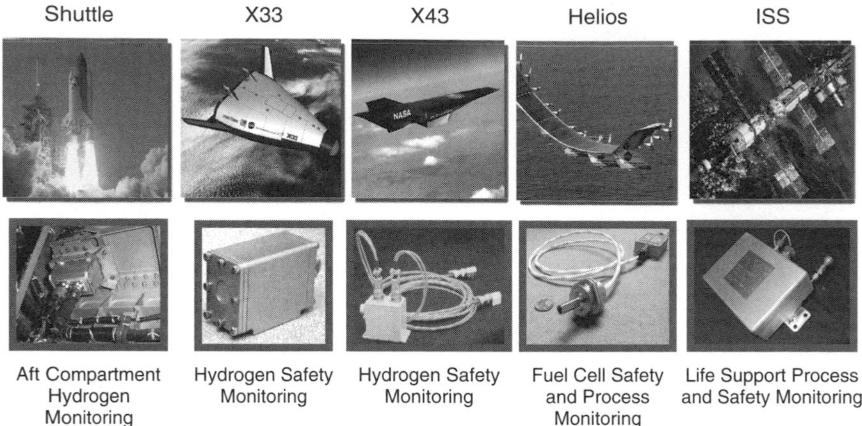

Fig. 8.5 The range of applications to which the hydrogen sensor has been adapted. The sensors as well as associated hardware and software had to be tailored in each case for the needs of the application

higher altitudes. The Shuttle application during launch into flight is not a Criticality 1 function but rather the data is meant for informational purposes. Thus, issues such as mandatory triple redundancy of a measuring system with supporting software features do not come into play. These operational environments are very different and highlight the need to design a system for the needs of the application.

A vivid example of operating a sensor outside its designed mode of operation is in Fig. 8.6, which shows a picture of the hydrogen sensor operating under water. This condition is possible if the sensor is exposed to an environment that results in condensation of water on the sensor surface, such as one with very high humidity without a method to dry the sensor before operation. However, the hydrogen sensor is not designed for underwater operation. Bubbles are seen rising from the sensor surface due to the fact that the electrodes are open to the environment. The condensed water across the electrodes in effect allows the sensor to become an electrochemical cell which dissociates water producing hydrogen and oxygen resulting in the observed bubbles. Operation of the sensor in this manner has been observed to cause drift in calibration and sensor damage over time. Thus, operating the sensor outside of its design range has significant effects on sensor operation and stability.

In summary, this section has detailed the development of a hydrogen sensor for leak detection applications. This development includes the design of the basic sensor mechanism, testing in the specific environment, and inclusion of the sensor with hardware and software. While a sensor may meet the needs of one application, it may not meet the needs of another. Tailoring the sensor structure, as well as supporting hardware and software, to assure proper operation within the application is necessary.

Fig. 8.6 Hydrogen sensor operation when covered with water. The bubbles being evolved are due to electrolytic processes occurring on the sensor surface which can damage the sensor. This example illustrates the consequences of operating a sensor in an application for which it was not designed

8.2.2 Sensor Structure Determines the Technical Challenges Part 1, Importance of Surface Interface Control: Case Study of SiC Based Hydrogen and Hydrocarbon Sensors

Silicon carbide (SiC) has high potential as the electronic semiconductor material for a new family of high temperature sensors and electronics. This is due to the ability of SiC to operate as a semiconductor in conditions under which silicon cannot sufficiently perform, such as at temperatures above 400°C [18]. One area where SiC semiconductor technology can be applied is in chemical sensing. Silicon carbide gas sensors have been in development for a number of years employing a range of designs including capacitors [19], transistors [20], and Schottky diodes [21–23]. These sensors have been shown to be responsive to several gases, including hydrogen and hydrocarbons, making them useful for a range of applications over a broad range of environments [1].

In particular, SiC Schottky diodes have been developed due to their high sensitivity and wide temperature range of operation. As described in the previous subsection, a gas-sensitive Schottky diode is composed of a metal (often catalytic) in direct contact with a semiconductor (MS) or a metal in contact with a very thin insulator or oxide on a semiconductor (MIS or MOS). The detection mechanism for H_2 involves the dissociation of H_2 on the surface of a catalytic metal leading to the formation of a dipole layer at the interface of the metal and the insulator (or metal-semiconductor interface depending on the structure). This dipole layer affects the Schottky barrier height of the diode resulting in an exponential change in the forward current while the diode is under fixed bias [9, 24]. The detection of hydrocarbons is possible if the sensor is operated at a high enough temperature to dissociate the hydrocarbon and produce atomic hydrogen. The resulting atomic hydrogen affects the sensor output in the same way as molecular hydrogen [21, 25, 26]. Predominately, the temperature for sensitive hydrocarbon detection is beyond the upper limit for silicon-based Schottky diode functionality and thus SiC enables high temperature detection of hydrogen and hydrocarbons with high sensitivity.

The use of a SiC Schottky diode structure for gas sensing allows high sensitivity but then introduces its own technical issues. In particular, the successful use of a SiC Schottky diode structure as a gas sensor depends on strict control of the metal-semiconductor interface that makes up the diode and determines its electrical properties [24]. One complicating factor in control of this interface is high temperature operation of these gas sensors. Higher temperature operation implies possible reactions between the catalytic sensing metal and the SiC. While one can decrease the metal/SiC reactivity, if this occurs in such a manner so as to "pin" the interface potential barrier, then sensor sensitivity can be significantly decreased [23], effectively defeating the purpose of using a Schottky diode structure. Thus, the choice of surface treatment or barrier layer(s) between the catalytic metal and the SiC substrate is complicated by simultaneous requirements of high sensor stability during high temperature operation while maintaining high sensitivity.

A second major complicating factor in the control of the SiC interface is the nature of the present-day SiC semiconductor substrates. Compared to silicon wafer standards, present-day SiC wafers are small, expensive, and of inferior crystalline quality. In addition to high densities of crystalline defects such as micropipes and closed-core screw dislocations, commercial SiC wafers also exhibit significantly rougher surfaces and larger warpage than is typical for silicon wafers [27]. The highly variable SiC surface itself significantly complicates efforts to control the catalytic metal/SiC interface.

This section will describe the evolution of the NASA Glenn Research Center (GRC) based SiC gas sensor development and approaches used to improve surface interface control. Overall, this section makes the fundamental point that for some gas sensing structures, control of the interfaces is necessary for adequate sensor operation.

NASA GRC's SiC-based Schottky diode development began with Pd on SiC (Pd/SiC) MS structures without a barrier layer between the Pd metal and semiconductor. Direct contact between the gas-sensitive catalytic metal and the semiconductor was thought to allow changes in the catalytic metal to have maximum effect on the semiconductor. Studies of this baseline system helped determine limits of diode sensitivity, potential material interactions between Pd and SiC, and whether a barrier layer between the Pd and SiC was necessary for long-term sensor stability. The details of this work are reviewed in reference [21]. The sensor detects hydrogen and hydrocarbons in inert or oxygen-containing environments with high sensitivity. However, the sensor response is adversely degraded by extended high temperature heating. For example, prolonged heating at 425°C has been shown to change the sensor properties and to decrease sensor sensitivity [21]. The reason for this change in diode properties is thought to be due to reactions between the Pd and SiC at the interface upon heating causing disruption of the metal structure due to oxygen (O_2) and Si [28, 29]. Overall, massive amounts of silicide and oxide formation have been noted in Pd/SiC diodes with heating that significantly decrease their performance.

One structure which has shown improved stability over that of Pd/SiC is PdCr directly deposited on SiC (PdCr/SiC) [30, 31]. The advantages of PdCr as a high temperature alloy have been explored extensively in strain gage applications [32]. It is a stable high temperature material which is able to provide static strain measurements at temperatures up to 1100°C. However, its use in a gas-sensing SiC-based structure depends not only on its inherent stability upon exposure at high temperatures, but also on such factors as the alloy's reactivity to SiC and the catalytic interactions of PdCr alloy with the gases to be measured.

Initial results related to the PdCr/SiC Schottky diode performance were highly encouraging. Monitoring of the sensor performance periodically during long-term heating of 250 hours at 450°C, Fig. 8.7, suggests the PdCr/SiC diode has significantly improved stability over the Pd/SiC structure. After an initial break-in period of near 40 hours, the results show that the sensor response in hydrogen is relatively constant [30].

However, despite these promising results, the ability to systematically produce PdCr/SiC gas sensors with combined stability and sensitivity has been problematic. Later results have shown that PdCr films of different Pd/Cr ratios have varying levels of reactivity with the SiC surface, as well as potential sensor drift after extended durations at high temperature. A representative Auger Electron Spetroscopy (AES) depth profile is shown in Fig. 8.8 of a PdCr film on SiC heated at 450°C for 100 hours and periodically exposed to 0.5% hydrogen in nitrogen [33]. Two major trends have been found to correspond to a decline in sensor response: Pd migration into the SiC interface to form palladium silicides ($PdSi_x$) and the subsequent migration of elemental silicon to the surface from the SiC. Palladium silicides are present throughout the film and the Si on and near the surface oxidizes to form silicon dioxide. Accompanying this silicide and oxide formation are the migration of carbon (C) into the PdCr from the SiC and of Cr into the SiC. Overall, as with Pd/SiC, significant migration of the components of the PdCr/SiC interface can occur with high temperature heating. Thus, the use of the PdCr alloy alone on SiC is not sufficient to produce stable and sensitive sensors. Control of the migration of the constiuents of the PdCr/SiC including Pd, Si, and Cr as well as the subsequent formation of silicides and oxides is necessary to achieve improved stability.

One possible explanation for the stability observed in some of the PdCr/SiC samples was that chrome carbide (Cr_3C_2) formed naturally with heating from PdCr

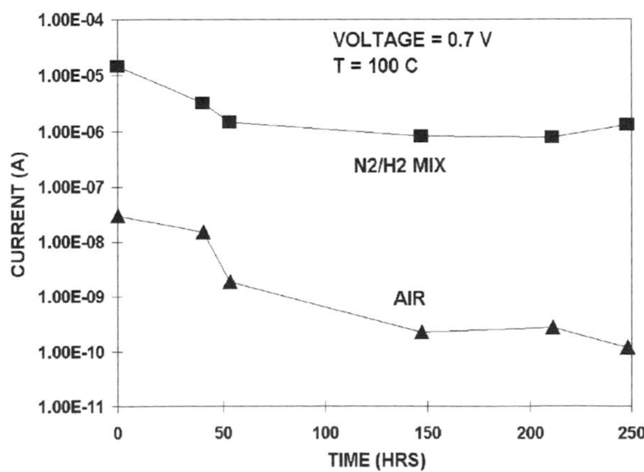

Fig. 8.7 The forward current at 100°C versus heating time at 425°C in air (▲) and in 120 ppm H_2 in N_2 (■) of a PdCr/SiC Schottky diode. The sensor is characterized at 100°C after heating at 425°C. After an initial break-in period, the diode shows a stable response to hydrogen and overall improved sensitivity (change in response from the air baseline to 120 ppm H_2 in N_2) over the heating period

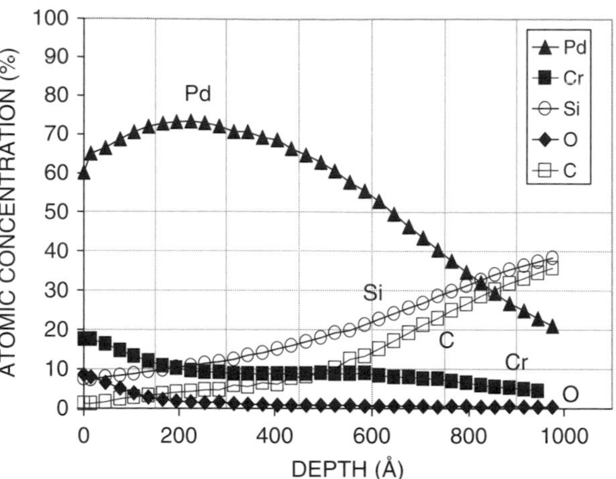

Fig. 8.8 AES depth profile of a representative PdCr/SiC diode after periodic 0.5% hydrogen in nitrogen gas exposure for 100 hours at 450°C. Despite early results showing a stable sensor structure, significant migration of multiple chemical species is observed

and SiC and then provided a barrier layer between the catalytic metal and the SiC [33]. This barrier layer may then have been responsible for the observed stability by inhibiting the formation of the oxides and silicides noted above. The lack of reproducibility in the sensor response may then be explained if this protective barrier layer is formed in an irreproducible or non-uniform way. Thus, the sensor repeatability could be improved significantly if the interface control of the barrier layer could be improved. Rather than relying on a natural formation of Cr_3C_2 from PdCr and SiC, the next step in controlling the interface layer of the Schottky diode was to directly deposit a Cr_3C_2 barrier layer onto the SiC surface [33].

A thin layer (600 Å) of Cr_3C_2 covered by 300 Å of Pt was deposited onto a SiC substrate forming a Pt/Cr_3C_2/SiC Schottky diode structure [33]. This sensor structure was tested at 450°C for 70 hours and then 580°C for 600 hours for a total of 670 hours. The gain in the sensor response to propylene versus air at 580°C is shown over time in the inset of Fig. 8.9. While the sensor loses some sensitivity over the first 200 hours at 580°C, the sensor seems to stabilize after this break-in period. The sensor shows a consistently strong response to propylene with an average gain of over two thousand for at least 270 hours. Figure 8.9 shows representative data comparing the sensor response at 3.2 V at 500 hours and 670 hours. The gas exposure period in Fig. 8.9 is somewhat extended for the 670 hour data to allow better comparison between the two sets of data. The data shows good repeatability of the signal in form and magnitude. The data suggests a sensor with very good sensitivity to propylene and reasonable stability at 580°C.

However, the sensor did change response over time from the onset of sensor testing. AES analysis was performed on Pt/Cr_3C_2/SiC samples both before and after testing in order to gain insight into the reason for the change. The AES results from the as-deposited sample are shown in Fig. 8.10a. The results show well-defined interfaces with no indication of Pt silicide ($PtSi_x$) formation throughout the sample. Figure 8.10b shows the AES analysis of the sensor after 670 hours of testing. The data did not show any evidence of carbide or carbon remaining from the original

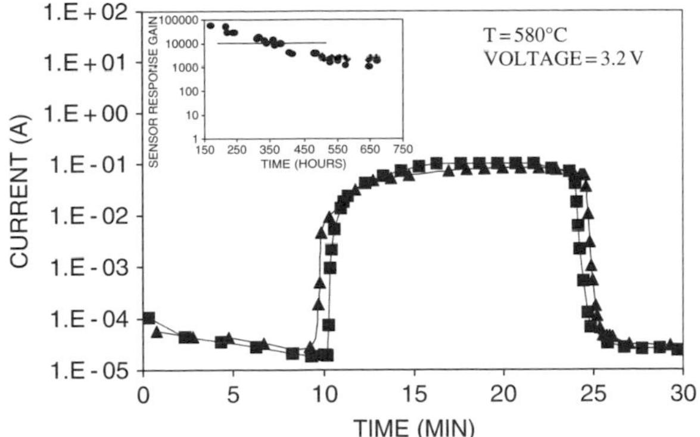

Fig. 8.9 The Pt/Cr_3C_2/SiC sensor tested at 580°C at 500 hours (■) and 670 hours (▲) in 0.5% propylene. The sensor was tested by first being exposed to air for 5 minutes, N_2 for 5 minutes, 0.5% propylene in N_2 for 10 minutes, pure N_2 for 5 minutes, and then air. Inset: The sensor gain over time at 580°C measured at 1.0 V (•) and 3.2 V (♦). The sensor shows good stability and sensitivity at these high temperatures [33]

Cr_3C_2 layer. Oxygen had diffused into the layers, replacing the carbon to form chromium oxide, which migrated toward the surface. This migration likely allowed the Pt to diffuse toward the SiC interface. However, the major point is the massive formation of metal silicides, a likely cause of sensor failure in other sensor structures, was not observed in this sample.

Although the reaction mechanisms for the Pt layer on Cr_3C_2/SiC are still not completely understood, several observations can be made. The Cr_3C_2 layer appears to be effective in preventing immediate reaction of the Pt layers with silicon from the crystal substrate to form metal silicides. The presence of carbide in the chromium layer slows down migration of the chromium to the surface, and allows formation of an oxygen-poor chromium oxide, which gradually diffuses throughout the metal layer. This chrome sub-oxide likely prevents metal silicide formation at the SiC interface. Subsequent work has shown repeatable fabrication ability and stable operation of this sensor structure up to 950 hours at 580°C.

Thus, the reaction mechanism of Cr_3C_2 with SiC significantly changes the dynamics of the sensor interface and improves the stability and performance of the sensor. Other work has shown directly depositing palladium oxide (PdO) as the interface layer also has significant effects in producing a stable high temperature gas-sensing structure [34]. Further work has also shown that moving from a standard commercially available surface to atomically flat SiC has a significant effect on the performance and stability of even a Pt/SiC MS Schottky diode gas sensor [33]. Overall, this work shows that the control of the interface for the catalytic metal/SiC structure is necessary for their operation as sensitive gas sensors and details an evolving process to control that interface.

In summary, for some sensors such as a resistor whose response depends on the bulk material properties, the surface interface may not be dominant and need for its control may be limited. However, for sensors such as Schottky diodes, the sensing

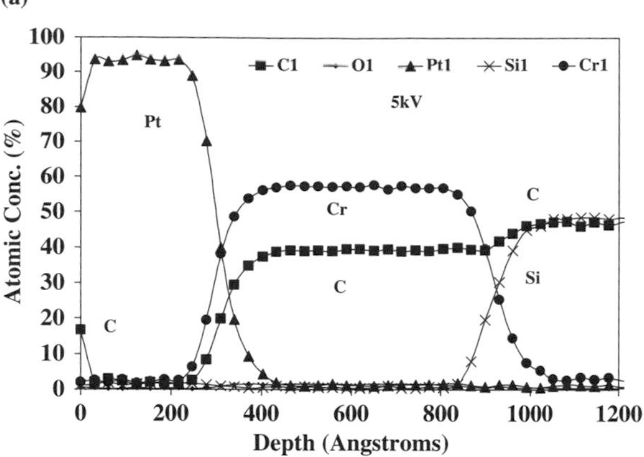

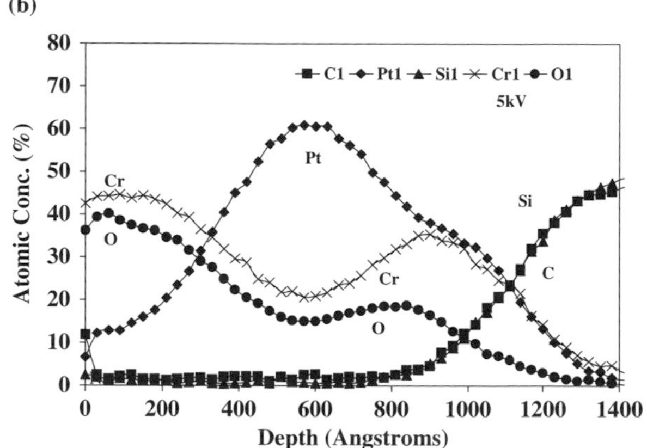

Fig. 8.10 AES depth profile of the $Pt/Cr_3C_2/SiC$ sensor: a) as-deposited. b) after annealing at 450°C for 70 hours and then 580°C for 600 hours and periodic exposure to hydrogen-bearing gases

mechanism is determined by the interface. Since a limited surface interface layer can then determine the sensor response, this allows the sensor to have high sensitivity. Correspondingly, however, this mandates that interface control is fundamental to this type of sensor's stability and performance. This case study demonstrates the evolving design of a sensor structure resulting in a more stable interface and thus a more stable sensor but one which maintains it sensitivity.

The lessons suggested by this case study also apply to other sensor structures that strongly depend on surface and interface effects. This includes the developing field of chemical sensors based on nanotechnology e.g., nanotubes, nanorods, etc. The structure of these sensors typically have a much higher surface to volume ratio than micro- or macro-based sensors and thus can be viewed as predominately interface; thus, these sensors too will face significant challenges in the control of these interfaces in order to be utilized as stable and functional systems.

8.2.3 Sensor Structure Determines Technical Challenges Part 2, Microfabrication is Not Just Making Something Smaller: Case Study of Carbon Dioxide Sensor Development

A significant direction in chemical sensor research, development, and application is the use of microfabrication and micromachining technology or MicroElectroMechanical Systems (MEMS) technology to produce sensors with minimal size, weight, and power consumption. These attributes allow sensor implementation in some applications, especially aerospace applications, where the implementation of traditional sensor system technology would be limited due to factors such as:

- The burden a larger sensor system would place on vehicle weight or power.
- Inability to smoothly integrate a larger sensor system into a vehicle without affecting vehicle operation or the measurement.
- Reproducibility or cost issues associated with sensor systems which are handmade or not mass produced.
- Even if a larger sensor system could be included into a vehicle, that sensor system alone may have limited capabilities. This implies the need for multiple larger sensor systems exacerbating the size, weight, and power issues discussed in the first two bullets.

Thus, miniaturization of instrument systems using microfabrication and micromachining technology can enable new capabilities and improve overall system performance. Microfabrication and micromachining technology is derived from advances in the semiconductor industry [1–6]. A significant number of silicon-based microfabrication processes were developed for the integrated circuit (IC) industry to produce small devices. Techniques such as lithographic reduction, thin film metallization, photoresist patterning, and chemical etching have found extensive use in chemical sensor development allowing the fabrication of microscopic sensor structures. The ability to mass-fabricate many of these sensors on a single wafer (batch processing) using semiconductor processing techniques significantly decreases the fabrication costs per sensor. A large number of sensors are fabricated simultaneously resulting in high reproducibility between the sensors in batch, and thus reproducibility in the sensor responses. The smaller sensor size also may enable better sensor performance especially in low concentration measurements where the small surface area of the sensor allows a smaller number of molecules to have a larger relative effect on the sensor output.

These microfabrication processes produce mainly two-dimensional planar structures. By combining these processes with micromachining technology, three-dimensional structures can be formed that can be applied to chemical sensing technology. These three-dimensional structures can have a wider range of properties, such as reduced thermal mass and power consumption or diffusion control of gas species coming into contact with a reactive surface.

While MEMS processes allow the mass production of sensors that have minimal size, weight and power consumption, the microfabrication of chemical sensors involves much more than just making a macrosized sensor smaller [1]. The processing used to produce a sensor material as a macroscopic bulk pellet can change considerably when it is desired to fabricate the material as part of a miniaturized system. For example, a chemical sensor in the form of a macroscopic bulk pellet can be fabricated from the powder of a starting material, pressed into a pellet containing lead wires, and then sintered at high temperatures to form the resulting sensor. However, using the same starting material, pressing the material onto a substrate to form a smaller or even microscopic sensor is often not a viable option. The pressing process itself may not form a uniform thin or thick film material. The underlying substrate with sensor film may not survive the applied pressure or sintering that typically is done with the pellet sensor material. Rather, a thin or thick film of the starting sensor material must be deposited onto a substrate which can be processed using MEMS techniques, e.g., sputtering or evaporation. This substrate must, at a minimum, support the sensor, enable connections to the outside world, and allow further sensor material processing if necessary.

Further, given the surface sensitive nature of many chemical sensors, the effects of miniaturization can be dramatic and include significant changes in sensor sensitivity and response time. This is in part due to the fact that the sensor film is often produced by techniques such as sputtering which may result in different material properties than those of bulk materials. The resulting surface to volume ratio of a thin film is larger than that of a bulk material: surface effects that may affect only a small percentage of a sensor in the bulk form may occur within a significantly larger percentage volume of a thin film sensor. This can strongly affect the sensor's response. For example, oxidation may occur on the surface of a sensor exposed to high temperatures. In a bulk material, this oxidation may only be a small percentage of the sensor's volume while in a thin film material the same oxidation thickness may account for a sizable percentage of the sensor's volume. If the sensor's detection mechanism relies on bulk conduction, this oxidation could significantly affect the sensor response by changing the nature of the volume of the sensor. In addition, stresses in sensor thin films that degrade sensor response or catastrophically damage the sensor structure may be less of a factor in bulk materials. Therefore, new technical challenges often must be overcome as sensor technology is miniaturized.

An example of the challenges of microfabrication of a chemical sensor is the miniaturization of a carbon dioxide (CO_2) sensor [35]. Previously, CO_2 sensors have been based on liquid electrochemical cells that often use a corrosive liquid as the electrolyte. This means that the cell needs to be sealed to prevent possible electrolyte leaks. Overall, the approach to miniaturization of these sensors is to move from the traditional liquid electrolytes to solid state devices. These attempts have included the use of solid state electrolytes such as bulk or thick film materials. The basis behind these materials choices is due to the idea that increasing miniaturization is not always conducive to liquid based system but is more conducive to solid state devices. However, these solid state devices do not operate like a standard liquid electrochemical cell. Some major differences include:

- Liquid devices operate through the flow of constituents through the liquid electrolyte; corresponding electrolyte constituent flow in solid state devices can be problematic and require higher temperatures depending on the electrolyte. Such operation at higher temperatures can bring its own problems related to sensor degradation and stability.
- An important factor in the operation of an electrochemical cell is the three phase boundary, i.e., the surface in the electrochemical cell where the gas interacts with both the electrode and the dielectric. Miniaturizing a sensor decreases the surface area of such boundaries by decreasing the overall size of the sensor.
- Liquid devices can maintain chemical equilibrium of the constituents through mixing motion of the liquid. A solid state device that uses a solid electrolyte by its nature may not have ready flow of constituents to enable easy mixing.

In recent years, there has been a significant effort to develop solid state miniaturized electrochemical CO_2 sensors. However, most bulk material or thick film based electrochemical sensors [36–41] consume high power, and the fabrication processes of these sensors are complicated because of the involvement of hot pressing and screen printing [39]. These sensor systems are often still at a comparatively early stage of development [40–43].

The NASA GRC/Case Western Reserve University (CWRU) work in miniaturized CO_2 sensors has concentrated on the development of a solid state electrochemical cell. One of solid state electrochemical CO_2 sensor miniaturization efforts uses super ion conductors such as NASICON (sodium super ionic conductor, $Na_3Si_2Zr_2PO_{12}$) as solid electrolyte. and sodium carbonate and barium carbonate ($Na_2CO_3/BaCO_3$, 1:1.7 molar ratio) as auxiliary electrolytes [36,41,44].

The resulting design is shown in Fig. 8.11 [35]. The following features of the sensor are generated by the drive for miniaturization and to address the changes in operation resulting from this solid state design. NASICON needs higher temperatures to act as an electrolyte. Thus, the base substrate on which the sensor is fabricated is alumina, which is durable at high temperatures. Integrated into the sensor design are a temperature detector and heater to maintain a device temperature at which the electrolyte can operate. To increase the number of three phase sites and the ability of the constituents of the sensor to maintain equilibrium, an interdigitated finger pattern was used with selective deposition of the electrolyte/auxiliary electrolytes. The working and reference electrodes for this solid state electrochemical cell are Pt, and the NASICON electrolyte is deposited between these Pt electrodes. This working electrode/NASICON/reference electrode pattern is repeated multiple times on the substrate using an interdigitated finger electrode structure as shown in Fig. 8.11a. The auxiliary electrolyte, $Na_2CO_3/BaCO_3$, is deposited over the whole surface of the interdigitated fingers. This approach of interdigitated fingers combined with uniform deposition of electrolyte and auxiliary electrolytes increases the number of sites where the three phase boundary of electrode, electrolyte, and auxiliary electrolyte exist. This increases the number of active electrochemical sites

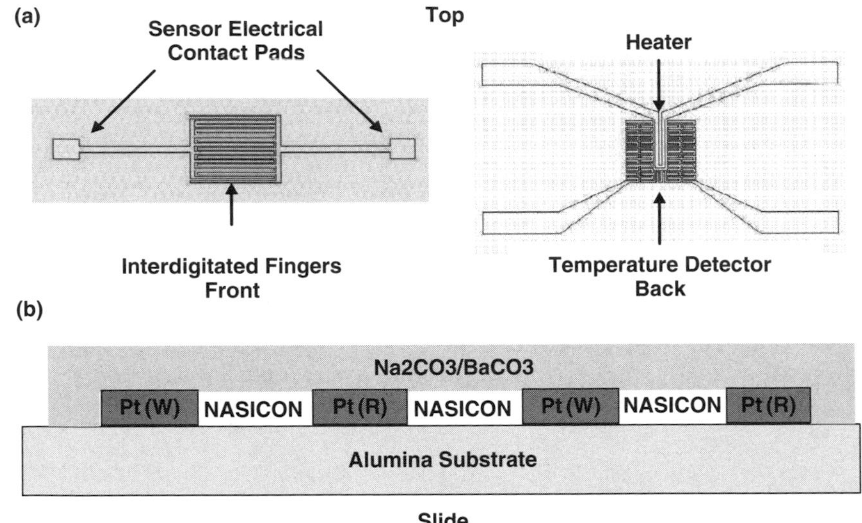

Fig. 8.11 (a) Sensor with the interdigitated finger electrodes on the frontside (left), heater and temperature detector on the backside (right). (b) Partial side view of the CO_2 sensor showing the NASICON and $Na_2CO_3/BaCO_3$ electrolytes, the Pt interdigitated electrodes (both working (W) and reference (R)), and the alumina substrate [35]

for detection in this microstructure and improves sensitivity over a standard two electrode design.

Testing has taken place to determine the optimum operational conditions for this device [35]. The sensor was operated in an amperometric mode and the current was measured at a given voltage applied across the interdigitated fingers. The approach was to determine which temperatures and voltages allowed the sensor to produce a large signal but still maintain a constant baseline without long-term drift. It was found that the optimum operating conditions are at higher temperatures (600°C) and at 1V yielding results such as those seen in Fig. 8.12. A range of CO_2 gas concentrations can be measured and the results are linear with ln[CO_2 Concentration]. The sensor demonstrated stable operation for weeks using these parameters.

The present understanding of the reaction mechanisms of this sensor is as follows: The repeatable operation without sensor drift significantly relies upon the current flow of Na^+ ions and regeneration of Na_2O from Na_2CO_3 [35]. These processes rely on both the NASICON and $Na_2CO_3/BaCO_3$ electrolytes. In previous work, the auxiliary electrolyte $Na_2CO_3/BaCO_3$ (1:1.7) was deposited on only the working electrode [39]. In the design in Fig. 8.11, the auxiliary electrolyte was deposited homogeneously on the entire sensing area of the sensor, including both working (sensing) and reference electrodes. This allows improved flow of species within the auxiliary electrolyte. For example, at the working electrode, the depleted Na^+ concentration could be recovered by the transfer of Na^+ from NASICON through the three-phase boundary. In turn, the decreased concentration of Na^+ in NASICON could be supplemented from the $Na_2CO_3/BaCO_3$ through the reference

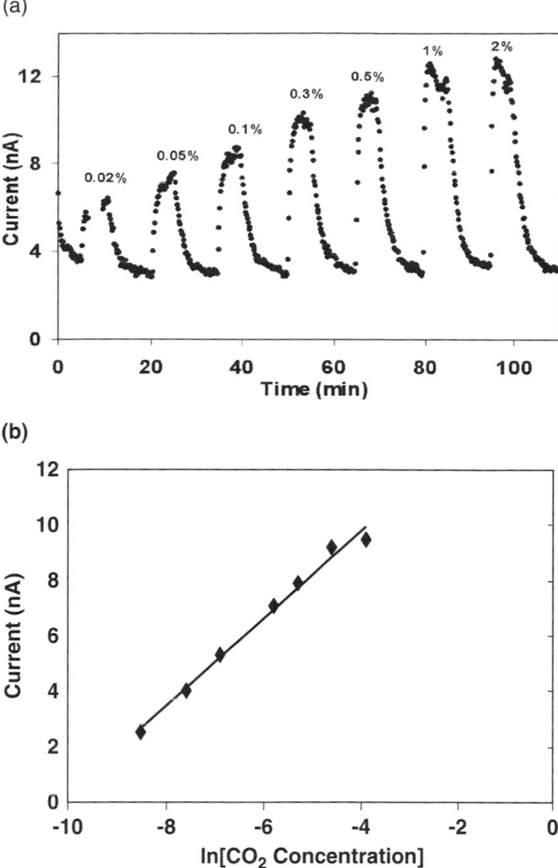

Fig. 8.12 (a) Sensor was tested in CO_2 gases with concentrations of 0.02%, 0.05%, 0.1%, 0.3%, 0.5%, 1% and 2% at 1V and 600°C. (b) Sensor current output versus ln[CO_2 Concentration] [35]

electrode three-phase boundary. The Na_2CO_3 deposited at the working electrode could be transferred to the reference electrode through the $Na_2CO_3/BaCO_3$ if the temperatures are high enough to allow equilibration. These mechanisms allow the sensor to measure CO_2 but recover back to its initial state easily. This process would be made more difficult if the $Na_2CO_3/BaCO_3$ was not distributed across both the working and the reference electrodes. The $Na_2CO_3/BaCO_3$ also likely serves as a diffusion barrier to CO_2 migration to the electrodes resulting in the linear response with ln[CO_2 Concentration].

Thus, the miniaturization of this CO_2 sensor design involved specific steps that needed to be taken to provide the functionality available with larger structures. One step involved changing the materials to a solid state design that would be stable at the higher temperatures required for operation. An interdigitated fingers design was used to improve the response of the electrochemical cell by maximizing the reactive (three-phase) interface. The design of the electrolyte/auxiliary electrolyte structure combined with the operating temperature was chosen to maintain equilibrium and stable operation within the electrochemical cell.

In summary, while the design of this sensor is evolving and other CO_2 electrochemical cell designs are possible, this case study describes the migration of a liquid electrochemical cell design to a solid state microsensor structure. The overall approach is to provide the functionality of the traditional system with materials which can be used in microfabrication processes. This was not just a matter of making a traditional system smaller; rather it involved addressing a wide range of technical issues, from material choice to operational parameters, resulting from the attempt to microfabricate a system.

8.2.4 One Sensor or Even One Type of Sensor Often will Not Solve the Problem, The Need for Sensor Arrays: Case Study of Multifunctional Fire Detection Sensor Array

Applications that require the use of chemical sensors often involve complex chemical environments or require the measurement of multiple parameters. The use of one chemical sensor or sensing technique may not be adequate to meet the needs of these applications. Thus, the use of sensor arrays is necessary to simultaneously measure the multiple parameters needed for some applications.

An array of sensors used to measure multiple chemical species is often called an electronic nose [44]. Standard commercial electronic noses are often composed of multiple versions of the same sensor type, e.g. metal oxide semiconductors such as tin oxide. Each element of the nose is modified slightly to provide slightly different responses to the environment. The resulting signal is then analyzed by software to understand the environment.

In contrast, a different approach is a sensor array that is composed of multiple sensors that are not of the same type, have different sensing approaches, and combine together to give complementary information [1,45]. This can be achieved by using different sensor platforms or sensing mechanisms within the array, such as electrochemical cells and Schottky diodes, or significantly different sensing approaches such as point contact sensors and optical techniques. Analogous to the five senses where, for example, the visual inputs are fundamentally different but complementary to auditory inputs, the overall approach is that each sensor in the array gives different types of information on the environment. That is, the suggestion is that the elements in the array, as much as feasible, be orthogonal to each other.

An example of the application of a sensor array designed for orthogonality is the detection of fires on-board commercial aircraft for safety applications. Fire detection systems, for example, presently in cargo hold fire detection equipment have been shown to be susceptible to false alarms at a rate, depending on the study, as high as 200-1 [1,46–49]. The standard method for fire detection in these cargo bay aircraft applications is the detection of smoke particulates. However, these sensors also can be set off by other particulates beside smoke, such as dust or water molecules in higher humidity conditions.

A second, independent method of fire detection to complement the conventional smoke detection techniques, such as the measurement of chemical species indicative of a fire, was proposed to reduce these false alarms. These chemical sensors have a very different false alarm mechanism and measure very different quantities (chemical species) than the particulate detector (mass or decrease in transmitted light depending on the sensor). Although many chemical species are fire indicators, two species of particular interest are carbon monoxide (CO) and CO_2 [1, 46–49]. Further, miniaturization of the fire detection equipment, both chemical species and particulate detectors, will allow distribution of sensors at a wider variety of locations and improve early detection and location of a fire [1].

Such a microsystem based fire detection system has been fabricated and tested [50]. This system, the Multifunctional, MultiParameter Fire Detection System (MMFDS), combines micro- and nano-based sensors with signal processing hardware and software to interpret the data. Development and inclusion of a miniaturized chemical sensor array and miniaturized particulate detectors was central to the system. The chemical sensor array includes CO and CO_2 sensors as well as the ability to measure other species of interest such as humidity and hydrogen/hydrocarbons. The particulate detector has also been miniaturized and has the potential for particulate size classification.

The overall MMFDS is designed so each sensor response is different and is reflective of a different aspect of the environment. While achieving complete orthogonality is difficult, the approach is to attempt orthogonality using different sensor platforms such as electrochemical cells, Schottky diodes, resistors, and mass measurements. Different fires have distinctive signatures and thus the array measures a range of parameters in order to determine the presence of a fire. These include:

- Particulate detection with an Ion Mobility Spectrometer (IMS): Indicative of smoke;
- Relative Humidity (RH): Indicative of a source of false alarms for the particulate detector;
- Carbon Monoxide and Carbon dioxide (CO/CO_2) production: Indicative of the presence of a fire when combined with particulate detection;
- Hydrogen/hydrocarbons (H_2/C_xH_y): Indicative of the presence of a smoldering fire or prevalent in some types of fires.

The software examines the input from the different sensors and determines which combinations are indicative of the presence of a fire. For example, if the particulate level increases as well as the humidity level increases, but there is no increase in chemical species, then it is likely that the particulate detector is seeing a false alarm due to a rise in humidity. However, if humidity level is constant while both the particulate level and the CO level increase, then it is likely a fire. The approach is to set the alarm levels so as to minimize false alarms while still seeing every fire that occurs.

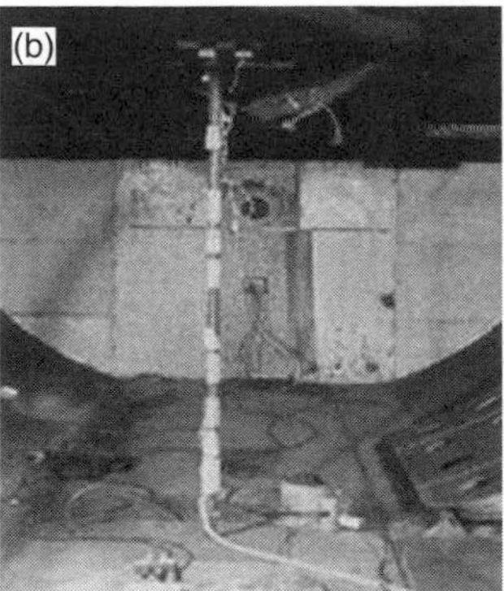

Fig. 8.13 FAA Cargo Compartment Fire Detection Testing a) Boeing 707 used for testing, b) Sensors and facility's overhead instrumentation, c) Flaming "Biscuit" resin block fire source of known and consistent smoke and gas output representative of fire

This MMFDS has been tested at the Federal Aviation Administration (FAA) in the cargo bay of a Boeing 707 aircraft as shown in Fig. 8.13. Within this cargo bay, fires are intentionally set using a known fire source (a polymer "biscuit") [51]. The emissions from these fires are monitored with standard fire detectors and a reference monitoring system, and then compared to prototype fire detection systems such as the MMFDS. Simplified fire detection algorithms were used with the MMFDS sensor array for this testing to show the basic system operation. More complete algorithms are available and can be tuned given knowledge of the application environment.

The results of this comparative FAA testing are dramatic [46–50]. Over a series of exposures to both dust and humidity, the MMFDS had a zero false alarm rate. The commercial system had a 100% false alarm rate. Over the entire test series with real fires, the MMFDS sensed the onset of actual fire nearly equally as well as the conventional smoke detectors if not better, depending on how the MMFDS software was set. Figure 8.14 shows the response of all the sensors with two algorithms (simply listed as fast or slow) as well as the response of the commercial sensor. The standard FAA requirement for aircraft fire detectors in cargo bays is the detection

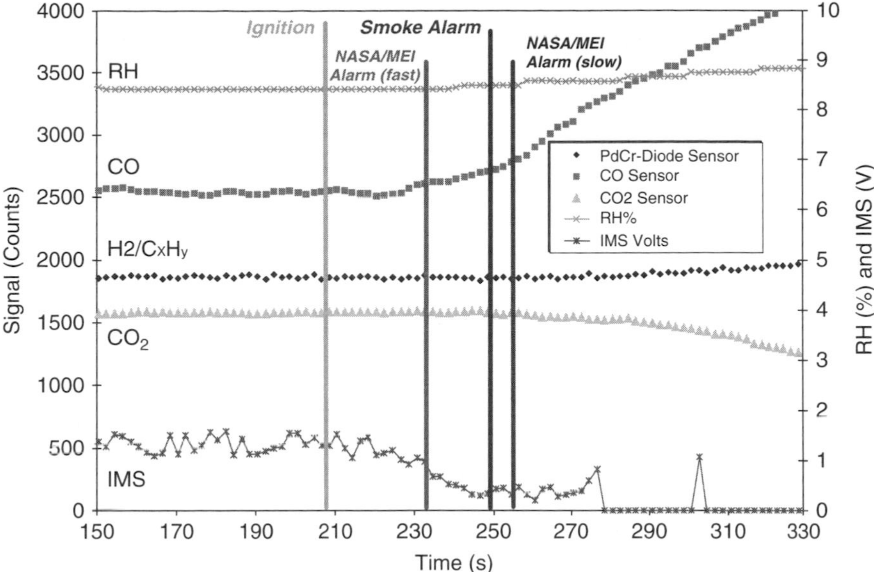

Fig. 8.14 Response for several different sensors including both chemical and particulate detection to the ignition of a resin block. The time of ignition of the fire, detection of a fire by the commercial system, and detection of the fire by the MMFDS using two different algorithms are indicated. The MMFDS responds comparably to the commercial sensor and within the 1 minute limit set by the FAA

of a fire within 1 minute. In all cases, the MMFDS system met the FAA standard of 1 minute. Figure 8.14 also shows the very different response of each sensor; each giving different pieces of information regarding the environment. Subsequent testing has shown that other fires behave differently with different chemical species concentration profiles being produced, but that the general approach of a sensor array providing different information on the environment is effective in determining the presence of those fires.

Thus, the approach of a multiple sensor system designed to maximize orthogonality was shown to significantly reduce false alarms (to the point of eliminating them in these FAA tests) while still consistently enabling rapid detection of fires. These tests demonstrate that the combination of these very different types of sensing technologies is significantly more effective in understanding a fire event than an individual sensor technology alone. This work was recognized with an R&D 100 Award in 2005 as one of the 100 most technologically significant new products of the year.

In summary, this case study shows a significant advantage of using multiple sensors in a sensor array to improve system reliability and to better understand the environment. The range of information needed to eliminate false alarms was not obtained by using a single sensor, but rather a combination of both chemical and particulate sensors was needed for improved results. The chemical sensors involved are designed to measure orthogonally with respect to each other (although there may be some cross-sensitivities) while the particulate detector is strongly orthogonal to the chemical sensors. In this case, the combination of the technologies has allowed drastic improvements in the reliable detection of fires without false alarms. It is suggested that this same approach of orthogonal, multi-parameter detection can be used to gain a better understanding of the environment in a range of applications.

8.2.5 Supporting Technologies Often Determine Success in a Sensor Application: Case Study of Smart Leak Detection Sensor Array

The ability of a sensor system to operate in a given environment often depends as much on the technologies supporting the sensor element as the element itself. If the supporting technology cannot handle the application, then no matter how good the sensor is itself, the sensor system will fail. This supporting technology includes the ability to mount the sensor in the environment, power the sensor operation, communicate with the sensor, and process the information provided by the sensor. Loss of connection to communication or power wires, or packaging that degrades and fails over time can leave an operational sensor element with no means for the user to get information from it. No matter how good the sensor element, if one cannot communicate with it, then it will not provide the information it was designed to deliver. Therefore, technologies that support sensor operation must be able to do so reliably within the operating environment or the sensor system will not be effective.

Further, sensors and their supporting technologies can be a burden on the vehicle or operational system. For every sensor going into an environment, i.e., for every new piece of measurement hardware that improves the awareness of the condition or operational state of the components or overall system, communication and power wires almost always must follow. These wires may be within the sensor or between the sensor to an operating system. As more hardware is added, more wires, weight, complexity, and potential for unreliability is also introduced. While adding sensors may be desired, adding all the wires that come with the sensors may yield a prohibitive weight gain and increased complexity. Further, depending on the sensor, processing of sensor outputs is better done at or very close to the sensor element itself. This is for reasons such as decreasing noise of the sensor signal by localized processing, decreasing the burden that transmitting the sensor information might place on the bandwidth of the vehicle communication system, or even enabling a distributed system which does not depend on a centralized processing unit for continued operation.

Thus, there are often significant advantages to integrating as much of the supporting technology, including packaging, in-situ data processing, power, and communication, with the sensor to form a smart system. Smart sensor systems are defined here as basic sensing elements with embedded intelligence and capable of a range of processing, communication, and self-control/monitoring functions [45, 50]. The ability to easily place smart sensors (complete with supporting technology) where they are needed without changes to the overall vehicle or system architecture would significantly improve the ability to include sensors into applications. A long-term vision for an intelligent vehicle system is a system that is self-monitoring, self-correcting and repairing, and self-modifying. In order for this vision to become a reality, the capabilities afforded by smart sensors integrated into the system are necessary [45, 50]

However, supporting technologies for a sensor are not always available or reliable. An example where supporting technology for smart sensors is not presently available are harsh environment applications such as engine or Venus environments where the supporting technologies are often not capable of operation in the ambient high temperature conditions. These supporting technologies include device contacts [52], packaging and interconnects [53], wireless communication systems, and on-board signal processing and power [54, 55]. Even material processing techniques for high temperature semiconductors such as SiC need to be matured in order to fully enable high temperature smart sensor systems [56].

Nonetheless, in other applications the supporting technology is available. One specific area of smart sensor system development is an integrated Smart "Lick and Stick" Leak Detection System for propulsion systems [15, 33, 50]. The objective of this work is to provide a stand-alone leak detection system that can be placed where needed in a launch vehicle system to detect hazardous conditions. The measurement of fuel and oxygen is included, as well as all the supporting technology in a compact package meant for easy integration into propulsion systems or wherever one would like to monitor hazardous leak conditions.

The components of the Smart "Lick and Stick" Leak Detection System include a microsensor array of hydrogen, hydrocarbon, and oxygen sensors fabricated by MEMS based technology. The development of the hydrogen sensor and SiC hydrocarbon sensors for this leak detection system was discussed in the Case Studies above and the O_2 sensor development is discussed elsewhere [1, 15]. The approach is to minimize the cross-sensitivity between the hydrogen, hydrocarbons, and oxygen measurements, i.e., maximize orthogonality as discussed above. Thus, a range of potential launch vehicle fuels (hydrogen or hydrocarbons) and oxygen can be measured simultaneously to determine if there is a hazardous condition. The array is being incorporated with signal conditioning electronics, power, data storage, and telemetry. The temperature of the sensors is controlled individually as needed and the data is processed for easy interpretation by the user. Parameters related to sensor element and overall sensor system health can be monitored if needed. The parts are chosen for operation in aerospace environments and include a microprocessor. The system is self-contained with the surface area slightly larger than a postage stamp.

Thus, this postage stamp sized "Lick and Stick" type gas sensor technology can enable a matrix of leak detection sensors placed throughout a region with minimal size and weight as well as with no power consumption from the vehicle. The sensors can detect a fuel leak from the vehicle, and combine that measurement with a determination of the oxygen concentration to ascertain if an explosive condition exists. A pressure sensor is mounted in the system to provide pressure compensation as, for example, a vehicle is launched from the ground into the vacuum of space. The electronics hold calibration tables and sensor history with built-in test capability. They can be programmed to provide the user with certain information required on a regular basis, but much further diagnostic information when needed. Wireless communication and battery power are features of the system to allow the unit to be stand-alone. However, a hardwire connection for communication and power is also an option integrated into the sensor system. Sensor outputs can be fed to a data processing station, enabling real-time visual images of leaks thereby enhancing vehicle safety. This leak detection system is an example of a smart microsensor system that is also multifunctional and designed with orthogonal sensor elements.

A prototype model of the "Lick and Stick" sensor system has been fabricated and is shown in Fig. 8.15a [33]. The complete system has signal conditioning electronics, power, data storage, and telemetry with hydrogen, hydrocarbon, and oxygen sensors. Figure 8.15b shows the operation of the electronics with the three sensor system simultaneously. The data highlights the response of the SiC-based gas sensor at various hydrocarbon fuel (RP-1) concentrations. The oxygen concentration is held constant at 21% and the hydrogen sensor signal shows no response, suggesting a lack of cross-sensitivity between the hydrogen and hydrocarbon sensors to the detection of this hydrocarbon. The hydrocarbon sensor is able to detect fuel concentrations from 300 ppm to 3000 ppm although lower concentrations are possible. This data is transmitted by telemetry and viewed on an exterior computer monitoring system.

This example demonstrates the combination of multiple sensor types into a complete, self-contained system with supporting technologies that gives more full-field information than would be available individually. The modular "Lick and Stick" approach allows sensors to be placed where they are needed without the addition of lead wires for power and communication. The "smart" nature of the system means that built-in self tests can be performed to determine parameters related to the health of the sensor system and if necessary recalibration of the sensors can be done easily and stored on-board. While further system development is still necessary, this is an example of a complete "Lick and Stick" smart, multi-parameter sensor microsystem that is usable wherever and whenever needed thus opening a range of monitoring applications. This basic "Lick and Stick" architecture is being considered for possible Shuttle and Crew Launch Vehicle implementation, in part, due to the smart, self-contained nature of the technology enabled by the integrated supporting technologies.

In summary, this case study discussed the importance of supporting technologies to implementation of a sensor system. The ability to mount the sensor in the environment, power the sensor operation, communicate with the sensor, and process the information provided by the sensor in a way that has minimal impact on vehicle operation are significant factors in enabling its application. While the sensor element and its operation are core to being able to make a measurement, successful implementation of the sensor mandates that consideration be made regarding integration of the sensor element, enabled by the supporting technology, into the overall application system. This case study has shown an example of a "Lick and

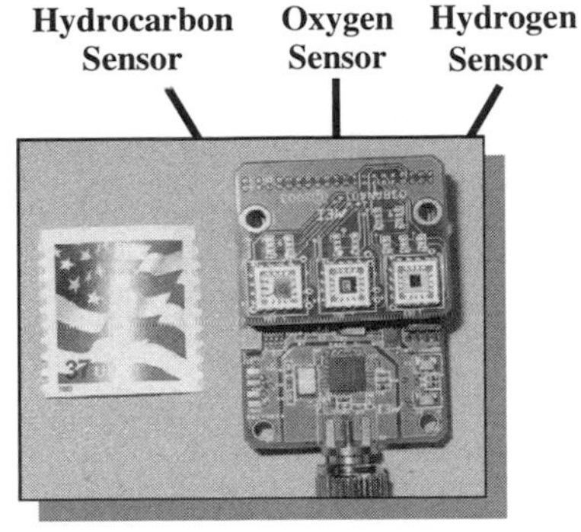

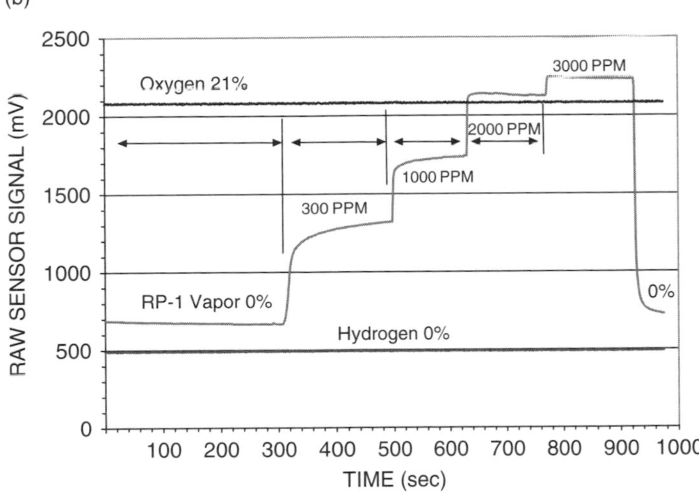

Fig. 8.15 a) A prototype version of a "Lick and Stick" leak sensor system with hydrogen, hydrocarbon, and oxygen detection capabilities combined with supporting electronics. b) Response of the three sensors of this system to a constant oxygen environment and varying hydrocarbon (RP-1) concentrations. The sensor signal shown is the output from the signal conditioning electronics which processes the measured sensor current at a constant voltage and transmits the data by telemetry

Stick" sensor system which has included the sensor elements and the necessary supporting technologies into a near postage stamp sized system. The integration of supporting technologies into this complete, compact, and stand-alone nature of this system enables use of the sensors in applications in which their use may have been problematic otherwise.

8.3 Summary and Sensor Technology Application Approaches

These Case Studies have discussed a range of sensor technology and application issues. The development of sensor systems for different applications has been discussed in light of the theme of the given case study. Each case study illustrated the development of a chemical sensor or sensor system by generally describing the history of the sensor system development and measurement approach related to a given system or application challenge. The case studies described in detail specific sensor system development challenges, approaches to those challenges, and lessons that might be taken from this work. Simply summarized, the themes are:

- Sensors and supporting hardware need to be tailored for the application.
- The sensor structure determines the technical challenges related to development.
- Sensor arrays with multi-parameter input are often necessary to meet the needs of an application.
- Supporting technologies often determine success of a sensor application.

These themes cover a wide range of sensors and sensor systems development. For a sensor user, implementation issues related to a sensor system have a common thread of technology attributes that enables the sensor technology to be useful. To a degree, these go beyond the case studies but reflect many of the case study themes. These include the following [45, 50]:

- INTERACT WITH THE DEVELOPER TO TAILOR THE SENSOR FOR THE APPLICATION:

Sensor technology is best applied with strong interaction with the user. If at all possible, sensor implementation should be considered in the design phase of the vehicle or application system. A sensor developer should strongly interact with the user to make sure the sensor system meets the needs of the application and to tailor the sensor development in order to meet those needs. This includes understanding the application needs in the design of the individual sensor elements as well as how these elements form complete sensors system. The sensor system should not just provide raw parameters to the user, but as a whole tell the user the information they need to know.

- EASE OF APPLICATION:

Sensor system development, including the use of micro/nano fabrication, optical techniques, spray-on technology, etc., should enable multipoint inclusion of complete sensor systems throughout the vehicle without significantly increasing size, weight, and power consumption. If adding vehicle intelligence becomes as easy as "licking and sticking", like postage stamps, smart sensor systems that are self-contained, self-powered, and do not require significant vehicle integration, one major barrier to inclusion of sensors for intelligence is significantly lessened.

- RELIABILITY:

Sensor systems have to be reliable and rugged. Users must be able to believe the data reported by these systems and have trust in the ability of the sensor system to respond to changing situations. Presently, removing a sensor may be viewed as a way to improve reliability and decrease weight. In contrast, removing sensors should be viewed as decreasing the available information flow about a vehicle. Broad use of intelligence in a system will also have a much better chance of occurring if the inclusion of intelligence is achieved with highly reliable systems that users want to have on the vehicle. Further, reliable sensor systems enable the vehicle as whole to be more reliable.

- REDUNDANCY AND CROSS-CORRELATION:

If the sensor systems are reliable and easy to install, while minimally increasing the weight or complexity of a vehicle subsystem, the application of a large number of sensor systems is not problematic. This allows redundant systems, for example, sensors spread throughout the vehicle. Multiparameter sensor systems, i.e., those that can measure multiple measurands related to system health at the same time, can be combined together to give full-field coverage of the system parameters while also allowing cross-correlation between the systems to improve reliability of both the sensor data and the vehicle system information.

- ORTHOGONALITY:

The information provided by the various sensory systems should be orthogonal, that is, each provides a different piece of information on the state of the vehicle system. A single measurement is often not enough to give situational awareness. Thus, the mixture of different techniques to "see, feel, smell, and hear" can combine to give complete information on the vehicle system and improve the capability to respond to the environment.

While not exhaustive, this list of attributes combined together significantly addresses a range of sensor system shortcomings to enable intelligent vehicle systems [45, 50]. A long-term vision for an intelligent vehicle system is a system that is self-monitoring, self-correcting and repairing, and self-modifying. One approach is to build the system bottom up from smart components. These smart components are independently self-monitoring, self-correcting, and self-modifying. In biological terms, the smart component will know its environment (see, feel, hear and smell), think (process information), communicate, and adapt to the environment (move and self-reconfigure). Overall, the approach is self-aware components integrated together to yield a "self-aware" vehicle system. While like a biological system, the requirements for the components of an intelligent system in, e.g., an engine environment, are far beyond those of biological systems would be exposed to at room temperature. The realization of such a vision depends, in part, on developing the sensor system technology to enable improved vehicle system awareness as well as the successful implementation of that technology. This book chapter suggested some possible approaches toward enabling improved sensor development and implementation, which in the long-term are hoped to help enable the vision of an intelligent vehicle as described above.

Acknowledgments The authors would like to acknowledge the invaluable contributions of Professor C.C. Liu of Case Western Reserve University who has been central to this development. The authors would also like to acknowledge the contributions of Dr. L. Matus, R. McKnight,

Dr. G. Beheim, Dr. P. Neudeck, P. Greenberg and Dr. R. Okojie of NASA GRC; Dr. C. Chang and D. Lukco of ASRC Aerospace/NASA GRC; A. Trunek, D. Spry, and Dr. L. Chen of OAI; S. Yu and Q. Wu of Case Western Reserve University, Dr. B. Ward and S. Carrazana of Makel Engineering, Inc.; J. Perotti of NASA Kennedy Space Center; T. Hong of NASA Johnson Space Center; Prof. P. Dutta of Ohio State University; D. Blake of FAA; J. A. Powell of SEST; J. Hunter for manuscript preparation, and the technical assistance of C. Blaha, J. Gonzalez, D. Androjna, M. Artale, B. Osborn, P. Lampard, K. Laster, and M. Mrdenovich of Sierra Lobo/NASA GRC

References

1. Hunter GW, Liu CC, and Makel D (2006) *Microfabricated Chemical Sensors for Aerospace Applications.* In: Gad-el-Hak, M ed. MEMS Handbook 2 edn, Design and Fabrication, CRC Press LLC, Boca Raton, Chapter 11.
2. Liu CC, O'Connor E, Strohl KP, Klann KP, Ghiurca GA, Hunter G, Dudik L, and Shao MJ (2002) *An Assessment of Microfabrication to Sensor Development and the Integration of the Sensor Microsystem.* In: Hesketh P (ed.) Proceedings – Electrochemical Society Microfabricated Systems and MEMS VI, Electrochemical Society Inc., Pennington, pp. 1–8.
3. Wu QH, Lee KM, and Liu CC (1993) *Development of Chemical Sensors Using Microfabrication and Micromachining Techniques.* Sensors Actuators B, vol. 13–14, pp. 1–6.
4. Stetter J, Hesketh P, and Hunter G (2006) *Sensors: Engineering Structure and Materials from Micro to Nano*, Interface Magazine, Electrochemical Society Inc., Pennington, vol. 15, no. 1, Spring, 66–69.
5. Liu CC, Hesketh P, and Hunter GW (2004) *Chemical Microsensors*, Interface Magazine, Electrochemical Society Inc., Pennington, vol. 13, no. 2, Summer, 22–29.
6. Madou M (1997) *Fundamentals of Microfabrication.* CRC Press, Boca Raton.
7. Hunter GW (1992) *A Survey and Analysis of Commercially Available Hydrogen Sensors.* NASA Technical Memorandum 105878.
8. Hunter GW (1992) *A Survey and Analysis of Experimental Hydrogen Sensors.* NASA Technical Memorandum 106300.
9. Lundstrom I (1989) *Physics with Catalytic Metal Gate Chemical Sensor.* CRC Crit. Rev. Solid State Mater. Sci., vol. 15, pp. 201–278.
10. Hughes RC, Schubert WK, Zipperian TE, Rodriguez JL, and Plut TA (1987) *Thin Film Palladium and Silver Alloys and Layers for Metal-Insulator-Semiconductor Sensors.* J. Appl. Phys., vol. 62, pp. 1074–1083.
11. Hunter GW, Liu CC, Wu QH, and Neudeck PG (1994) *Advances in Hydrogen Sensor Technology for Aerospace Applications.* Advanced Earth-to-Orbit Propulsion Technology, Huntsville, AL, May 17–19, 1994, NASA Conference Publication 3282, vol. I, pp. 224–233.
12. Hunter GW, Neudeck PG, Chen LY, Liu CC, Wu QH Makel DB, and Jansa E (1995) A *Hydrogen Leak Detection System for Aerospace and Commercial Applications.* 31st AIAA/ASME/SAE/ASEE Joint Propulsion Conference and Exhibit, San Diego, CA, July 10–12, 1995, American Institute of Aeronautics and Astronautics, Washington, DC, Tech. Rep AIAA-95-2645.
13. Barnes HL and Makel DB (1995) *Quantitative Leak Detection Using Microelectronic Hydrogen Sensors.* 31st AIAA/ASME/SAE/ASEE Joint Propulsion Conference and Exhibit, San Diego, CA, July 10–12, 1995, American Institute of Aeronautics and Astronautics, Washington, DC, Tech. Rep AIAA Paper 95-2648.
14. Hunter GW, Chen L, Neudeck PG, Makel D, Liu CC, Wu QH, and Knight D (1998) *A Hazardous Gas Detection System for Aerospace and Commercial Applications.* 34th IAA/ASME/SAE/ASEE Joint Propulsion Conference and Exhibit, Cleveland, OH, July 1998 Tech. Rep AIAA-98-3614.

15. Hunter GW, Xu J, Neudeck PG, Makel DB, Ward B, and Liu CC (2006) *Intelligent Chemical Sensor Systems for In-Space Safety Applications.* 42nd AIAA/ASME/SAE/ASEE Joint Propulsion Conference & Exhibit, July 10–12, 2006, Sacramento, California, Tech. Rep AIAA-06-58419.
16. Hunter GW, Neudeck PG, Fralick G, Liu CC, Wu QH, Sawayda S, Jin Z, Makel DB, Liu M, Rauch WA, and Hall G (2000) *Chemical Microsensors for Aerospace Applications.* Microfabricated Systems and MEMS V, Proceedings of the International Symposium 198th Meeting of the Electrochemical Society, Oct. 22–27, Phoenix, AZ, P. J. Hesketh et al. Editors, Electrochemical Society Inc., pp. 126–141.
17. Hunter GW, Neudeck PG, Chen LY, Liu CC, Wu QH, Sawayda S, Jin Z, Hammond JD, Makel DB, Liu M, Rauch WA, and Hall G (1999) *Chemical Sensors for Aeronautic and Space Applications III* a short course presented at Sensors Expo 99, Cleveland, OH, NASA TM 1999-209450.
18. Neudeck PG (2000) *SiC Technology.* In: Chen WK, ed. The VLSI Handbook, The Electrical Engineering Handbook Series, CRC Press and IEEE Press, Boca Raton, Chapter .
19. Spetz AL, Baranzahi A, Tobias P, and Lundstrom I (1997) *High Temperature Sensors Based on Metal-Insulator-Silicon Carbide Devices.* Phys. Status Solidi A, vol. 162, pp.493–511.
20. Spetz AL, Tobias P, Uneus L, Svenningstorp H, Ekedahl L, and Lundstrom I (2000) *High Temperature Catalytic Metal Field Effect Transistors for Industrial Applications.* Sensors Actuators B, vol. 70, pp. 67–76.
21. Chen LY, Hunter GW, Neudeck PG, Knight DL, Liu CC, and Wu QH (1996) *Silicon Carbide-Based Gas Sensors.* In: Proceeding of the Third International Symposium on Ceramic Sensors, Anderson H U et al. eds, Electrochemical Society Inc., Pennington, pp. 92–105.
22. Hunter GW, Neudeck PG, Gray M, Androjna D, Chen LY, Hoffman RW Jr., Liu CC, and Wu QH (2000) *SiC-Based Gas Sensor Development.* Materials Science Forum, Silicon Carbide and Related Materials 1999, Carter CH et al. eds, Switzerland: Trans Tech Publications, vol. 469–472, pp.1439–1442.
23. Hunter GW, Neudeck P, Okojie R, Thomas V, Chen L, Liu CC, Ward B, and Makel D (2002) *Development of SiC Gas Sensor Systems.* In: Proceeding of the State-of-the-Art Program on Compound Semiconductors XXXVI/Wide Bandgap Semiconductors for Photonic and Electronic Devices and Sensors III, 201st Meeting of The Electrochemical Society, Philadelphia, Pennsylvania, May, 2002, F. Ren, et al. eds, Electrochemical Society Inc., Pennington, pp. 93–111.
24. Sze SM (1981) *Physics of Semiconductor Devices*, 2nd edn. New York: John Wiley & Sons.
25. Baranzahi A, Spetz AL, Glavmo M, Nytomt J, and Lundstrom I (1995) *Influence of the Interaction Between Molecules on the Response of a Metal-Oxide-Silicon Carbide, MOSiC, Sensor.* In: Tech. Digest Transducers'95 and Eurosensors IX Stockholm, vol. 1, pp. 722–725.
26. Hunter GW, Neudeck PG, Chen LY, Knight D, Liu CC, and Wu QH (1995) *Silicon Carbide-Based Hydrogen and Hydrocarbon Gas Detection.* 31st AIAA/ASME/SAE/ASEE Joint Propulsion Conference and Exhibit, San Diego, CA, July 1995, AIAA paper 95-2647.
27. Powell JA and Larkin D (1997) *Process-Induced Morphological Defects in Epitaxial CVD Silicon Carbide.* J. Phys. Status Solidi (b), vol. 202, pp. 529–548.
28. Chen L, Hunter GW, and Neudeck PG (1997) *Comparison of Interfacial and Electronic Properties of Annealed Pd/SiC and Pd/SiO2/SiC Schottky Diode Sensors.* J. Vacuum Sci. Technol. A, vol. 15, pp. 1228–1234.
29. Chen L, Hunter GW, and Neudeck PG (1998) *X-ray Photoelectron Spectroscopy Study of the Heating Effects on Pd/6H-SiC Schottky Structure.* J. Vacuum Sci. Technol. A, vol. 16, pp. 2890–2895.
30. Hunter GW, Neudeck PG, Chen LY, Knight D, Liu CC, and Wu QH (1998) *SiC-Based Schottky Diode Gas Sensors.* In: Silicon Carbide, III-Nitrides and Related Materials in Proceedings of International Conference on SiC and Related Materials, Stockholm, Sweden, Sep., 1997, Pensl G et al. eds, pp. 1093–1096.

31. Chen L, Hunter GW, and Neudeck PG (1998) *Surface and Interface Properties of PdCr/SiC Schottky Diode Gas Sensor Annealed at 425°C*. Solid-State Electron., vol. 42, pp. 2209–2214.
32. Lei JF (1991) *Palladium–Chromium Strain Gauges, Static Strain Measurable at High Temperatures*. Plat. Met. Rev., vol. 35, pp. 65–69.
33. Hunter GW, Neudeck PG, Xu J, Lukcol D, Trunek A, Artale M, Lampard P, Androjna D, Makel D, Ward B, and Liu CC (2004) *Development of SiC-based Gas Sensors for Aerospace Applications*. In: Dudley M et al. eds., Materials Research Society Symposium Proceedings, Silicon Carbide 2004 – Materials, Processing, and Devices, Materials Research Society, Warrendale, pp. 287–297.
34. Hunter GW, Xu J, Neudeck PG, Trunek A, Chen L, Spry D, Lukco D, Artale M, Lampard P, Androjna D, Makel D, Ward B, and Liu CC (2005) *An Overview of Relevant Technology Development for SiC-based Based Gas Sensors Systems*. In: Presented at the 207th Meeting of the Electrochemical Society, May 15–20, 2005, Quebec City, Quebec.
35. Hunter GW, Xu JC, Liu CC, Hammond JW, Ward B, Lukco D, Lampard P, Artale M, and Androjna D (2006) *Miniaturized Amperometric Solid Electrolyte Carbon Dioxide Sensors*, ECS Transactions, vol. 3, Chemical Sensors 7 and MEMS/NEMS 7, P. Hesketh et al. eds, The Electrochemical Society Inc., Pennington, pp. 203–214.
36. Yao S, Shimizu Y, Miura N, and Yamazoe N (1990) *Solid Electrolyte CO_2 Sensor Using Binary Carbonate Electrode*. Chem. Lett., vol. 19, pp. 2033–2036.
37. Holzinger M, Maier J, and Sitte W (1997) *Potentiometric Detection of Complex Gases: Application to CO_2*. Solid State Ionics, vol. 94, pp. 217–225.
38. Chu WF, Fischer D, Erdmann H, Ilgenstein M, Koppen H, and Leohard V (1992) *Thin and Thick Film Electrochemical CO_2 Sensors*. Solid State Ionics, vol. 53–56, pp. 80–84.
39. Lee J.-S., Lee J.-H., and Hong S.-H. (2003) *NASICON-Based Amperometric CO_2 Sensor Using Na_2CO_3–$BaCO_3$ Auxiliary Phase*. Sensors *Actuators,* B, vol. 96, pp. 663–668.
40. Yang Y and Liu CC (2000) *Development of a NASICON-Based Amperometric Carbon Dioxide Sensor.* Sensors Actuators B, vol. 62, pp. 30–34.
41. Steudel E, Birke P, and Weppner W (1997) *Miniaturized Solid State Electrochemical CO_2 Sensors*. Electrochim. Acta, vol. 42, pp. 3147–3153.
42. Bang Y, Son K, Huh J, Choi S, and Lee D (2003) *Thin Film Micro Carbon Dioxide Sensor Using MEMS Process*. In: The 12th International Conference on Solid State Sensors, Actuators and Microsystems, Boston, June 8–12, Transducers '03, pp. 532–535.
43. Ward BJ, Liu CC, and Hunter GW (2003) *Novel Processing of NASICON and Sodium Carbonate/Barium Carbonate Thin and Thick Films for a CO_2 Microsensor*. J. Mater. Sci., vol. 38, pp. 4289–4292.
44. Gardner JW and Bartlett PN (1994) *A Brief History of Electronic Noses*. Sensors Actuators B, vol. 18, pp. 211–220.
45. Hunter GW (2003) *Morphing, Self-Repairing Engines: A Vision for the Intelligent Engine of the Future*. In: AIAA/ICAS International Air & Space Symposium, 100th anniversary of Flight, 14–17 July 2003, Dayton, OH, AIAA paper 2003–3045.
46. Nuisance Alarms in Aircraft Cargo Areas and Critical Telecommunications Systems: Proceeding of the Third NIST Fire Detector Workshop, Grosshandler WL ed., Gaithersburg, MD, NISTIR 6146, Dec. 1997.
47. Blake D (2000) *Aircraft Cargo Compartment Smoke Detector Alarm Incidents on U.S.-Registered Aircraft, 1974–1999*, FAA Report DOT/FAA/AR-TN00/29.
48. Grosshandler WL (1995) *A Review of Measurements and Candidate Signatures for Early Fire Detection*, Nat. Inst. of Stand. And Tech., Gaithersburg, MD, January 1995, NISTIR 555.
49. Hunter GW, Xu JC, Greenberg P, Ward B, Carranza S, Makel D, Liu CC, Dutta P, Lee C, Akbar S, Blake D (2004) *Miniaturized Sensor Systems for Aerospace Fire Detection Applications*, presented at the Fourth Aircraft Fire and Cabin Safety Research Conference in Lisbon, Portugal.
50. Hunter GW, Oberle LG, Baakalini G, Perotti J, and Hong T (2005) *Intelligent Sensor Systems for Integrated System Health Management in Exploration Applications*, First International Forum on Integrated System Health Engineering and Management in Aerospace, Napa, CA.

51. Blake D (2006) *Development of a Standardized Fire Source for Aircraft Cargo Compartment Fire Detection Systems*, FAA Report DOT/FAA/AR-06/21, May 2006.
52. Okojie RS, Spry D, Krotine J, Salupo C, and Wheeler DR (2000) *Stable Ti/TaSi2/Pt Ohmic Contacts on N-Type 6H-SiC Epilayer at 600C in Air*, Materials Research Society Symposia Proceedings, MRS, Warrandale, PA: vol. 622, pp. H7.5.1–H7.5.6.
53. Liang-Yu C and Jih-Fen L (2002) *Packaging of Harsh-Environment MEMS Devices*, The MEMS Handbook, Mohamed Gad-el-Hak ed., CRC Press, Boca Raton, 2002, Chapter 23.
54. Hunter GW, Okojie RS, Neudeck PG, Beheim GM,Ponchak GE, Fralick G, Wrbanek J, and Chen LY (2006) *High Temperature Electronics, Communications, and Supporting Technologies for Venus Missions*, Proceedings of the Fourth Annual International Planetary Probe Workshop, June 27–30, Pasadena, California.
55. Spry D, Neudeck P, Okojie R, Chen L, Beheim G, Meredith R, Mueller W, and Ferrier T (2004) *Electrical Operation of 6H-SiC MESFET at 500°C for 500 Hours in Air*, Presented at International Conference on High Temperature Electronics Conference 2004, Santa Fe, NM.
56. Powell JA, Neudeck PG, Trunek AJ, Beheim GM, Matus LG, Hoffman RW Jr., and Keys LJ (2000) *Growth of Step-Free Surfaces on Device-Size –(0001)-SiC Mesas*, Appl. Phys. Lett. vol. 77, pp. 1449–1451.

Glossary

Ambient: The gas composition which is dominant in the surrounding environment (e.g. air, pure nitrogen, pure helium etc.).
Criticality 1: A safety term used to refer to operations whose failure can result in loss of life/vehicle.
Detection Mechanism: The chemical and/or physical reaction by which a sensor responds to a given chemical species.
Interfering gases: Gases which can cause a competing response in a sensor and can thus mask or interfere in the sensor's response to a given chemical species. For example, many sensors that respond to hydrogen can also respond to carbon monoxide; thus carbon monoxide is an interfering gas in the measurement of hydrogen.
Lower explosive limit (LEL): The lowest concentration at which a flammable gas becomes explosive. This limit depends on the flammable gas and the corresponding amount of oxidant. For hydrogen in air, this limit is a hydrogen concentration of 4%.
MEMS: MicroElectroMechanical Systems, often used to refer to miniature systems produced by microfabrication and micromachining technology.
Micromachining technology: The fabrication of three dimensional miniature structures using processing techniques such as etching.
Orthogonal: As used in this context, orthogonal sensors are sensors which provide information on the state of system using different sensing mechanisms with no cross-sensitivity.
Response time: The time it takes for a sensor to respond to the environment. Since some sensors never completely stabilize and reach a stable maximum value, often a value equal to 90% the steady state value is cited.
Sensitivity: The amount of change in a sensor's output from baseline to a given chemical species.
Stability: The reproducibility and repeatability of a sensor signal and baseline over time.
Thin Film: Typically, a film whose thickness is less than 1 micron.

Chapter 9
Engineered Nanopores

Amir G. Ahmadi and Sankar Nair

Abstract We review current progress in the science and technology of engineered nanopore devices (ENDs), which are individually addressable nanoporous channels embedded in a thin film substrate. These biologically-inspired devices can be constructed by nanofabrication approaches employing solid-state materials (e.g., oxides and nitrides), or from soft matter (e.g., proteins and lipids) by self-assembly. Coupled with single-molecule electrical and optical techniques, ENDs are a potential 'next-generation' platform for developing high-resolution, ultra-high throughput nanoscale biomolecule analysis systems. We discuss current END fabrication methods and some seminal demonstrations of their potential use in DNA sequencing and protein analysis. We also discuss theoretical and experimental aspects of biomolecule transport in ENDs, and the development of END simulation, control, and operation methodologies. The concluding section discusses future prospects and challenges for ENDs, including the long-term goal of approaching 'single-nucleotide resolution' as well as system integration challenges in constructing technologically viable biosensing END arrays and chips.

Keywords: Nanopore · Ion channel · DNA sequencing · Biomolecule analysis · Nanotechnology · Microfabrication · Biosensors · Microfluidics · Molecular simulation

9.1 Nanopores in Biology and Technology

Nanopores are ubiquitous in biology, in the form of ion channels embedded in cell membranes and nuclear membranes. A biological ion channel is formed from one or more proteins self-assembled into a nanoporous channel-like structure, with a nominal pore size of less than 1 nm and length of ~20 nm spanning the lipid membrane. Functional ion channels are often characterized by the presence of a 'gate'

S. Nair
School of Chemical & Biomolecular Engineering, Georgia Institute of Technology, Atlanta GA 30332-0100

P. J. Hesketh (ed.), *BioNanoFluidic MEMS.*

(that opens and closes in response to an external chemical or electrical stimulus) as well as a 'selectivity filter' that preferentially allows permeation of particular ions (such as sodium, potassium, calcium, or chloride) while excluding others [1]. Ion channels are involved in diverse biological processes such as the transmission of nerve impulses, metabolic pathway regulation (e.g., insulin release from pancreatic β-cells), and muscle (e.g., cardiac) function. Ion channels are thus an important therapeutic target in the case of diseases like diabetes and cystic fibrosis.

The functional characteristics of biological nanopores (high permeation rates and selectivity) are also highly desirable for replication in synthetic systems. Functional nanopores (e.g., those in nanoporous zeolites [2] or nanotubes [3, 4]) are already important in many technological areas including energy-efficient separations, energy conversion, and chemical or biomolecule sensing. In these applications, it is not required to address individual nanopores, and the collective behavior of the nanoporous material or thin film is of main interest. However, nanopores that function as individually addressable devices have recently assumed importance as a platform for 'next-generation' methods for ultra-rapid performance of basic biotechnological operations such as DNA/RNA sequencing and protein analysis [5]. These operations form the foundation of genomics and proteomics, which hold the promise of ultimately providing a complete understanding of biological systems from a genetic perspective as well as cures for diseases that are influenced by genetic factors [6]. In particular, 'engineered nanopore devices' (ENDs) that can analyze the properties of individual DNA or protein strands, offer a promising route towards applications such as ultra-rapid DNA sequencing. In this chapter we discuss the current knowledge of several aspects of END science and technology. We also provide a perspective regarding future developments that have potential to create technological applications of these devices, and the corresponding scientific and engineering challenges that must be overcome.

9.2 Nanopores from Soft Matter

The first synthetic ENDs were produced from 'soft matter' embodied by channel-forming bacterial proteins reconstituted in synthetic lipid bilayers. The general principle behind ENDs is as follows: the analyte (e.g., DNA) is dissolved in a conductive salt solution and driven through the nanopore with an applied electric field. Sensing is accomplished in several possible ways, the most popular being the measurement of modulations in ionic current through the nanopore during transport (translocation) of the analyte. The duration and degree of these modulations can be correlated with parameters such as the biomolecule length, and (in principle) its sequence.

One prevailing example is the protein α-hemolysin (α-HL), which is a naturally occurring compound secreted by the bacterium *Staphylococcus aureus*. When exposed to a synthetically prepared lipid bilayer, each α-HL molecule acts as a monomer in a self assembly process at the surface of the lipid substrate. Seven such monomers fold into a unique quaternary structure that forms a nanoporous

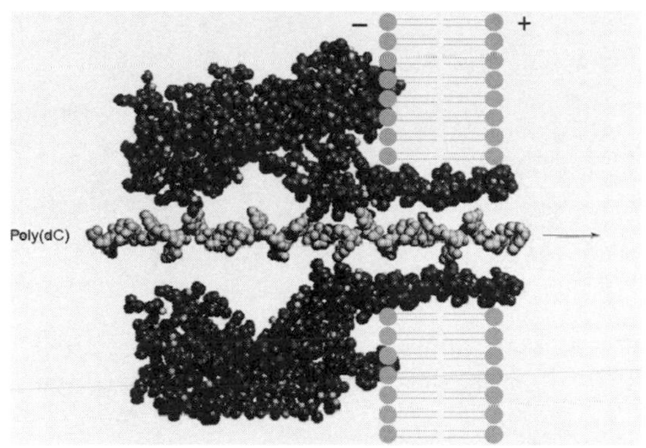

Fig. 9.1 Cross-section of assembled heptameric structure of α-HL in a lipid bilayer [7]. A double-strand of DNA is shown traversing the nanopore

transmembrane channel (Fig. 9.1). The nanopore opening is ~2.6 nm (on the left), leading into a wider vestibule which then narrows to a limiting diameter of 1.5 nm followed by an elongated cylindrical channel embedded in the lipid membrane. As discussed later, biomolecule translocation has been found to be affected by both the channel and vestibular regions

The apparatus for a typical α-HL nanopore experiment consists of two compartments filled with aqueous salt solutions separated by a lipid bilayer membrane. Since the lipid bilayer is impermeable at this time, there is no measured ionic current. α-HL is added to one compartment, and formation of the first nanopore can be observed in about 5–30 minutes as evinced by a sudden rise in the measured ionic current. At this point the compartment is flushed with fresh solution to prevent further pore formation, and the analyte may be introduced for sensing measurements. Figure 9.2 illustrates this setup and typical ionic current readings for different DNA species. Sensing applications with α-HL nanopores have been developed for single-molecule detection, identification, and quantification of a wide range of analytes ranging from TNT to divalent metal ions to single stranded DNA and RNA [8–11]. In the first two cases, selective binding sites for the analytes are created (e.g., by genetic engineering) so that the ionic current modulation is specific only to the desired analyte and long-lived enough to be reliably observed. Measurements using α-HL also demonstrate identification and quantification of unknown analytes as well as the ability to distinguish between related species through duration and amplitude of current blockage.

In the case of DNA sensing, there are two potential ways in which nanopores can be employed. The ultimate (and more difficult) objective is direct sequencing by recognizing and distinguishing individual bases on a single DNA strand. This method requires high enough sensitivity in measuring the ionic current change in response to each base passing through the pore, such that the relative changes in amplitude can be used to separate and identify every base in the sequence. The α-HL nanopore is approximately 25 DNA bases in length. Although there is a limiting constriction in the α-HL nanopore that has a length comparable to a single base, the measurement noise (~1 pA) is too high to detect individual bases as they pass through the constriction (with a time scale of ~1 μs per base). Hence, the initial

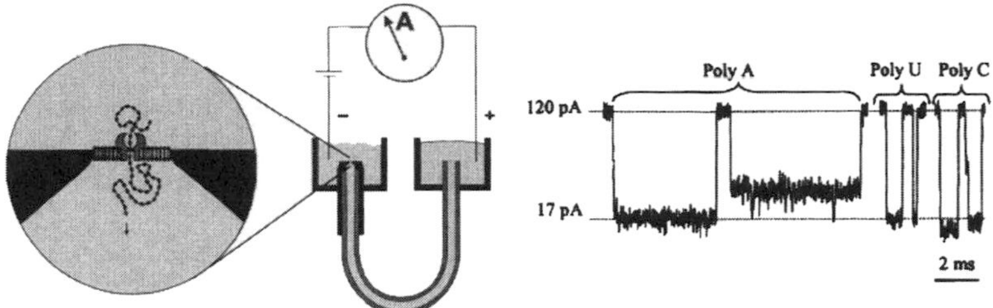

Fig. 9.2 Prototype setup for a-HL DNA sensor [12]. An example measurement of ionic current amplitude and blockage duration for polyA, polyU, polyC 100-mers

focus has been on using the nanopore to determine the length of DNA strands. This would allow coupling of nanopore detectors to the well established technique of polymerase chain reaction (PCR). Thus, nanopore detectors could replace the current processing of PCR-generated samples by gel or capillary electrophoresis. Accordingly, nanopore detectors will require the capability to distinguish a DNA strand of length N bases from a strand with $N+1$ bases.

Work in the latter direction has taken multiple approaches; important steps include demonstrating the ability to distinguish between DNA molecules of slightly different lengths and/or compositions [12]. In one study using α-HL nanopores, six DNA samples 100 bases in length but with varying compositions were shown to have distinguishable statistical translocation properties, even in cases where the overall compositions of two DNA molecules were identical while the base sequences differed (Fig. 9.3). The ionic current data from repeated translocation events is collected and interpreted in the form of a histogram or event diagram. A translocation event is indicated by a drop in current, the duration of which is assumed to be directly proportional to the length of the strand. The varying amplitude of the events may also be useful in identifying specific bases or sets of bases along the strand.

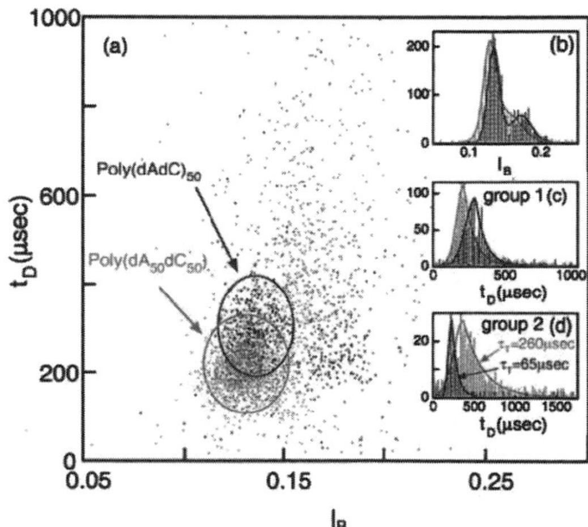

Fig. 9.3 Event diagram for translocation events of poly-(dAdC)50 and poly-(dA50dC50) 100-mers in two separate experiments, indicating differentiability on basis of the sequence, despite identical overall composition [13]

Other experiments have used strategies to slow down the translocation of DNA through the pore, thus allowing data collection from a larger number of ions as each bases passes the pore constriction. This would presumably increase the signal-to-noise ratio to allow DNA strand lengths to be distinguished with higher resolution, or even allow individual bases to be distinguished. For example, DNA 'hairpins' are single strands with a small portion of the sequence (at the end of the strand) being composed of two blocks of complementary bases. Thus, the end of the strand curls back on itself to create a 'hairpin'. The hairpin portion is too large to travel through the constriction, and hence is trapped in the vestibule while the single-stranded portion partially traverses the pore. Considering the time taken by the hairpin portion to dissociate and follow the rest of the strand through the pore, there is an, overall 5- to 10-fold decrease in the translocation rate. Research has shown that with the hairpin approach, the ionic current readouts from strands varying only by a single base in length are distinguishable from each other [14, 15]. Additionally, strands identical in length and composition except for a single base mismatch in the duplex, can also be distinguished.

These advances show promise for α-HL in nanopore sensing applications. However, there are intrinsic disadvantages in working with nanopores made from soft matter [7, 16, 17]. The α-HL nanopore is not very robust and cannot be maintained for extended periods (greater than 1 day). Although advances towards single-base resolution have been made, the limiting signal-to-noise ratio issues (caused by fast translocation and low ionic conductivity) continue to hinder their applications to DNA sequencing. Additionally, computer simulations of DNA transport through the α-HL nanopore have revealed their intrinsic structural limitations. Brownian dynamics simulations of DNA translocation through α-HL pores correctly predicted the presence of multiple peaks and long tails in translocation time distribution curves commonly observed in experiments (e.g., insets in Fig. 9.3). These features are undesirable since it then becomes impossible to distinguish strands differing slightly in length. The simulation studies indicate that the pore geometry, particularly the vestibule (Fig. 9.1), is the primary source of the low length-resolution of the nanopore. Due to the large volume available in the vestibule, the translocating strand can adopt a large number of configurations as it passes through the pore, thus resulting in a wide translocation time distribution. This finding was further supported by carrying out simulations of DNA translocation through a smooth cylindrical nanotube of comparable size, in which case a much narrower translocation time distribution was observed [18]. In conclusion, degradability and intrinsic limits on resolution due to complex pore geometry are the main problems in DNA sizing or sequencing with α-HL nanopores. These issues have led to a shift towards fabrication of solid-state inorganic nanopores that allow greater robustness and better control over pore geometry.

9.3 Solid-State Nanopore Devices

Development of solid-state nanopores for DNA sequencing and other applications is an approach rather different from using α-HL nanopores that rely on biological self assembly and genetic engineering. Nanopores formed in solid state materials are subject to completely different design considerations that resemble those found in semiconductor and microelectronic device manufacturing. The first critical design consideration is the pore size, which should ideally be narrow enough to accommodate one strand of DNA at a time and short enough to instantaneously attribute the blockage in ionic current to a single base during translocation. The pore diameter is thus chosen on the order of the diameter of single stranded DNA, in the range of 2–10 nm. The nominal pore length is essentially determined by the thickness of the solid-state thin film (e.g., silicon nitride) through which the nanopore is fabricated. However, the effective pore length may be shorter than the film thickness if the pore has a strongly tapered or conical shape. Thus, the pore length and diameter are ultimately a function of various empirical parameters, such as the material in which the pore is being fabricated (e.g., silicon nitride, silicon dioxide), the diameter of the beam (ionic or electronic) used to create the pore, the intensity of the beam (i.e., the ionic or electronic flux contained within the beam), and the exposure time. Mechanical, chemical, and electrical properties of the thin film material can all be of importance. In general, the materials should be rigid, mechanically robust, impermeable, and tolerant to processing under wide ranges of temperature and pressure. Silicon wafers are the substrates of choice, while silicon dioxide (SiO_2) and/or silicon nitride (Si_3N_4) thin films can be deposited or grown on the substrate and used to produce a nanopore. Silicon is used widely in micro- and nano-fabrication due to its abundance, crystallinity, availability of 'planarized' manufacturing processes, and desirable properties as a semiconductor. Silicon nitride is robust and non-reactive whereas silicon oxide is hydrophilic and electrically insulating; both can be deposited to nanometer precision. However, there is currently no systematic information in the open literature about the effects of these variables on the properties of fabricated solid-state nanopores.

The processing strategy usually starts with fabrication of relatively large holes and reduction of pore size (usually by at least an order of magnitude) in each material removal step, until a nanopore of desired size is formed. In a typical process, the silicon substrate (300–550 μm in thickness) is used to deposit layers of silicon nitride and/or oxide. The silicon is etched from behind (using an appropriate photolithographic mask and an etchant such as potassium hydroxide or tetramethylammonium hydroxide) to expose a square window of size 10–50 μm below the nitride/oxide layers. Next, these layers are attacked from the top using combinations of etchants (such as reactive ion etching for silicon nitride and buffered oxide etch for silicon dioxide) until a free-standing membrane of the desired thickness (< 50 nm) is exposed. Finally, the nanopore is produced in this free-standing membrane. Many of the steps leading up to the formation of the nanopore (such as deposition, photolithography, and wet and dry chemical etching) are well characterized and commonly practiced. However, the nanopore reproducibility, nanopore size control, and stability of the thin membrane are much less understood.

The earliest reported solid-state END fabrication method [19,20] used controlled ion beam milling to generate a nanopore. This process begins with deposition of a 500 nm low-stress silicon nitride film on the front-side of a (100) silicon substrate, followed by patterning of a hole of diameter 500 μm in the backside of the wafer by photolithographic techniques. The silicon is then etched in solution at a 54.7° angle following the (111) crystal planes, until the nitride layer is exposed and free-standing, creating a 25 μm square window. At this point, two approaches can be taken. In the first, a bowl-shaped cavity about 100 nm wide is fashioned in the newly revealed nitride surface using reactive ion etching (RIE) or focused ion beam (FIB) milling. Next, material is removed from the cavity until the remaining thickness of nitride at the center is 5–10 nm. This surface provides the desired length of thc porc, and the pore itself is created by milling the surface with a focused ion beam tool. The second approach uses RIE or FIB to open a large through-hole of size 60 nm which penetrates the silicon nitride. The ion beam is then used to laterally deposit material to the pore, filling it in to the desired size with nanometer precision. This technique of manipulating matter at the nanoscale by bombarding a sample with ions in a controlled manner is introduced as 'ion beam sculpting'. The basic idea is that each incident ion removes an atom of material at the surface. However, depending on processing conditions of temperature and flux, ion exposure can lead to lateral deposition of material.

One reported experimental setup of this mechanism consists of a dual-beam tool which erodes the surface with Ar^+ ions while using a transmission electron microscope (TEM) for imaging of the surface. The apparatus contains a feedback mechanism which monitors the number of ions transmitted through the sample and is used to regulate the milling rate. The beam is focused on the surface and material is sputtered until formation of a nanopore is observed. TEM images combined with the ion rate counter characterize pore formation. Figure 9.4 demonstrates the strategy and apparatus used in ion beam sculpting of a silicon nitride layer by material sputtering.

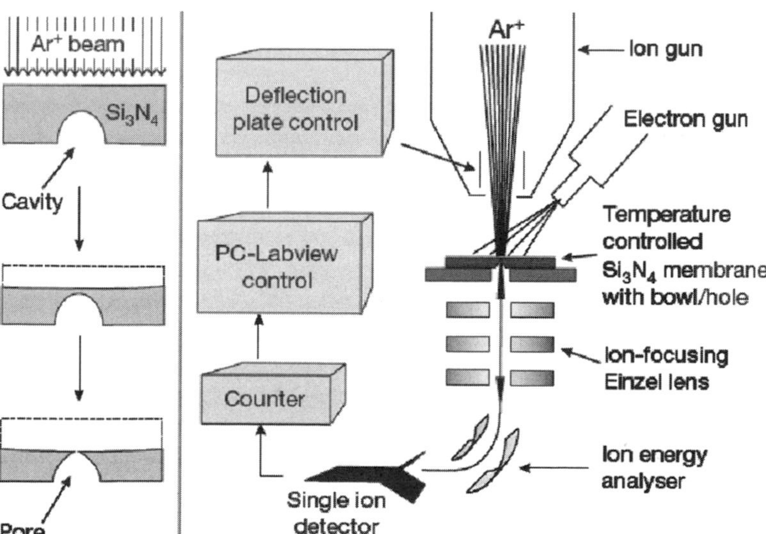

Fig. 9.4 Opening up a nanopore in a free-standing silicon nitride membrane using a focused ion beam, and the dual beam setup with detection and feedback mechanism [19]

Pore sizes of 3 and 10 nm have been demonstrated in 5–10 nm free-standing silicon nitride using ion beam sculpting.

Another method used to produce nanopores is the application of a tightly focused, high-voltage (200–300 keV) electron beam (generated in a transmission electron microscope) to a thin free-standing membrane supported on a silicon wafer [21–27]. The film is usually silicon nitride or oxide (20–50 nm thin), deposited and exposed using sequential photolithography and etching steps as described earlier. An example of the fabrication in such a configuration is shown in Fig. 9.5. In this case, the purpose of the top layers is to provide mechanical support and reduced capacitance for the free surface. We note that the solid-state fabrication methods discussed here are capable of mass producing nanopores in arrays on a single wafer relatively quickly and cheaply. Once a nanopore has been generated in a free-standing surface by one of these methods, the nanopore can be directly used in sensing applications or undergo further processing for a different application, such as a nanoscale electrode [21]. Several current efforts are underway to use these pores to achieve single base resolution DNA sequencing. In a translocation experiment parallel to the a-HL detector, the pores are immersed in ionic solution joining two chambers. The DNA sample is usually double-stranded (which is too large to traverse the α-HL pore).

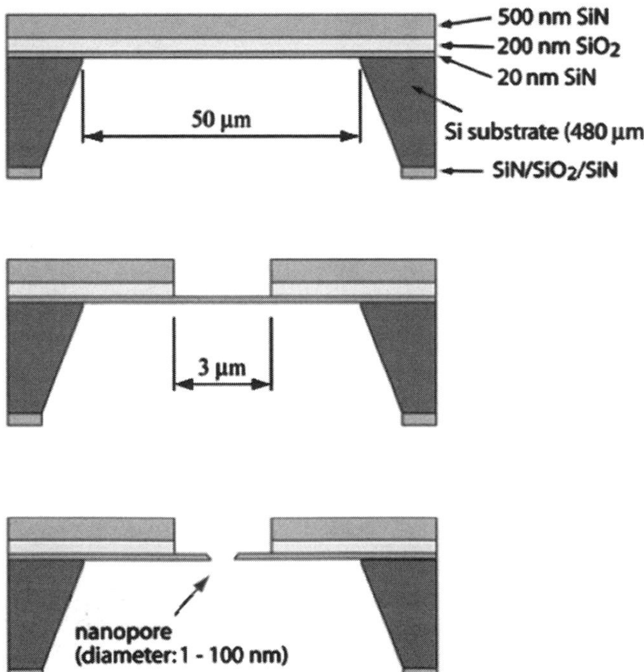

Fig. 9.5 Basic fabrication steps, opening a large window in silicon to expose the nitride from the bottom, followed by patterning and etching of the topside, and using high voltage TEM to make the nanopore [21]

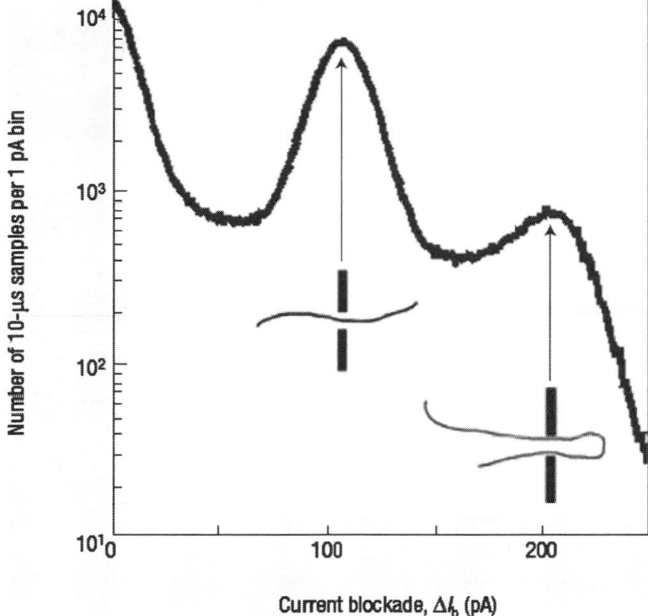

Fig. 9.6 Time distribution of current blockades in 10 nm pore indicative of simple translocation and folding during translocation respectively [20]

Preliminary DNA sensing experiments with solid-state ENDs have verified that translocation is actually taking place. Data is analyzed in event diagrams similar to those obtained with α-HL nanopores. Effects of variation of important factors such as temperature, pH, voltage, and salt concentration on translocation have been studied (note that biological nanopores have much lower tolerance for changes in pH and temperature). For example, experiments in which pore sizes of 3 nm and 10 nm were compared [20] demonstrate that a portion of the DNA traversing the larger pore experienced folding during translocation. As might be expected, the current blockade is approximately doubled since the pore accommodates twice the number of bases at a given time. The time distribution plot of this phenomenon is shown in Fig. 9.6. Experiments carried out at varying pH have demonstrated capabilities of solid-state ENDs to discriminate between single-stranded and double-stranded DNA [24].

9.4 Nanopore Simulation and Control Techniques

The use of nanoscale transport theory and simulation techniques can provide an important tool in understanding and optimizing both biological and solid-state ENDs. At a first level, modeling and simulation can assist in the quantitative explanation of experimental results and probing the effects of parameters that may be difficult to control experimentally. However, theory and simulation, coupled with principles of device operation and control, can assist in ultimately addressing the issue of achieving single-base resolution. One aspect in which simulation can have an immediate impact is in providing accurate correlations between translocation time and biomolecule (e.g. DNA) properties for wide ranges of operational parameters (e.g., field strength, pore size, temperature) For this purpose, the translocation process is split into steps which can be modeled separately. A biopolymer chain must first 'find' the nanopore and partially enter it. It can then either traverse the nanopore or retract back into the solution, depending upon whether the applied electric field can overcome the loss of entropy in the polymer associated with entering the pore as an extended chain. There is also an unavoidable random thermal force superimposed on the system that affects the statistics of the translocation process. If the polymer is successfully captured or trapped by the pore, it then traverses the nanopore driven by the applied field. Finally, the lagging end of the chain passes through the nanopore and the analyte escapes on the other side of the nanopore, completing the translocation.

These complex molecular-level transport processes can be modeled at different levels of detail. Brownian Dynamics (BD) simulations are a popular method of simulating translocation of a polymer chain across nanopores [18, 28–30]. In this method, the polymers are modeled as a chain of monomer 'beads', each of which experiences van der Waals, Coulomb, and nearest–neighbor-bonding interactions with other beads on the same chain, as well as with the nanopore. The Newtonian equation of motion is integrated numerically for each bead, thereby simulating the motion of the chain as a whole. In addition to 'coarse-graining' the structure of the monomers, Brownian Dynamics also permits a further saving in computational costs by replacing explicit consideration of solvent molecules by an effective medium characterized by a dielectric constant and electrolyte concentration. Thus, the screened electrostatic interactions are modeled in a continuum framework. Another result of omitting explicit solvent molecules is that the random collisions of the solvent molecules with the polymer can be modeled by the Langevin approach, viz. the addition of a random force to the Newtonian equation of motion whose magnitude is sampled from a Gaussian distribution [29].

The simulation environment typically consists of compartments separated by a solid wall containing the nanopore, as shown in Fig. 9.7. In that study, translocation was modeled for different solid-state nanopore sizes of 1.6, 2.5, and 3 nm. These case studies examined effects of both attractive and repulsive polymer–nanopore interactions, and indicate that attractive nanopore–poymer interactions can significantly improve the probability of polymer capture. In the case of the 2.5 nm pore, the simulation predicted two peaks and non-Gaussian behavior in the translocation time distribution curve, which is also a common empirical observation in α-hemolysin ENDs. The latter ENDs have also been studied with Brownian Dynamics simulations [18, 29]. In this case, an important finding was that the complex shape of the α-hemolysin nanopore plays an important role in the observed low resolution and non-Gaussian behavior in the translocation time distributions. In particular, comparative simulations of α-hemolysin and hypothetical nanotubes clearly showed the

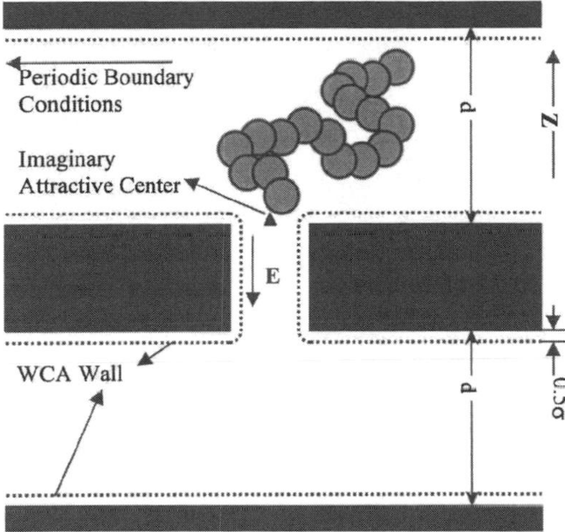

Fig. 9.7 Schematic of Brownian Dynamics simulation setup [28]

benefits (mainly increased resolution and better reproducibility) of the shift towards cylindrical solid-state nanopores (Fig. 9.8). This work showed that the vestibule of the α-hemolysin nanopore in fact creates an entropic trap for the traversing chain, allowing a number of different chain configurations in successive translocation events and leading to a wide translocation time distribution. The cylindrical nanotube leads to a much lower entropy for the translocating chain.

Molecular Dynamics (MD) is another simulation tool used to study translocation dynamics in ENDs. Here the system is modeled at a detailed atomistic level with the interatomic forces being described by detailed force field models [31]. Thus,

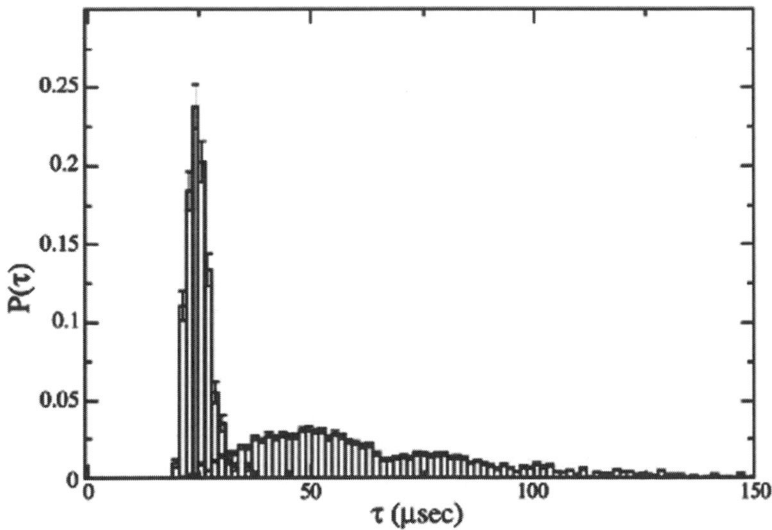

Fig. 9.8 Simulated translocation time distributions for a-HL and a nanotube [29]

MD permits much more accurate study of translocation phenomena, but at much shorter time scales (picoseconds to nanoseconds) than BD (up to milliseconds) due to the much larger computational cost. A recent MD study investigated translocation of double-stranded DNA through a nanopore in a silicon nitride membrane, with a potassium chloride electrolyte solution [32]. DNA bases were found to exhibit strong affinity for the silicon nitride surface during translocation. The magnitude of the interaction tended to cause double stranded DNA to 'unzip' within the pore. It was also shown that DNA trapped in the nanopore (without actually translocating through it) can cause a blockage in ionic current comparable to that experienced during translocation. These effects and their implications are important in understanding translocation in solid-state nanopores and in avoiding detrimental phenomena such as 'false positive' signals from DNA trapped in the nanopore without translocation. There have also been theoretical studies on predicting translocation times in nanopores, using the theory of stochastic polymer dynamics. These works consider a simplified system [33–37], viz. a Gaussian polymer chain traversing a nanopore. Explicit formulae were developed for the dependence of translocation duration on chain and pore length, pore size, polymer–nanopore interactions, and applied chemical gradients. Again, a competitive mechanism was observed between attractive polymer–nanopore interactions and the entropic barrier to translocation.

A potential next step in modeling and simulation of nanopore translocation could be in finding ways to control and optimize transport in the nanopore and achieve single base resolution. One significant problem in current ENDs is that translocation occurs too quickly to resolve the identity of single nucleotides or even the length of each strand to a single-nucleotide level. Control of environmental parameters such as temperature, viscosity, and salt concentration have not been enough to solve this problem, although ongoing efforts have been able to decrease translocation time by up to an order of magnitude [38]. Furthermore, the action of random thermal forces imposes a fundamental limit on the resolution of ENDs if operated in a conventional manner. A recent computational study employing Monte Carlo methods, it was shown that a rotating transverse (i.e., in the plane of the nanopore substrate) electric field could help to control random fluctuations in DNA dynamics in the transverse plane, thus assisting in a more reproducible capture and translocation events [39,40]. A recent study investigated the use of optimized AC stimuli coupled with the conventional DC stimulus to drive DNA across nanopores [41, 42]. The main idea was that by choosing an appropriate stimulus (e.g., AC, pulse-train, or multisine) that is 'tuned' to the characteristic time scale of translocation, one can make the translocation time a strong function of the strand length and thus greatly increase the resolution with which strands of different lengths can be distinguished. This appropriate stimulus is obtained from a predictive simulation of the translocation process. In addition, the optimal stimulus is different for different length domains, so that the simulation engine must be interfaced with a control algorithm that computes optimal parameters for different length regions of the sample and hence allows a high-resolution analysis starting from a coarse DC-driven analysis of the sample (Fig. 9.9). In these studies, a simple rigid-rod model was parameterized to reproduce existing experimental data and used to demonstrate the above

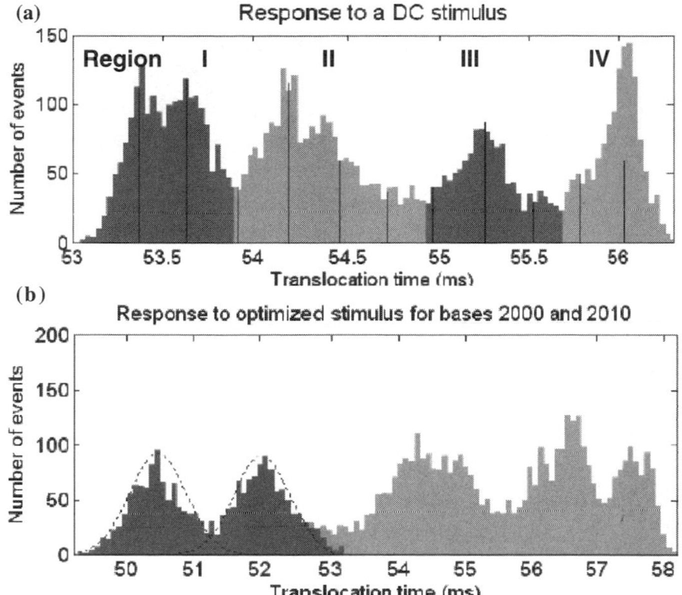

Fig. 9.9 (a) Simulated DC translocation time distribution of a sample containing eleven base lengths between 2000 and 2100, showing a coarse resolution. (b) Simulated translocation time distribution for an optimal stimulus that clearly resolves the two base lengths (2000 and 2010) in Region I. Optimal stimuli are generated for each of the four regions, allowing a complete analysis of the sample [41]

concepts. Detailed validation with BD and MD simulations, as well as experiment, would establish these combined simulation and optimization techniques as tools for operating ENDs at high resolution.

9.5 Prospects and Challenges in END Science and Technology

The potential advantages of a single-molecule method based upon direct reading of the sequence from unamplified DNA, are increasingly accepted. Conceptually, it is natural to pursue the development of a technology that directly reads sequence information from individual genome-length DNA, rather than the much more complex sample preparation and analysis methods necessitated by current approaches. The development of such a technology is extremely challenging and is projected to take at least a decade of science and engineering research targeted at device fabrication, detection methods, and the dynamics of biomolecule transport in confined nanopore spaces. It is, however, clear that even without 'single-base' resolution, the development of ENDs is of high importance in biomolecule analysis. For example, the capability of the END to accurately size DNA strands several orders of magnitude faster than electrophoresis offers the possibility of replacing an important component of current DNA sequencing technology. There are also a number of other applications wherein single-nucleotide resolution is not required. For example, ENDs with functionalized sites can be employed to detect extremely small levels of toxic agents such as the anthrax lethal factor (LF) and edema factor (EF) [43], based upon modulation of END properties (such as ionic current) by specific binding events occurring inside (or in the vicinity of) the nanopore. The use of α-hemolysin nanopores with various molecular 'adapters' has been shown to lead to new biomolecule sensors that offer much higher sensitivity, much faster response, single-molecule resolution, and potentially lower cost, than conventional biomolecule sensors based on measuring the collective response from a macroscopic ensemble of sensing/detection sites [5, 44, 45]. Similarly, solid-state nanopores fabricated from materials such as silicon dioxide, could be functionalized [46] with biomolecules of various types and used to analyze proteins [47–49], viruses, and other biomolecular analytes with all the advantages (e.g., speed, sensitivity) offered by ENDs. Furthermore, future ENDs need not be restricted to measurements of ionic current modulation. Other proposed detection methods, such as transverse electron tunneling between metal electrodes on the nanopore walls [50] and fluorescence resonance electron transfer (FRET) measurements [51] between quantum dots embedded in the nanopore walls, may lead to entirely new capabilities that cannot be achieved by ionic current measurements alone. Concurrently, advances in operation and control methodologies for nanopore operation [41, 42] can allow optimal operation of the END. Figure 9.10 is a schematic of the possible appearance of a future END device.

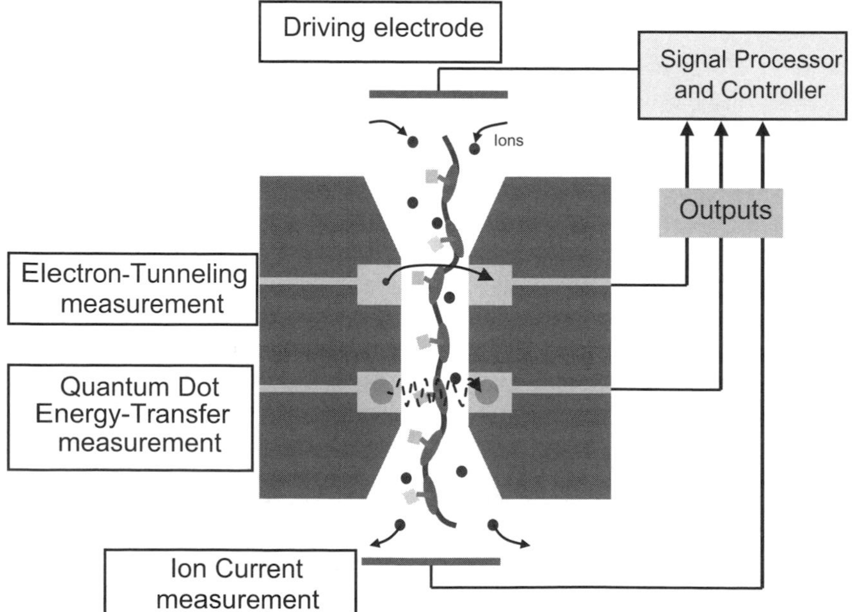

Fig. 9.10 Conceptual schematic of a future Engineered Nanopore Device (END)

In order for ENDs to be successfully commercialized for both 'short-term' applications (that do not require single-nucleotide resolution) as well as 'long-term' applications (such as genome sequencing with single-nucleotide resolution), two important challenges must be overcome that are common to all the variations of ENDs. Firstly, there is currently no strategy for reliable and reproducible fabrication of large arrays of solid-state nanopores that can be operated in parallel. Previous reports have focused on fabricating prototype single nanopores using ion track etching, ion milling, or electron microscopy [19, 21, 24, 52]. Encouraging results have been obtained, and it is clear that each technique involves a number of experimental parameters (e.g., the nature of the substrate in which the nanopore is being created, the energy and diameter of the ion/electron beam, the exposure time) that strongly influence the quality of the nanopores produced. Future ENDs may be operated in massively parallel arrays to provide bioanalytical information at a small fraction of the time and cost of present-day technology. In the fabrication of large arrays of nanopores, these characteristics cannot be monitored manually, and hence they must be known to a degree of precision similar to that achieved in the fabrication of microelectronic devices. However, there is currently very little systematic knowledge of the relationship of nanopore fabrication parameters to the structure of the resulting nanopores [53]. It is essential to develop a fabrication strategy that is based on the knowledge of 'process-product' relationships of nanomanufacturing tools as pertaining to nanopore fabrication, and then to demonstrate the application of this strategy to create and characterize arrays of high-quality ENDs. This capability can then be interfaced with existing methods of integrated circuit and microfluidic fabrication to build completely integrated systems containing on-chip detection electronics and microfluidic sample handling systems. The second challenge has already been discussed extensively in this work, i.e. the issue of controlling the DNA dynamics and developing methods to achieve single-nucleotide resolution in ENDs. Current directions being pursued in multiple research laboratories create considerable optimism that these challenges will be overcome. It is likely that the commercial appearance of END systems will require close collaborations of scientists and engineers to ultimately deliver robust, high-speed, low-cost bioanalytical platforms that retain all the intrinsic advantages offered by ENDs. Such a development would have a truly revolutionary impact on innumerable areas in biotechnology and medicine that require the ability to analyze complex biomolecular mixtures.

References

1. Aidley, D.J. and P.R. Stanfield, *Ion Channels: Molecules in Action*. 1996: Cambridge, MA: Cambridge University Press.
2. Davis, M.E., *Ordered porous materials for emerging applications*. Nature, 2002. **417**(6891): pp. 813–821.
3. Baughman, R.H., A.A. Zakhidov, and W.A. de Heer, *Carbon nanotubes – The route toward applications*. Science, 2002. **297**(5582): pp. 787–792.
4. Goldberger, J., R. Fan, and P.D. Yang, *Inorganic nanotubes: A novel platform for nanofluidics*. Accounts of Chemical Research, 2006. **39**(4): pp. 239–248.

5. Kasianowicz, J.J., *Nanometer, scale pores: Potential applications for analyte detection and DNA characterization*. Disease Markers, 2002. **18**(4): pp. 185–191.
6. Collins, F.S., E.D. Green, A.E. Guttmacher, and M.S. Guyer, *A vision for the future of genomics research*. Nature, 2003. **422**(6934): pp. 835–847.
7. Deamer, D.W. and M. Akeson, *Nanopores and nucleic acids: prospects for ultrarapid sequencing*. Trends in Biotechnology, 2000. **18**(4): pp. 147–151.
8. Braha, O., B. Walker, S. Cheley, J.J. Kasianowicz, L.Z. Song, J.E. Gouaux, and H. Bayley, *Designed protein pores as components for biosensors*. Chemistry & Biology, 1997. **4**(7): pp. 497–505.
9. Guan, X.Y., L.Q. Gu, S. Cheley, O. Braha, and H. Bayley, *Stochastic sensing of TNT with a genetically engineered pore*. Chembiochemistry, 2005. **6**(10): pp. 1875–1881.
10. Butler, T.Z., J.H. Gundlach, and M. Troll, *Translocation of RNA block-copolymers through the alpha hemolysin nanopore*. Biophysical Journal, 2005. **88**(1): pp. 347A–347A.
11. Butler, T.Z., J.H. Gundlach, and M.A. Troll, *Determination of RNA orientation during translocation through a biological nanopore*. Biophysical Journal, 2006. **90**(1): pp. 190–199.
12. Akeson, M., D. Branton, J.J. Kasianowicz, E. Brandin, and D.W. Deamer, *Microsecond timescale discrimination among polycytidylic acid, polyadenylic acid, and polyuridylic acid as homopolymers or as segments within single RNA molecules*. Biophysical Journal, 1999. **77**(6): pp. 3227–3233.
13. Meller, A., L. Nivon, E. Brandin, J. Golovchenko, and D. Branton, *Rapid nanopore discrimination between single polynucleotide molecules*. Proceedings of the National Academy of Sciences of the United States of America, 2000. **97**(3): pp. 1079–1084.
14. Vercoutere, W., S. Winters-Hilt, H. Olsen, D. Deamer, D. Haussler, and M. Akeson, *Rapid discrimination among individual DNA hairpin molecules at single-nucleotide resolution using an ion channel*. Nature Biotechnology, 2001. **19**(3): pp. 248–252.
15. Vercoutere, W.A., M. Akeson, H. Olsen, and D.W. Deamer, *Analysis of hairpin structures within single DNA molecules using a nanopore detector*. Biophysical Journal, 2000. **78**(1): pp. 402A–402A.
16. Marziali, A. and M. Akeson, *New DNA sequencing methods*. Annual Review of Biomedical Engineering, 2001. **3**: pp. 195–223.
17. Vercoutere, W. and M. Akeson, *Biosensors for DNA sequence detection*. Current Opinion in Chemical Biology, 2002. **6**(6): pp. 816–822.
18. Kong, C.Y. and M. Muthukumar, *Modeling of polynucleotide translocation through protein pores and nanotubes*. Electrophoresis, 2002. **23**(16): pp. 2697–2703.
19. Li, J., D. Stein, C. McMullan, D. Branton, M.J. Aziz, and J.A. Golovchenko, *Ion-beam sculpting at nanometre length scales*. Nature, 2001. **412**(6843): pp. 166–169.
20. Li, J.L., M. Gershow, D. Stein, E. Brandin, and J.A. Golovchenko, *DNA molecules and configurations in a solid-state nanopore microscope*. Nature Materials, 2003. **2**(9): pp. 611–615.
21. Krapf, D., M.Y. Wu, R.M.M. Smeets, H.W. Zandbergen, C. Dekker, and S.G. Lemay, *Fabrication and characterization of nanopore-based electrodes with radii down to 2 nm*. Nano Letters, 2006. **6**(1): pp. 105–109.
22. Heng, J.B., A. Aksimentiev, V. Dimitrov, Y. Grinkova, C. Ho, P. Marks, K. Schulten, S. Sligar, and G. Timp, *Stretching DNA using an artificial nanopore*. Biophysical Journal, 2005. **88**(1): pp. 659A–659A.
23. Heng, J.B., A. Aksimentiev, C. Ho, V. Dimitrov, T.W. Sorsch, J.F. Miner, W.M. Mansfield, K. Schulten, and G. Timp, *Beyond the gene chip*. Bell Labs Technical Journal, 2005. **10**(3): pp. 5–22.
24. Heng, J.B., C. Ho, T. Kim, R. Timp, A. Aksimentiev, Y.V. Grinkova, S. Sligar, K. Schulten, and G. Timp, *Sizing DNA using a nanometer-diameter pore*. Biophysical Journal, 2004. **87**(4): pp. 2905–2911.
25. Keyser, U.F., B.N. Koeleman, S. Van Dorp, D. Krapf, R.M.M. Smeets, S.G. Lemay, N.H. Dekker, and C. Dekker, *Direct force measurements on DNA in a solid-state nanopore*. Nature Physics, 2006. **2**(7): pp. 473–477.

26. Mara, A. and Z. Siwy, *An asymetric nanopore for biomolecular sensing*. Biophysical Journal, 2004. **86**(1): pp. 603A–603A.
27. Yan, H. and B.Q. Xu, *Towards rapid DNA sequencing: Detecting single-stranded DNA with a solid-state nanopore*. Small, 2006. **2**(3): pp. 310–312.
28. Tian, P. and G.D. Smith, *Translocation of a polymer chain across a nanopore: A Brownian dynamics simulation study*. Journal of Chemical Physics, 2003. **119**(21): pp. 11475–11483.
29. Muthukumar, M. and C.Y. Kong, *Simulation of polymer translocation through protein channels*. Proceedings of the National Academy of Sciences of the United States of America, 2006. **103**(14): pp. 5273–5278.
30. Kong, C.Y. and M. Muthukumar, *Simulations of stochastic sensing of proteins*. Journal of the American Chemical Society, 2005. **127**(51): pp. 18252–18261.
31. Randel, R., H.C. Loebl, and C.C. Matthai, *Molecular dynamics simulations of polymer translocations*. Macromolecular Theory and Simulations, 2004. **13**(5): pp. 387–391.
32. Aksimentiev, A., K. Schulten, J. Heng, C. Ho, and G. Timp, *Molecular dynamics simulations of a nanopore device for DNA sequencing*. Biophysical Journal, 2004. **86**(1): pp. 480A–480A.
33. Muthukumar, M., *Polymer escape through a nanopore*. Journal of Chemical Physics, 2003. **118**(11): pp. 5174–5184.
34. Slonkina, E. and A.B. Kolomeisky, *Polymer translocation through a long nanopore*. Journal of Chemical Physics, 2003. **118**(15): pp. 7112–7118.
35. Flomenbom, O. and J. Klafter, *Single stranded DNA translocation through a nanopore: A master equation approach*. Physical Review E, 2003. **68**(4): art. no. 041910.
36. Grosberg, A.Y., S. Nechaev, M. Tamm, and O. Vasilyev, *How long does it take to pull an ideal polymer into a small hole?* Physical Review Letters, 2006. **96**(22): art. no. 228105.
37. Wolterink, J.K., G.T. Barkema, and D. Panja, *Passage times for unbiased polymer translocation through a narrow pore*. Physical Review Letters, 2006. **96**(20): art. no. 208301.
38. Fologea, D., J. Uplinger, B. Thomas, D.S. McNabb, and J.L. Li, *Slowing DNA translocation in a solid-state nanopore*. Nano Letters, 2005. **5**(9): pp. 1734–1737.
39. Chen, C.M., *Driven translocation dynamics of polynucleotides through a nanopore: Off-lattice Monte-Carlo simulations*. Physica A-Statistical Mechanics and its Applications, 2005. **350**(1): pp. 95–107.
40. Chen, C.M. and E.H. Peng, *Nanopore sequencing of polynucleotides assisted by a rotating electric field*. Applied Physics Letters, 2003. **82**(8): pp. 1308–1310.
41. Bhattacharya, S., S. Nair, and A. Chatterjee, *An Accurate DNA Sensing and Diagnosis Methodology Using Fabricated Silicon Nanopores*. IEEE Transactions on Circuits & Systems – I: Special Issue on Advances in Life Science Systems and Applications, 2006. **53**(11): pp. 2377–2383.
42. Bhattacharya, S., V. Natarajan, A. Chatterjee, and S. Nair, *Efficient DNA sensing with fabricated silicon nanopores: diagnosis methodology and algorithms*. Proceedings of the 19th International Conference on VLSI Design, Hyderabad, India 2006.
43. Halverson, K.M., R.G. Panchal, T.L. Nguyen, R. Gussio, S.F. Little, M. Misakian, S. Bavari, and J.J. Kasianowicz, *Anthrax biosensor, protective antigen ion channel asymmetric blockade*. Journal of Biological Chemistry, 2005. **280**(40): pp. 34056–34062.
44. Bayley, H. and P.S. Cremer, *Stochastic sensors inspired by biology*. Nature, 2001. **413**(6852): pp. 226–230.
45. Kasianowicz, J.J., *Nanopores – Flossing with DNA*. Nature Materials, 2004. **3**(6): pp. 355–356.
46. Nilsson, J., J.R.I. Lee, T.V. Ratto, and S.E. Letant, *Localized functionalization of single nanopores*. Advanced Materials, 2006. **18**(4): pp. 427–431.
47. Han, A.P., G. Schurmann, G. Mondin, R.A. Bitterli, N.G. Hegelbach, N.F. de Rooij, and U. Staufer, *Sensing protein molecules using nanofabricated poresn*. Applied Physics Letters, 2006. **88**(9): art. no. 093901.
48. Harrell, C.C., Z.S. Siwy, and C.R. Martin, *Conical nanopore membranes: Controlling the nanopore shape*. Small, 2006. **2**(2): pp. 194–198.

49. Harrison, O., B. Ledden, J. Uplinger, B. Thomas, T. Mitsui, D.S. McNabb, J. Golovchenko, and J.L. Li, *Probing single polypeptides with a solid state nanopore sensor*. Biophysical Journal, 2004. **86**(1): pp. 480A–480A.
50. Lagerqvist, J., M. Zwolak, and M. Di Ventra, *Fast DNA sequencing via transverse electronic transport*. Nano Letters, 2006. **6**(4): pp. 779–782.
51. Chan, E.Y., *Advances in sequencing technology*. Mutation Research-Fundamental and Molecular Mechanisms of Mutagenesis, 2005. **573**(1–2): pp. 13–40.
52. Li, J.L., D. Stein, C. Qun, E. Brandin, A. Huang, H. Wang, D. Branton, and J. Golovchenko, *Solid state nanopore as a single DNA molecule detector*. Biophysical Journal, 2003. **84**(2): pp. 134A–135A.
53. Chen, L.M., P.C. Li, X.L. Fu, H.Y. Zhang, L.H. Li, and W.H. Tang, *Fast fabrication of large-area nanopore arrays by FIB*. Acta Physica Sinica, 2005. **54**(2): pp. 582–586.

Chapter 10
Engineering Biomaterial Interfaces Through Micro and Nano-Patterning

Joseph L. Charest and William P. King

Abstract Patterning biomaterial surfaces with synthetic topographical and chemical features provides a means of engineering cell-biomaterial interfaces, thereby enabling the study of cellular response to specific external cues. Cleanroom-based fabrication techniques have created precise and consistent topographical and chemical patterns on cell substrates at the micro- and nano-scale, allowing characterization of cellular response to well-defined surface features. Techniques such as imprint lithography and micro-contact printing have advanced substrate fabrication by expanding material selection and increasing throughput. Independent combination of topographical and chemical patterns has provided sophisticated interfaces suitable for comparing the relative influence of and interplay between topographical and chemical patterns. Deliberately patterned topographical and chemical features have influenced cellular responses ranging from morphology and alignment through adhesion and differentiation. Enhanced patterning techniques will continue to lead cell substrate fabrication towards sophisticated, user-defined configurations of topographic and chemical patterns, providing a platform to establish mechanisms of cellular response to cell-material interfaces.

Keywords: Biomaterial · Interface · Surface · Pattern · Imprint lithography · Micro-contact printing · Morphology · Alignment · Proliferation · Differentiation

10.1 Introduction

Patterning biomaterial surfaces with synthetic topographical and chemical features provides a means of engineering cell-biomaterial interfaces. A precisely engineered biomaterial interface can provide controlled interaction with biological analytes in

W. P. King
Department of Mechanical Science and Engineering, University of Illinois Urbana-Champaign
Urbana, IL 61801
e-mail: wpk@uiuc.edu

P. J. Hesketh (ed.), *BioNanoFluidic MEMS.*

biosensors, cues for cellular growth in tissue engineering scaffolds, and largely determines the biological response to implanted devices.

Cells respond to external mechanical and chemical cues either within an in vivo environment via interactions with extracellular matrix (ECM) or with a biomaterial surface via mechanical and chemical features at the cell-biomaterial interface. Surface mechanical features can be classified either as *roughness* or *topography*. Surface *roughness* is comprised of 3-D features possessing randomness in size, shape, and periodicity, whereas surface *topography* possesses well-defined 3-D features of deliberately designed size, shape, and organization with a regular periodicity. Surface *chemical patterns* are defined by their chemical composition, as well as their feature size, shape, and periodicity. Various patterning techniques can produce surface topography on cell substrates with a wide variety of feature shapes and sizes [1,2] and a variety of chemical patterns that influence cellular function [3]. Although roughness, topography, and chemistry all affect cellular response [4, 5], topography and chemical patterns applied to cell substrate surfaces provide user-defined and well-characterized substrates for the investigation of specific cell responses to surfaces.

This chapter reviews both cell substrate surface patterning techniques and cellular responses to substrate surface patterns. The review focuses on top–down patterning methods for cell culture substrates, as they provide methodologies for deliberate and user-configurable feature geometries in well-controlled models for cellular study. The techniques section discusses traditional cleanroom microfabrication methods, such as photolithography and electron beam lithography, for patterning both topography and chemistry. Additional topographical patterning techniques include molding methods such as injection molding, casting, and imprint lithography. Discussion of non-cleanroom chemical patterning techniques includes various methods with an emphasis on micro-contact printing. Methods of independently patterning chemistry and topography are also discussed.

Additionally this review discusses the response of cells to synthetic surface patterns. For response of cells to topography, the review focuses on 'contact guidance' of cells to surface features and includes effects of topography on higher-order responses such as proliferation and differentiation. For response of cells to chemical patterns, the review discusses restriction of location and shape, and consequent influence on adhesion and cell–cell contact, as well as modulation of apoptosis, proliferation, and differentiation. Evaluation of relative influence of and interplay between topographical and chemical patterns is also discussed.

10.2 Techniques for Surface Patterning Cell Substrates

The evaluation of cellular response to surface patterns requires substrate fabrication techniques that provide feature consistency, high-resolution patterning, and high-throughput production of substrates. Feature consistency leads to substrates inducing repeatable cellular response, enabling robust, quantitative analysis. High-resolution patterning results in features that appropriately mimic the sub-micron and nanoscale feature sizes present in cellular components and ECM. High throughput fabrication processes provide sufficient numbers of samples to provide statistically, and potentially clinically, relevant sample sizes for biological assays. Patterning techniques for both topography and chemical patterns fall into two main sub-categories: 1) cleanroom methods based on traditional microfabrication or 2) non-traditional techniques that do not depend on cleanroom methods.

10.2.1 Topographical Patterning Methods

Cleanroom techniques provided initial approaches to pattern micro and nano-scale resolution topographies on cell substrates of materials such as silicon and glass with good consistency [1]. Although silicon and glass are not necessarily ideal biomatcrials, cleanroom techniques provided a means to attain consistent micron and sub-micron resolution features to demonstrate cellular response to the material interface. Cleanroom manufacturing techniques of cell substrates generally pattern features using masked ion beam lithography, photolithography, or electron beam lithography, then transfer the topography from the resist to the base substrate with an etching step.

10.2.1.1 Masked Ion Beam Lithography

Although not a common technique, masked ion beam lithography (MIBL) has created topography in non-standard microfabrication materials such as polymethylmethacrylate (PMMA) [6] allowing additional material selection over silicon-based manufacturing techniques. A nickel mesh placed onto the PMMA film served as a mask while an ion beam rastered over the entire sample. Shown in Fig. 10.1, the resulting substrate, possessed topography about 400 nm deep, with horizontal dimensions similar to the nickel screen pattern. The consequential implantation of ions into the material resulted in a chemical modification where the ion beam etched the material.

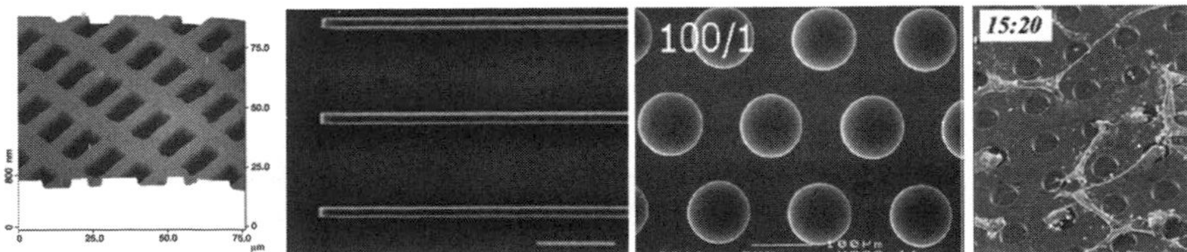

Fig. 10.1 Masked cleanroom-based methods of topographic patterns resulted in cell substrates of various materials. Masked ion beam etching of PMMA [6], direct photopatterning of polyimide on glass [9], photolithography and subsequent chemical etching of titanium [10], and photolithography and subsequent reactive ion etching of quartz [11], are various approaches. Images from [6] reprinted with kind permission of Springer Science and Business Media. Images from [9, 11] reprinted from Biomaterials with permission from Elsevier

10.2.1.2 Photolithography

Photolithography creates microscale patterns on a substrate by selectively exposing areas of a photo-active polymer resist coating. The exposed material is then removed chemically or thermally to produce the pattern. Typically, the photopatterning is followed by a subsequent etching step to transfer the polymer pattern into the substrate material, resulting in surface topographic features on the substrate.

Photolithography has patterned cell substrates possessing features of square grooves, V-grooves, and pits ranging in size from .5 μm through several hundred μm [1]. Early photopatterned cell substrates possessed microscale grooves ranging from 70 to 165 μm etched into silicon with epithelial cells cultured on the microgrooves aligning to them [7,8]. Figure 10.1 shows examples where more recent photolithography has directly patterned polyimide channels [9], and patterned circular pits for subsequent chemical etching of titanium [10] reactive ion etching of quartz [11].

The resolution of photolithography has been extended to create features as small as 130 nm by substituting X-ray radiation for ultraviolet light and exposing the resist through a holographically produced mask [12]. As some cellular features possess length scales below 100 nm, it is critical to explore cellular response to features with nanoscale dimensions, requiring a technique with better resolution than photolithography. Photolithography is also limited by expensive cleanroom facilities and a subsequent etching step thereby slowing throughput and predominantly limiting material selection to silicon, glass, or quartz which are not readily applied to biomaterial applications.

10.2.1.3 Electron Beam Lithography

EBL is similar to photolithography, but exposes the resist through a finely-focused and precisely controlled beam of electrons. EBL patterning has regularly obtained feature sizes of 10 nm [13]. Figure 10.2 shows early cell culture substrates patterned through EBL which possessed 1, 2 and 4 μm wide grooves [14], and more recent substrates that possessed features ranging from 70 to 4000 nm [15, 16]. The serial processing nature of EBL limits the total patterned area of the substrate as well as the maximum feature size, resulting in low-throughput and substantial cost if patterns covering large areas are required. In addition, EBL is somewhat limited in material selection as it requires materials suitable for subsequent cleanroom etching techniques.

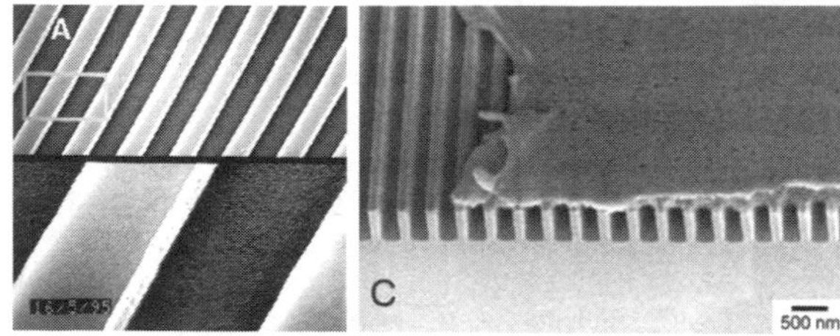

Fig. 10.2 Electron beam lithographically patterned topographical substrates consisting of 2 μm wide grooves in quartz [14] and 400 nm pitch grooves in silicon [16]. Reproduced with permission of the Company of Biologists

10.2.2 Molding Techniques

Molding techniques take advantage of the high resolution features created through traditional cleanroom techniques by replicating them in inexpensive polymer-based materials in a low-cost, high-throughput process. Molding techniques such as injection molding and imprint lithography require thermoplastic materials, while casting techniques require a material that can bc dissolved in a solvent or cured. Polymers provide an advantage in molding techniques as they have exhibited high resolution with the potential to replicate features of sub-nanometer size [17].

10.2.2.1 Injection Molding

Injection molding forces a polymer in melt form into a rigid mold to create 3-D structures of nearly arbitrary shape. The method has demonstrated the ability to replicate biomimetic features down to 4 nm [18]. A nickel mold, fabricated by electroplating a fibrillar collagen sample, served as tooling for an injection molding machine. Resulting substrates possessed replicas of the 3–4 nm collagen features, with replication fidelity dependent on polymer type and limited by the fidelity of the tooling rather than the injection molding process. Although injection molding shows great potential for mass production, its complex tooling and machinery carries high cost and inhibits substrate redesign, thereby limiting its application to research.

10.2.2.2 Casting: Solvent Casting and Cured Polymer Casting

Casting approaches replicate simple 2-D molds using a polymer in solution or a pre-polymer that is later cured. As no significant heat or pressure is used, casting approaches do not require complex machinery, making them conducive to production on an experimental scale for cell culture studies as well as on a mass-production scale.

In solvent casting, polymers are dissolved in a solvent and cast onto a mold prior to solvent evaporation. Solvent cast topographical cell substrates typically are polystyrene since it is a standard cell culture material. Solvent cast polystyrene substrates have possessed features as small as .5 μm wide grooves using a photolithographically patterned mold [19]. Figure 10.3 shows solvent cast polystyrene replicas of an etched silicon mold. The 2 μm wide grooves showed consistent replication of the mold, including nanoscale roughness inherent to the mold [20, 21]. One consequence of the solvent casting process was presence of residual solvent in the substrate after evaporation. Although cell growth was not significantly impacted by the residual solvent [20], solvent residue could potentially have unknown toxic effects.

In cured polymer casting, a cureable polymer or pre-polymer is loaded onto the mold, cured, and released to create a relief replica of the mold topography. An early example of a cured polymer topographical cell substrate possessed V-grooves cast in epoxy using an etched silicon mold [22]. Recent epoxy cast cell substrates have possessed complex patterns of discontinuous edges [23] as shown in Fig. 10.3.

Fig. 10.3 Molding techniques replicated molds in a low-cost, high-throughput fashion. Techniques employed various materials such as solvent casting 2 μm wide grooves in polystyrene [21], epoxy casting of 34 μm wide squares with discontinuous edges [78], casting 33 μm wide pyramids in PDMS [26], and hot-embossing 4 μm wide grooves into polyimide [31]. Image from [78] reprinted with kind permission of Springer Science and Business Media. Images from [26,31] reprinted with permission from Elsevier

Polydimethylsiloxane (PDMS), a cureable inorganic polymer, has been used extensively since its initial use for topographic cell substrates [24], due to its non-toxicity and inertness for most biological studies. In addition, the mechanical modulus of PDMS has been adjusted to investigate aspects of contractility in cells [25] adding further functionality to a cell culture substrate. Topographically patterned features on PDMS cell substrates have included 33 μm wide pyramids [26], as shown in Fig. 10.3, and 350 nm wide grooves [27]. Although both solvent casting and cured polymer casting are high-throughput techniques with excellent resolution, they inherently limit material selection to those that can be solvent cast or cured.

10.2.2.3 Imprint Lithography

Imprint lithography, referred to as hot-embossing if performed at elevated temperatures, forms a relief replica of a mold by pressing it into a thermoplastic material. In theory, virtually any thermoplastic material is suitable for imprint lithography, providing a wide substrate material selection. Imprint lithography has reproduced features as small as 10 nm with good fidelity in polymers [28, 29], making it ideal for mimicking nanoscale in vivo topography. In addition, polymers employed in cell substrate applications provide an inexpensive material with tuneable mechanical and chemical properties. Imprint lithography has fabricated sub-micron topography on 2-D polymer cell substrates [30] and hot-embossed substrates have displayed excellent replication fidelity and significant cellular response to the topography [31], as shown in Fig. 10.3. The high-throughput nature of imprint lithography as applied to cell substrates was demonstrated through replication of nanoscale features consistently over areas as large as 79 cm^2 in a non-cleanroom environment [32]. Recently, imprint lithography has emerged as a predominant method to fabricate cell substrates with sub-micron features [27, 33], microscale features in biodegradeable polymers [34], and multiple sets of features from sequential embossing steps [35].

10.2.3 Chemical Patterning Methods

Chemical surface patterning results in geometrically confined features composed of biologically interactive chemistries. Chemical patterning may consist of direct patterning of the biologically interactive chemical, or indirect patterning of the chemical through a patterned intermediate layer that selectively promotes or suppresses the adhesion of the biologically interactive chemical. Intermediate self-assembled monolayers (SAMs) have promoted or suppressed adsorption of protein and consequent adhesion of cells dependent on user-specified terminal groups of the SAMs [36]. Once the SAM was patterned, immersion of the substrate in the protein or cell solution resulted in geometric patterns due to the selective adsorption or adhesion. As reviewed here, chemical patterning serves to geometrically control cell attachment to substrates, resulting in influence of cells through spatial control.

10.2.4 Traditional Cleanroom Techniques

10.2.4.1 Photolithography

Chemical patterning through photolithography has produced substrates through both direct and indirect patterning approaches. Photopatterning of a protein has resulted in a substrate capable of a limited-interaction co-culture of cells [37]. Post-photopatterning liftoff resulted in lanes of collagen, surrounded by non-functionalized borosilicate. Cell adhesion was then modulated by seeding without serum, restricting strong cell adhesion to the collagen lanes, then seeding a second cell type with serum to allow adhesion to the non-functionalized areas. In this way, cell types were confined to specific areas thereby controlling heterotypic cell–cell interactions. Indirect photopatterning by liftoff of a polyethylene glycol (PEG) silane SAM from a glass substrate resulted in bare glass adhesive areas surrounded by PEG [38] as shown in Fig. 10.4. Since PEG typically suppresses attachment of cells, seeding of cells on the substrate resulted in restriction of cells to the bare glass. Simple patterns have been created through photopatterning and liftoff of metals, with circular patterns of aluminum on a niobium background [39] as illustrated in Fig. 10.4.

Beyond patterns that either suppress or promote cell adhesion, photopatterning of a specifically designed photo-active biotin resulted in precise geometric shapes of biotin-presenting SAMs that enabled further specific interaction [40]. After binding of avidin to the biotin layer, further biotin-conjugated antibodies were bound

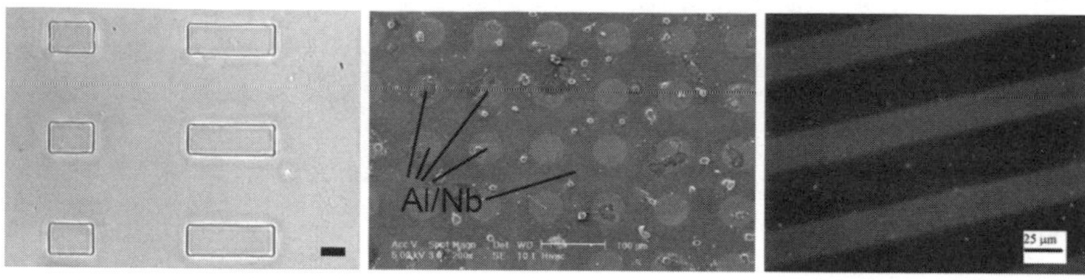

Fig. 10.4 Photolithography patterned chemistry through different subsequent steps for different chemistries such as liftoff of PEG SAMs yielding 25 μm wide rectangles of exposed glass [38], liftoff of metal to produce aluminum dots on a niobium background [39], and directly photolinkable biotin to create biospecifically adhesive lanes [40]. Image from [38, 40] reprinted with kind permission of Springer Science and Business Media. Image from [39] reprinted with permission from Elsevier

to the avidin layer resulting in geometric patterns with highly-specific preferential adhesion characteristics. The result, shown in Fig. 10.4, was the ability to restrict specific, fluorescently-labelled antibodies to lanes. To further the functionality of chemically patterned substrates, subsequent modification of photopatterned chemistry using layer-by-layer (LBL) assembly and multiple photolithography steps has resulted in multiple patterned chemistries on one substrate [41]. The LBL assembly enabled control of the thickness of the chemical features, as well as tuning of the physical–chemical properties. Although photolithography allows patterning of a variety of specific chemistries, resolution of the method is fundamentally limited.

10.2.4.2 Electron Beam Lithography

EBL has patterned resists to control SAM placement or ablated patterns directly into SAMs with reliable feature sizes in the range of 10 nm [13]. EBL has patterned gas-phase deposited SAMs with a minimum line width of 27 nm [42]. Utilizing this SAM patterning method has resulted in collagen patterned in 30–100 nm wide tracks [43] as shown through AFM in Fig. 10.5. Direct ablation of patterns into an existing SAM has resulted in patterning of biologically-active molecules with 250 nm linewidths [44]. Recent work has used EBL-ablated patterns with feature sizes as small as 40 nm in a protein resistant SAM [45]. The ablated patterns allowed selective backfilling of protein-coated spheres resulting in protein patterns of sub-100 nm dimensions, shown through an AFM image in Fig. 10.5. Although resolution of EBL chemical patterning is excellent, low-throughput and expense remain as limitations.

10.2.5 Non-traditional Techniques

Chemical patterning through non-cleanroom methods has been accomplished through a wide variety of methods, each with particular advantages and drawbacks.

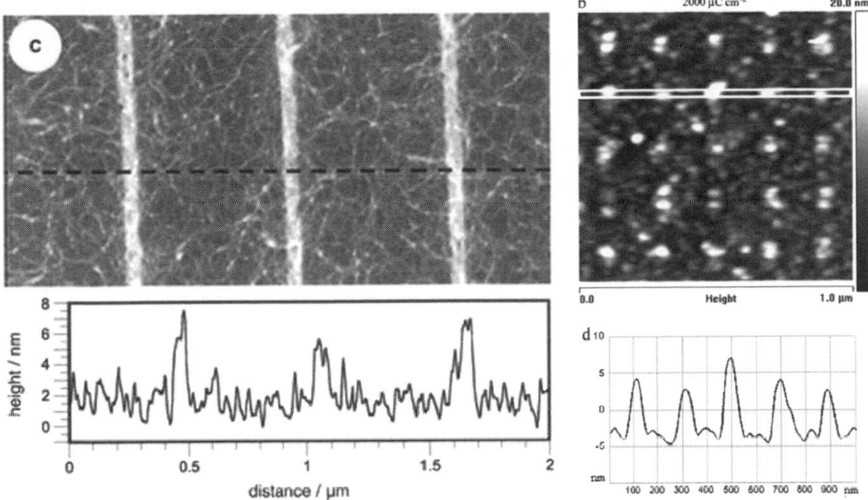

Fig. 10.5 EBL produced sub-micron chemical features. Collagen adsorbed to methyl-terminated SAMs patterned through EBL exposed resist [43], and protein-coated spheres adsorbed to areas where PEG SAMs were ablated by EBL [45]. Rightmost images reprinted from [45], copyright 2006 American Chemical Society

Several techniques have emerged that have specific advantages. Mechanical scraping of collagen has produced 50 µm wide lanes in a very inexpensive and simple manner [46]. Implantation of biologically relevant ions has been demonstrated for microscale patterns [6]. Stencil peeling has selectively removed cells or proteins from a substrate resulting in defined patterns of microscale dimensions [47]. The stencil, patterned through photolithography and subsequent etching, consisted of a thin layer of parylene adhered to a substrate before cell seeding or protein adsorption. Biologically-active lipid bilayers as small as 1.3 µm have been patterned through stencil peeling [48]. Since the patterning of the stencil occured before cell seeding, this technique provided a method to pattern live cells directly. Focused ion beams (FIB) have been used to induce localized topographical changes in gallium arsenide substrates that permit selective adsorption of protein into dot formations of approximately 100 nm diameter [49]. Similarly, microscale patterns of cell adhesive areas have been patterned using FIB ion implantation on polyhydroxymethylsiloxane [50]. Both processes required only one patterning step with the potential for nanoscale feature dimensions, however the resulting patterns were substrate material dependent and limited material selection. Dip-pen nanolithography (DPN), has created 100 nm patterns of mercaptohexadecanoic acid (MHA), with surrounding areas passivated by a PEG-terminated monolayer [51]. Specifically, 200 nm patterns of MHA coated in a fibronectin fragment served as patterning for cellular focal adhesions. While DPN produces nanoscale chemical patterns and is relatively substrate independent, the serial nature of the process limits its throughput.

10.2.5.1 Micro-contact Printing of Chemical Patterns

Micro-contact printing (μCP) represents the most often used method to create chemical patterns for cell substrates. Direct μCP prints an 'ink' of a biologically active compound, such as a protein, onto a substrate through contact transfer of the compound from an elastomeric stamp to the substrate. In a similar fashion, indirect μCP uses a SAM as an ink which is transferred from the stamp for initial chemical functionalization [52], with subsequent backfilling of a second background SAM creating a distinct difference in adhesive properties between the pattern and background. Protein adsorption or cell adhesion is restricted by the difference in adhesive properties. Cell substrates patterned through μCP exhibited the ability to distinctly restrict cell spreading and consequently control cell shape [53].

10.2.5.2 Direct μCP of Biologically Active Chemicals

Direct printing of proteins and biological macromolecules results in geometric patterns without the use of underlying SAMs. Directly printed protein patterns have survived long incubation times, as laminin lanes printed on a layer of bovine serum albumin (BSA) on polystyrene tissue culture dishes have remained stable in media or buffer for 4 weeks [54]. The 5–50 μm wide lanes permitted myoblast adhesion while the background BSA suppressed myoblast adhesion. Direct printing functions well for various proteins or biological macromolecule mixtures. For example, direct printing of an ECM-gel containing poly-D-lysine created grids of 4–6 μm wide lanes connecting circular nodes of 12–14 μm in diameter [55]. The printed areas served as adhesive sites for control of neuron placement. Direct printing has been expanded to include μCP of patterns onto biological tissues such as printing of 10 μm wide polyvinyl alcohol (PVA) lanes onto a human lens capsule [56]. Figure 10.6 shows a further enhancement of direct printing which used a flat PDMS stamp patterned with antigens through microwells [57]. The microwells localized delivery of multiple antigens, each to distinct locations on the stamp, enabling the spatially specific binding of antibodies from an antibody solution. The stamp could then print multiple antibodies in one step on a glass substrate. Resolution of μCP has been improved so that precise control of stamp aspect ratio and shape has resulted in μCP of features less than 100 nm using antibodies as an ink [58]. Pattern degradation of direct-μCP protein features surrounded by a PEG SAM showed minimal outgrowth of cells from the patterns after a 13 day culture and minimal degradation of patterns after 13 days in serum containing medium [59]. Direct μCP provides a non-cleanroom approach to chemical patterning that is high-throughput, stable for cell culture times, and patterns a variety of chemistries.

10.2.5.3 Indirect μCP of SAMs

Indirect μCP creates patterned SAMs that in turn selectively suppress or promote protein adsorption or cell adhesion, thereby geometrically restricting cell location,

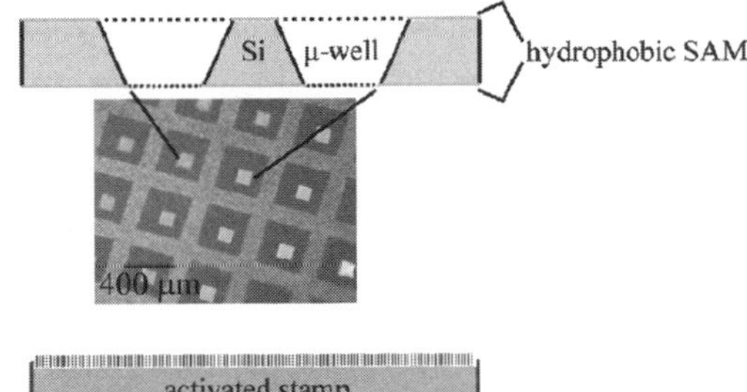

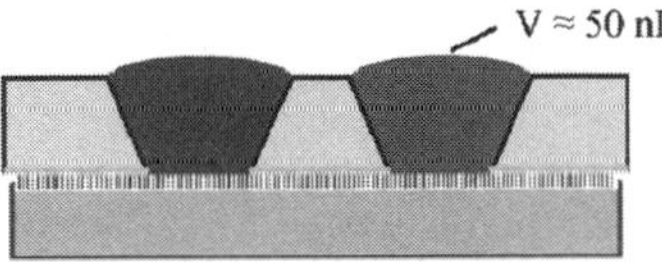

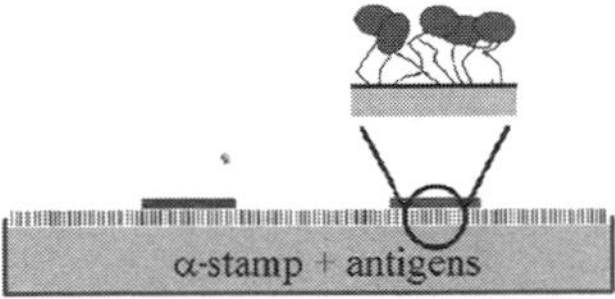

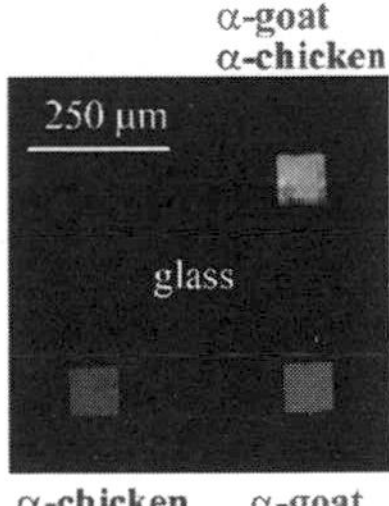

Fig. 10.6 Local delivery of antigens enabled μCP of multiple chemistries with one stamping step. After antigen patterning, the stamp was inked from a mixed solution of antibodies, and could then print the antibodies onto a substrate [57]

size, and shape. Since SAMs have been characterized for specific protein adsorption and activity [60, 61], indirect μCP can provide a well-controlled chemical model layer in addition to geometrical patterning. Indirect μCP has demonstrated pattern sizes as small as .3 μm squares by printing methyl-terminated SAMs [62]. Printing of an adhesive SAM, followed by backfilling with PEG-terminated SAM, resulted in control of adhesive island sizes. The adhesive island sizes in turn precisely controlled cell spread area to several designated increments in order to quantitatively study effects of cell spreading area on cell function [63]. Precise control of cell size and shape through indirect μCP of adhesive islands has resulted in providing consistency to adhesion studies [64] and modulating adhesion strength through controlling available cell spreading area [65]. Indirect μCP has also produced stable patterns, as μCP MHA surrounded by PEG areas showed good pattern fidelity over an 89 hour cell culture [66]. Like direct μCP, indirect μCP is a non-cleanroom, high-throughput, and stable chemical patterning method with the added feature of a possessing a well-defined underlying chemical model layer.

10.2.6 Combined Topographical and Chemical Patterning

Since cells in vivo respond to both topographical and chemical cues simultaneously, patterning both topography and chemistry leads to biomaterial interfaces with the potential to better mimic complex in vivo scenarios. In addition, studying the relative influence of and interplay between topographical and chemical cues necessitates the independent combination of both types of patterns. Although some chemical patterns have displayed shallow topographic features inherent to them [41, 67], and some topographic features have been composed of a functional chemistry [68], deliberate chemical patterning has the potential to add functionality to cell substrates possessing patterned topography. Specifically, chemical functionalization of spaces in between etched microwells has been demonstrated [69], as well as on plateaus between grooves [70]. However, the chemical patterns relied on and were spatially concurrent with the underlying topography, thus rendering the two patterning methods dependent on one another. Independent patterning of chemistry on topography has increased the sophistication of substrate interfaces and enabled new investigations into cellular response. Using photolithography, cell adhesive chemical lanes were patterned on a substrate possessing topographical grooves etched into fused silica [71, 72]. Since the chemical lanes were patterned independently of the topography, they could be user-specified to be oriented parallel or perpendicular to the grooves. The independent nature of the patterns enabled investigation of the relative influence of chemical and topographical patterns on cell response, however it required cleanroom fabrication to do so. Independent patterning of chemistry on topography without cleanroom techniques has been accomplished by combining imprint lithography with a subsequent μCP step. Chemical patterns of protein adhesive lanes of various spacing were printed on top of hot-embossed grooves [73]. The independent configurations of chemistry overlain onto the topography enabled comparison of chemical to topographical influence, while the non-cleanroom techniques permitted rapid substrate fabrication. The combination of imprint lithography with μCP has also created substrates with chemical dots and lanes overlaid onto nanoscale topographical grooves [32]. Figure 10.7 illustrates chemical patterns created through μCP independently of the underlying topography. Independent chemistry and topography on a cell substrate provide a platform for investigating cell response to user-defined scenarios of specific chemical and topographical patterns.

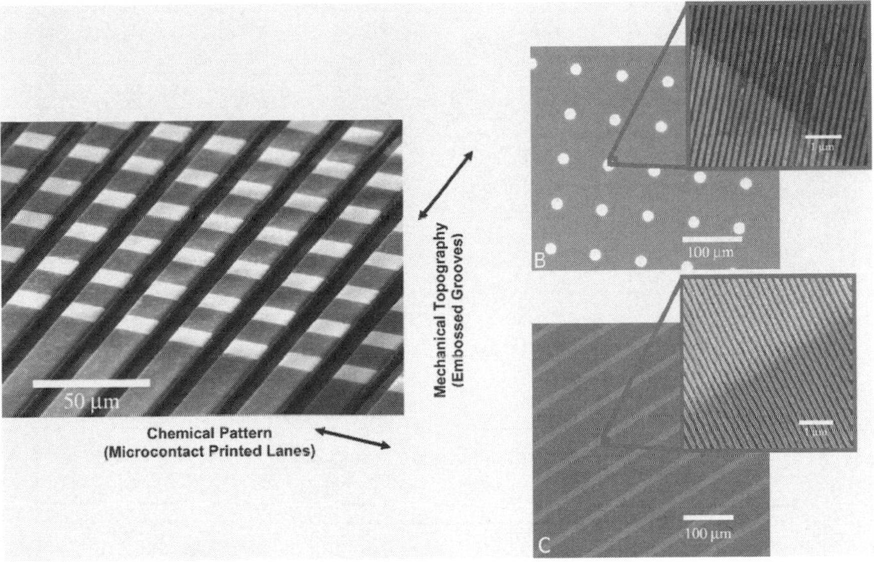

Fig. 10.7 Cell substrates with independently patterned chemistry and topography. Embossed microgrooves ran perpendicular to µCP lanes, and µCP dots and lanes were printed independently of the underlying embossed nanogrooves. Left image reprinted from [73] with permission from Elsevier, right image reused with permission from [32] Copyright 2005, AVS The Science & Technology Society

10.3 Cellular Response to Surface Patterns

10.3.1 Cellular Response to Topography

10.3.1.1 Morphological Response

Alignment, Orientation, Elongation

Cells respond morphologically to topography by alignment to and elongation along topographic features, termed 'contact guidance' [74]. More consistent microfabricated topography has advanced quantification of the extent of morphological changes, including alignment angles of cells and focal adhesions to microscale grooves [75]. Non-cleanroom microfabrication of topography has enabled more extensive quantification of morphological response. Hot-embossing polymer topographic substrates led to quantification of alignment and elongation of osteoblast cell bodies, nuclei, and focal adhesions to microscale grooves [31]. Although cell bodies, nuclei, and focal adhesions aligned to the topography, as shown in Fig. 10.8, only cell bodies significantly elongated. In addition, the polymer substrates demonstrated the effectiveness of a non-cleanroom technique in eliciting significant cellular response. The effects of contact guidance extend to neural cells as well. Embossed polymer substrates have elicited quantifiable angle and degree of alignment of axons from ganglia along topographic ridges in the substrate [76].

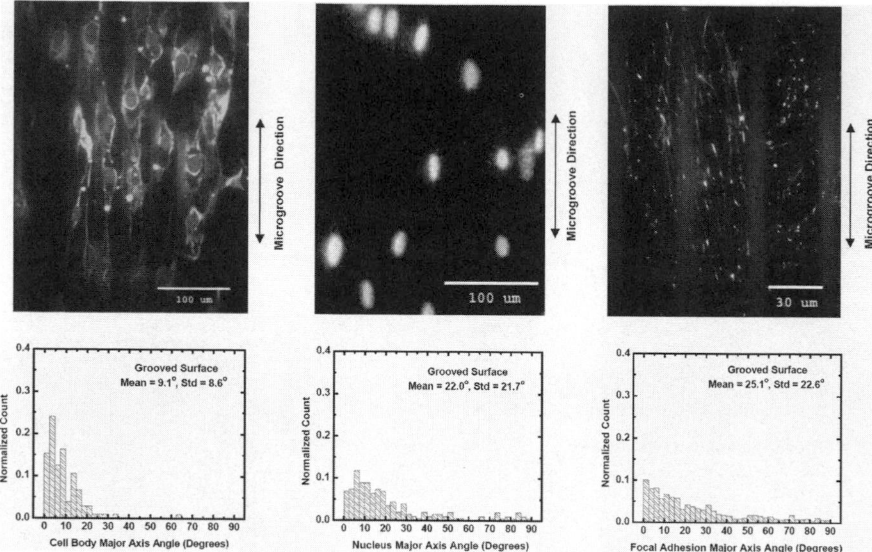

Fig. 10.8 Quantified alignment of cell components on embossed microscale grooves. Cell bodies were shown via membrane stain, nuclei via DNA stain, and focal adhesions through vinculin staining with accompanying histograms of alignment angles. Histograms were of uniform distribution for cells on smooth samples. Image reprinted from [31] with permission from Elsevier

Spatial distribution of proteins and other functional cellular components have been altered along with gross morphological changes of the cell. Human bone marrow stromal cells (HBMSCs) have shown not only alignment of the actin cytoskeleton, cell body, and focal adhesions to topography, but also concentration of tubulin protein to grooves [77]. Focal adhesions of osteoblasts have preferentially concentrated on raised topographical features, with consequent localization of focal adhesion kinase to the raised features [78].

Response of Cells to Discontinuities

The response of cells to topography may depend on the presence of surface discontinuities. A single step discontinuity in a cell substrate impeded cell migration across it according to step height, regardless of whether the cell was ascending or descending the step [79]. Gaps in discontinuous topography influenced alignment of fibroblasts in a similar fashion as contact guidance [23]. The alignment, or gap guidance, occurred for cells located between topographic features necessitating alignment of the cell body to accommodate the presence of the raised topography. A similar effect has also been observed due to recessed topography, with myoblasts aligning to rows of recessed circular pits (data accepted for publication in Biomaterials).

Cytoskeletal Involvement in Alignment to Topography

Examination of the sequence of alignment events has provided some insight into the involvement of cytoskeletal components in cellular alignment to topography. For fibroblasts on microscale grooves, myotubules aligned to the features first, followed by focal adhesions, actin filaments, then the overall cell body [80]. Further study using cytoskeletal inhibiting drugs has shown that cells with disrupted microtubules aligned to grooves wider than 1 μm wide, but did not align to smaller grooves whereas inhibition of actin filaments did not disrupt alignment on any groove sizes [81]. Disruption of actin filaments, microtubules, or both, did not significantly inhibit neurite alignment to grooves of 1, 2 and 4 μm widths [14]. Conversely, cellular alignment due to gap guidance has been inhibited by disruption of cytoskeletal components, as indicated by a reduction in the percentage of aligned cells due to the presence of either actin filament or microtubule inhibiting drugs [23]. Although cytoskeletal components have played some role in alignment of cells to topography, it is unclear exactly how large of an impact they have.

Parameters of Topography that Influence Cell Morphology

Topographical parameters, specifically dimensions of the topographic features, have impacted the extent of cellular alignment and elongation. Typically, depth of grooves has had more effect on cellular alignment than width or pitch of grooves. For example, the fraction of aligned cells to microscale grooves increased more with an increase in depth of 1.7 μm than a change in pitch of 20 μm [82]. Similarly, varying groove width from 1 to 10 μm did not significantly impact the extent of cellular alignment [75] whereas for a constant width and pitch, the fraction of aligned cells increased as groove depth increased from 200 nm to 1 μm [83]. Elongation of cells, as measured by ratio of cell major axis to minor axis, also increased with groove depth. Figure 10.9 shows data for epithelial cells which have aligned to grooves as

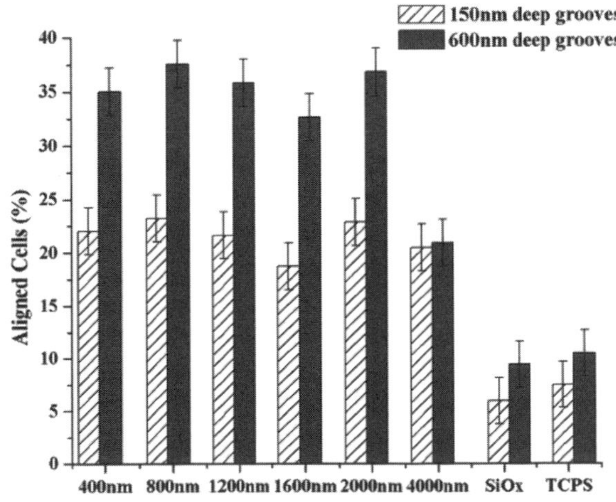

Fig. 10.9 Percentage of cells aligned within 10° of grooves. Cell alignment was constant for all pitches on 150 nm deep grooves, while cell alignment was constant only for groove pitches 200–2000 nm for 600 nm deep grooves [16]. Reproduced with permission of the Company of Biologists

narrow as 330 nm, with the fraction of aligned cells similar for grooves with pitches ranging from 400 to 2000 nm and a depth of 600 nm [16]. However, when the groove depth was decreased to 150 nm, the fraction of aligned cells remained similar for pitches ranging from 400 to 4000 nm indicating that sensitivity to depth and pitch may be interrelated. Using a similar substrate with 600 nm deep grooves, keratocytes aligned similarly for groove pitches ranging from 800 to 4000 nm, with significantly lower alignment levels on 400 nm pitch grooves, indicating the significance of feature dimensions may also depend on cell type [84]. Although cells display varying levels of sensitivity dependent upon topographic feature parameters, cells have responded to some extent on groove widths as small as 100 nm [32, 35]. Cellular alignment or response to groove widths less than 100 nm has yet to be established.

10.3.1.2 Higher-order Cellular Response to Topography

In addition to simple morphological changes, cells have exhibited some potential to modulate higher-order cell function in response to surface feature changes. Surface *roughness* has impacted differentiation in cell models such as bone marrow cells [85] and MG63 osteoblasts [86, 87], indicating that substrate mechanical features can impact higher-order cell function. Cells cultured on well-defined surface *topography* have also exhibited altered levels of bone-markers. Cells produced higher levels of alkaline-phosphatase (ALP) on topographically patterned pyramids than on smooth substrates [26]. Cells cultured on topographically patterned composite materials exhibited higher levels of ALP as compared to cells cultured on smooth composites [88]. However, the topographic patterning may have exposed varying amounts of the composite materials inducing a surface chemistry change concurrent with the topography. Bone markers have not only been altered by the presence of topography but modulated by surface topographical parameters such as circular pit size and spacing [10] as well as groove depth [89] with effects of topography extending to in vivo conditions [90].

Topographical influence of differentiation has not been limited to bone cell models. Neuritogenesis in PC12 neural cells has been modulated by varying substrate groove widths [91]. Neuron markers were upregulated in cells cultured on grooves as compared to cells cultured on smooth substrates, while glial markers were similar on both substrate types [92]. Conversely, several studies have shown a lack of influence of topography on differentiation [93] and proliferation [94] of osteoblasts, indicating topographical influence of higher-order effects may require specific cell-topography interactions. Although some effects have been documented, the overall effect of topography on differentiation remains uncharacterized.

10.3.2 Cellular Response to Chemical Patterns

10.3.2.1 Influence of Chemical Patterns on Location and Shape

Chemical patterns have influenced cell function through restriction of cell location and spreading with consequential control of cell shape. Patterns with dimensions similar to cells have controlled the shape of cells to rectangles [53], with precise control of cell shape restricted to teardrop shaped patterns [95] as well as squares, triangles, and other shapes [96]. Patterning adhesive islands of sub-cellular dimension has also influenced cell spreading, location, and shape. Although cells spanned several adhesive islands of sub-cellular dimension, the preferential adhesion to islands has controlled overall cell location [51], concentration of cell receptors [97] and even cell shape [62]. Cell location has been controlled by providing chemical patterns conducive to cell adhesion such as laminin lanes [98], irradiated areas of polymer with enhanced adhesive properties [50], and multiple chemistries with varying propensities for cell adhesion [41]. Selective removal of cells has also patterned cell location by using lift off techniques [37] and temperature responsive materials [99] to remove cells after seeding. Patterning of hexagonal adhesive islands onto a lens capsule demonstrated control of cell location on non-synthetic substrates of human tissue [100].

10.3.2.2 Influence of Chemical Patterns on Cell Extension, Adhesion, and Cell–Cell Contact

Chemical patterning has influenced the extension of various cellular processes. Cells on square-shaped adhesive islands, observed through time-lapse techniques, preferentially extended filopodia, lamellopodia, and microspikes at corners of square adhesive islands [101]. Cytoskeletons and focal adhesions were oriented such that tractional forces would be concentrated at the corners, with concentrations of fibronectin secreted at the corners [96]. Preferential cellular extension location and orientation of stress fibers may lead to directional migration of cells. Cells cultured on a teardrop shape exhibited preferential extension of lamellopodia at the blunt end, with actin filaments predominantly parallel to the long axis of the tear drop. When released from the pattern, the cell migrated in the direction of the blunt end along the long axis of the tear drop.

Chemical patterning has played a significant role in cell adhesion. Control of cell shape through the μCP of circular adhesive islands resulted in accurate quantification of cell adhesive strength by a spinning disk assay [64]. Both cellular adhesive shear strength and quantity of bound integrin $\alpha 5$ subunit increased with adhesive island size [65].

Chemical patterns have served to regulate cell–cell contact. Regulation of heterotypic cell–cell contact has been accomplished through patterning adhesive areas and plating hepatocytes, then rinsing non-adherent hepatocytes and plating fibroblasts [37]. The resulting co-culture possessed lanes of hepatocytes surrounded by spaces of fibroblasts, enabling limitation of the amount of contact between the two phenotype populations through modulation of lane width. The amount of cell–cell contact has also been regulated on an individual scale through chemical patterning. Non-adhesive agarose gel patterned on glass limited cell location to 'bowtie' shaped areas of bare glass, with either one cell on each half of a bowtie, or a single cell occupying one side of the bowtie [102]. The bowtie patterns provided a controlled cell spreading area and level of cell–cell contact between the two cells.

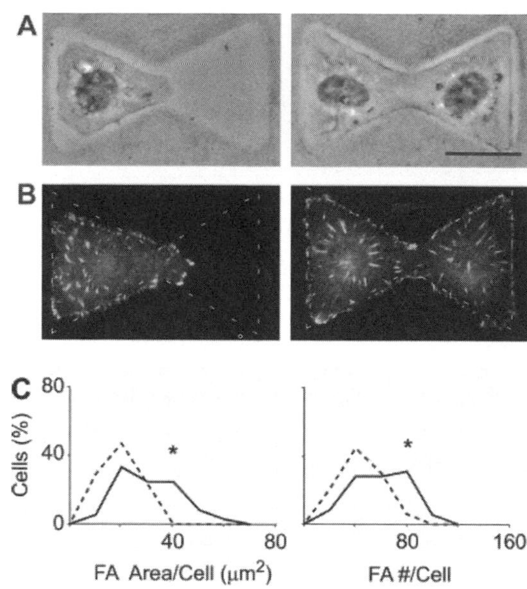

Fig. 10.10 Endothelial cells restricted to bowtie-shaped glass areas surrounded by agarose gel. Single and pairs of cells shown through phase-contrast (A), and vinculin staining (B). Overall focal adhesion area and focal adhesions per cell were significantly higher for paired cells than single cells (C) [102]. Reprinted from Developmental Cell, 6, McBeath R, Pirone DM, Nelson CM, Bhadriraju K, Chen CS, Cell Shape, Cytoskeletal Tension, and RhoA Regulate Stem Cell Lineage Commitment, 483–495, Copyright (2004), with permission from Elsevier

Figure 10.10 shows images and quantitative data that indicate more focal adhesions were observed on cells with cell–cell contact, as compared to single cells, with the effect abrogated by blocking VE-cadherin. Cell–cell contact in neurons has been achieved through μCP of ECM in a neural network formation [103]. Neurons preferentially adhered at patterned nodes, with axons and dendrites extended along patterned lanes, allowing connection of cells at adjacent nodes. The result was a prescribed network of interconnected neurons, controlled through the chemical patterns.

10.3.2.3 Chemical Pattern Influence of Apoptosis, Proliferation, Differentiation

Effects of chemical patterns on apoptosis, proliferation, and differentiation have largely been due to restriction of cell spreading area. Epithelial cells cultured on adhesive islands ranging in size from 25 to 1600 μm^2 exhibited more apoptotic markers on smaller islands and higher DNA synthesis on larger islands [104, 105]. Bone cells cultured on cell adhesive islands with areas varying from 75 to 10000 μm^2 synthesized DNA in proportion to projected cell area [106]. In addition, nuclear shape index varied with available cell spreading area, and collagen synthesis was highest for intermediate nuclear shape index. Restriction of cells to lanes has resulted in modulation from proliferative or apoptotic states to differentiative states. Epithelial cells restricted to lanes of 30 μm width aligned, spread and proliferated while cells restricted to lanes of 10 μm width down-regulated proliferation and expressed a differentiated epithelial morphology of tube-like structures. Cardiac myoblasts cultured on chemically patterned lanes have also exhibited

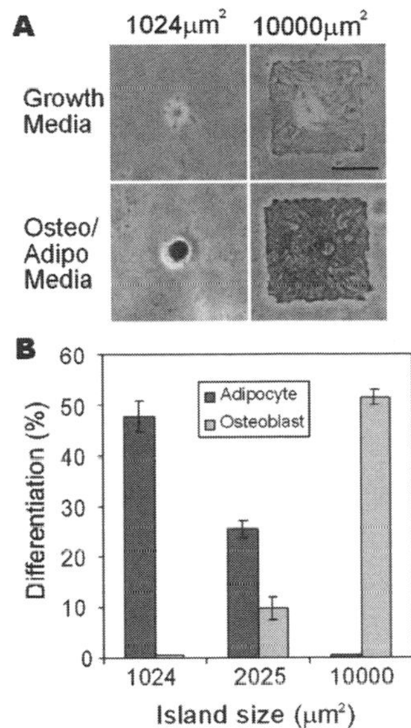

Fig. 10.11 Human mesenchymal stem cells exhibited lineage commitment dependent on adhesive island size. Brightfield images of cells on islands with lipids stained red and alkaline phosphatase stained blue (A). Percentage differentiation of cells according to island size (B) [107]

differentiation. As compared to unpatterned areas, cardiac myocytes on laminin lanes developed morphology similar to native heart tissue and began beating, with beating frequency synchronization dependent on lane spacing [54].

Precise control of spread cell area has resulted in modulation between different phenotype fates. A larger percentage of human mesenchymal stem cells cultured on adhesive islands underwent adipogenesis on small islands of 1024 μm^2 and osteogenesis on large islands of 10000 μm^2 [107]. Cells exhibited progression towards both lineages on intermediate islands of 2025 μm^2. Figure 10.11 provides data illustrat ing the effect of pattern size on cell lineage. Chemical pattern size and shape clearly influenced multiple cell functions.

10.3.3 Cellular Response to Combined Chemistry and Topography

10.3.3.1 Chemical Patterning to Restrict Cells to Topographic Features

Combining chemical and topographical patterns in some cases has biased cell location to user-specified topographic features. For example, patterning ridges with a cell adhesive domain resulted in restriction of cells to ridges, while patterning the ridges with an adhesion suppressing domain resulted in cells restricted to grooves [70]. In a similar fashion, patterning the plateau area in between microwells with a cell adhesion resistant SAM encouraged cell localization to the microwells [68,69]. Appropriate sizing of microwells resulted in restriction of single cells to the microwells. Chemical patterning concurrent with the topography provided selective cell adhesion to specific topographic features.

10.3.3.2 Chemical Patterning to Investigate Relative Influence of Patterns

Both chemical and topographical patterns have exhibited the ability to influence alignment of cells, however the relative influence of the two pattern types is not well understood. Independently patterning chemistry and topography has allowed investigation into the competitive and possibly synergistic effects of combined patterns. Photolithographic patterning of grooves in fused silica and adhesive lanes enabled independent placement of chemical lanes either parallel or orthogonal to the grooves. With lanes patterned parallel to grooves, fibroblast alignment increased beyond levels experienced with either type of pattern alone [71]. For lanes placed orthogonally the grooves of matched pitch, a larger fraction of fibroblasts aligned to the lanes rather than the grooves for all groove depths and widths. Neurite alignment of dorsal root ganglia behave similarly, with enhanced alignment for lanes and grooves parallel [72]. However, for lanes orthogonal to grooves, lanes dominated neurite alignment except for grooves with depths greater than or equal to 1 μm. Using a cleanroom approach produced substrates with independent chemical and topographical patterns, which enabled examination of relative pattern influence and interplay.

A non-cleanroom approach to independently pattern chemistry and topography has further aided investigation of the relative influence of pattern type by increasing substrate throughput and material selection. Independent μCP of cell adhesive lanes on embossed grooves enabled fabrication of a polymer cell substrate with lanes

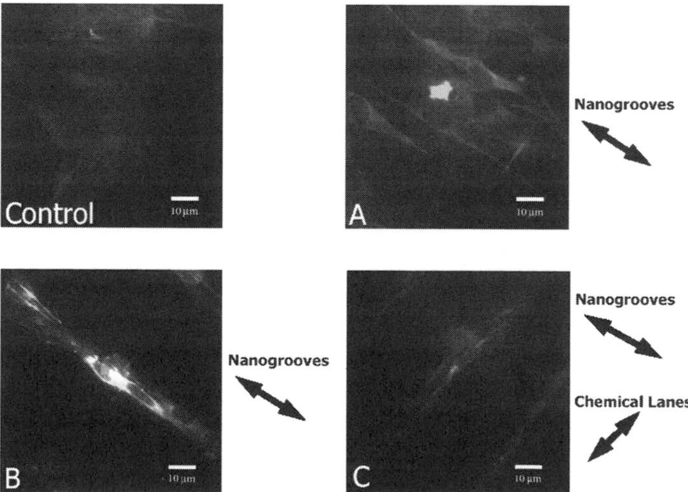

Fig. 10.12 Osteoblast response to independently patterned chemical and topographical patterns. Cells aligned to grooves on a grooved substrate with uniform chemistry (A) and to grooves on a grooved substrate with 10 μm chemical dots (B). However, on a grooved substrate with 10 μm lanes, alignment modulated to the chemical lanes [32]. Reused with permission from Joseph L. Charest, Marcus T. Eliason, Andrés J. García, William P. King, A. Alec Talin, and Blake A. Simmons, Journal of Vacuum Science & Technology B, 23, 3011 (2005). Copyright 2005, AVS The Science & Technology Society

of varying pitch placed orthogonally to the grooves [73]. Osteoblasts cultured on the substrates aligned predominantly to the grooves, although the lanes did serve to reduce the amount of alignment to the grooves. Grooves dominated osteoblast alignment as lane spacing was increased from 10 to 50 μm, with some osteoblasts bridging up to 50 μm of adhesion-resistant PEG SAMs while remaining aligned to the grooves. Although topography dominated the alignment, the chemical patterns were discontinuous and topographical feature size was not varied. A similar fabrication strategy created 100 nm wide by 100 nm deep grooves overlaid with either 10 μm wide orthogonal lanes or 10 μm diameter dots [32]. Figure 10.12 illustrates the response of cells cultured on the substrates. For the same underlying topography, continuous chemical lanes dominated alignment where the discontinuous pattern of dots did not. Modulation of alignment dominance from chemical to topographical patterns may have depended on a combination of parameters including length scale of feature sizes and continuity of patterns. As both topography and chemical patterns have significant influence on cells, independent control of both patterns results in substrates applicable to further investigations of the complex cell-material interface.

10.4 Summary and Conclusions

Micro and nanopatterned substrates have enabled quantitative analysis of the response of cells to well-defined biomaterial interfaces, furthering understanding of the complex communication between cell and material. Cleanroom microfabrication tools advanced the field to create patterns reliably with feature sizes on sub-cellular length scales. Non-cleanroom techniques, such as imprint lithography and micro-contact printing, have emerged to meet demands for processes with a wide material selection and the ability to fabricate numerous samples in a high-throughput fashion for robust biological assays. Current non-cleanroom methods show potential to achieve substrate surface features deep into the nanoscale. Specifically, imprint lithography has produced nanoscale topography in combination with independently patterned chemistry through μCP. Non-cleanroom methods will enhance cell-material interface research by enabling feature sizes in the nanoscale regime, widening material selection, and providing large sample sizes without costly equipment, while permitting independent combination of chemistry and topography that furthers the specificity and sophistication of cell-material interfaces.

Morphological response of cells to material surfaces has been characterized for various parameters of surface topography and chemical patterns. However, the mechanism for morphological response, in particular contact guidance, remains largely unknown. Although cytoskeletal involvement has been implicated for some cases, specific molecular mechanisms have not been established. While higher-order cellular response to chemical patterns has been characterized for some cases, topographic influence on proliferation and differentiation remains predominantly uncharacterized with much potential for future investigation. Although basic guidelines have been established through the use of independent chemical and topographical patterning, the relative influence of and complex interplay between topographic and chemical patterns remain as areas for further exploration. Enhanced patterning techniques will continue to lead cell substrate fabrication towards sophisticated, user-defined configurations of topographic and chemical patterns, providing a platform to establish mechanisms of cellular response to cell-material interfaces.

References

1. Flemming RG, Murphy CJ, Abrams GA, Goodman SL, Nealey PF. Effects of synthetic micro- and nano-structured surfaces on cell behavior. Biomaterials 1999;20(6): 573–588.
2. Curtis A, Wilkinson C. Topographical control of cells. Biomaterials 1997;18(24):1573–1583.
3. Falconnet D, Csucs G, Grandin HM, Textor M. Surface engineering approaches to micropattern surfaces for cell-based assays. Biomaterials 2006;27(16):3044–3063.
4. Schwartz Z, Boyan BD. Understanding mechanisms at the bone-biomaterial interface. Journal of Cellular Biochemistry 1994;56:340–347.
5. Allen LT, Fox EJP, Blute I, Kelly ZD, Rochev Y, Keenan AK, et al. Interaction of soft condensed materials with living cells: Phenotype/transcriptome correlations for the hydrophobic effect. PNAS 2003;100(11):6331–6336.
6. He W, Gonsalves KE, Batina N, Poker DB, Alexander E, Hudson M. Micro/nanomachining of polymer surface for promoting osteoblast cell adhesion. Biomedical Microdevices 2003;5(2):101–108.
7. Brunette DM, Kenner GS, Gould TRL. Grooved titanium surfaces orient growth and migration of cells from human gingival explants. Journal of Dental Research 1983;62(10):1045–1048.
8. Brunette DM. Fibroblasts on micromachined substrata orient hierarchically to grooves of different dimensions. Experimental Cell Research 1986;164(1):1–26.
9. Mahoney MJ, Chen RR, Tan J, Saltzman WM. The influence of microchannels on neurite growth and architecture. Biomaterials 2005;26:771–778.
10. Zinger O, Zhao G, Schwartz Z, Simpson J, Wieland M, Landolt D, et al. Differential regulation of osteoblasts by substrate microstructural features. Biomaterials 2005;26:1837–1847.
11. Berry CC, Campbell G, Spadiccino A, Robertson M, Curtis ASG. The influence of microscale topography on fibroblast attachment and motility. Biomaterials 2004;25:5781–5788.
12. Clark P, Connolly P, Curtis ASG, Dow JAT, Wilkinson CDW. Cell guidance by ultrafine topography in vitro. Journal of Cell Science 1991;99:73–77.
13. Broers AN, Hoole ACF, Ryan JM. Electron beam lithography – Resolution limits. Microelectronic Engineering 1996;32(1–4):131–142.
14. Rajnicek AM, Britland S, McCaig CD. Contact guidance of CNS neurites on grooved quartz: Influence of groove dimensions, neuronal age and cell type. Journal of Cell Science 1997;110:2905–2913.
15. Diehl KA, Foley JD, Nealey PF, Murphy CJ. Nanoscale topography modulates corneal epithelial cell migration. Journal of Biomedical Materials Research A 2005;75(3):603–611.
16. Teixeira AI, Abrams GA, Bertics PJ, Murphy CJ, Nealey PF. Epithelial contact guidance on well-defined micro- and nanostructured substrates. Journal of Cell Science 2003;116(10):1881–1892.
17. Xu Q, Mayers BT, Lahav M, Vezenov DV, Whitesides GM. Approaching zero: Using fractured crystals in metrology for replica molding. Journal of the American Chemical Society 2005;127(3):854–855.

18. Gadegaard N, Mosler S, Larsen NB. Biomimetic polymer nanostructures by injection moldings. Macromolecular and Materials Engineering 2005;288:76–83.
19. Chesmel KD, Black J. Cellular-responses to chemical and morphologic aspects of biomaterial surfaces. I. A novel in vitro model system. Journal of Biomedical Materials Research 1995;29(9):1089–1099.
20. Walboomers XF, Croes HJE, Ginsel LA, Jansen JA. Growth behavior of fibroblasts on microgrooved polystyrene. Biomaterials 1998;19(20):1861–1868.
21. Walboomers XF, Ginsel LA, Jansen JA. Early spreading events of fibroblasts on microgrooved substrates. Journal of Biomedical Materials Research 2000;51(3):529–534.
22. Brunette DM. Spreading and orientation of epithelial-cells on grooved substrata. Experimental Cell Research 1986;167(1):203–217.
23. Hamilton DW, Brunette DM. "Gap guidance" of fibroblasts and epithelial cells by discontinuous edged surfaces. Experimental Cell Research 2005;309(2):429–437.
24. Schmidt JA, von Recum AF. Texturing of polymer surfaces at the cellular level. Biomaterials 1991;12(4):385–389.
25. Chrzanowska-Wodnicka M, Burridge K. Rho-stimulated contractility drives the formation of stress fibers and focal adhesions. 1996;133(6):1403–1415.
26. Liao H, Andersson A-S, Sutherland D, Petronis S, Kasemo B, Thomsen P. Response of rat osteoblast-like cells to microstructured model surfaces in vitro. Biomaterials 2003; 24(4):649–654.
27. Yim KF, Reano RM, Pang SW, Yee AF, Chen CS, Leong KW. Nanopattern-induced changes in morphology and motility of smooth muscle cells. Biomaterials 2005;26:5405–5413.
28. Chou SY, Krauss PR, Renstrom PJ. Imprint lithography with 25-nanometer resolution. Science 1996;272(5258):85–87.
29. Chou SY, Krauss PR, Zhang W, Guo L, Zhuang L. Sub-10 nm imprint lithograpy and applications. Journal of Vacuum Science & Technology B 1997;15(6):2897–2904.
30. Casey BG, Cumming DRS, Khandaker II, Curtis ASG, Wilkinson CDW. Nanoscale embossing of polymers using a thermoplastic die. Microelectronic Engineering 1999;46(1–4): 125–128.
31. Charest JL, Bryant LE, Garcia AJ, King WP. Hot embossing for micropatterned cell substrates. Biomaterials 2004;25(19):4767–4775.
32. Charest J, Eliason M, Talin A, Simmons B, Garcia A, King W. Polymer cell culture substrates with combined nanotopographical patterns and micropatterned chemical domains. Journal of Vacuum Science & Technology B 2005;23(6):301–3014.
33. Johansson F, Carlberg P, Danielsen N, Montelius L, Kanje M. Axonal outgrowth on nanoimprinted patterns. Biomaterials 2006;27(8):1251–1258.
34. Sarkar S, Lee GY, Wong JY, Desai TA. Development and characterization of a porous micro-patterned scaffold for vascular tissue engineering applications. Biomaterials 2006;27(27):4775–4782.
35. Hu W, Yim EKF, Reano RM, Leong KW, Pang SW. Effects of nanoimprinted patterns in tissue-culture polystyrene on cell behavior. AVS 2005;2984–2989.
36. Prime KL, Whitesides GM. Self-assembled organic monolayers: Model systems for studying adsorption of proteins at surfaces. Science 1991;252(5010):1164–1167.
37. Bhatia SN, Yarmush ML, Toner M. Controlling cell interactions by micropatterning in co-cultures: hepatocytes and 3T3 fibroblasts. Journal of Biomedical Materials Research 1997;34(2):189–199.
38. Irimia D, Karlsson JOM. Development of a cell patterning technique using poly(ethylene glycol) disilane. Biomedical Microdevices 2003;5(3):185–194.
39. Scotchford CA, Ball M, Winkelmann M, Voros J, Csucs C, Brunette DM, et al. Chemically patterned, metal-oxide-based surfaces produced by photolithographic techniques for studying protein- and cell-interactions. II: Protein adsorption and early cell interactions. Biomaterials 2003;24(7):1147–1158.
40. Orth RN, Clark TG, Craighead HG. Avidin-biotin micropatterning methods for biosensor applications. Biomedical Microdevices 2003;5(1):29–34.

41. Mohammed JS, DeCoster MA, McShane MJ. Fabrication of interdigitated micropatterns of self-assembled polymer nanofilms containing cell-adhesive materials. Langmuir 2006;22(6):2738–2746.
42. Pallandre A, Glinel K, Jonas AM, Nysten B. Binary nanopatterned surfaces prepared from silane monolayers. Nano Letters 2004;4(2):365–371.
43. Denis FA, Pallandre A, Nysten B, Jonas AM, Dupont-Gillain CC. Alignment and assembly of adsorbed collagen molecules induced by anisotropic chemical nanopatterns. Small 2005;1(10):984–991.
44. Harnett CK, Satyalakshmi KM, Craighead HG. Bioactive templates fabricated by low-energy electron beam lithography of self-assembled monolayers. Langmuir 2001;17(1):178–182.
45. Rundqvist J, Hoh JH, Haviland DB. Directed immobilization of protein-coated nanospheres to nanometer-scale patterns fabricated by electron beam lithography of poly(ethylene glycol) self-assembled monolayers. Langmuir 2006;22(11):510–5107.
46. Ra HJ, Picart C, Feng HS, Sweeney HL, Discher DE. Muscle cell peeling from micropatterned collagen: Direct probing of focal and molecular properties of matrix adhesion. Journal of Cell Science 1999;112(10):1425–1436.
47. Ilic B, Craighead H. Topographical patterning of chemically sensitive biological materials using a polymer-based dry lift off. Biomedical Microdevices 2000;2(4):317–322.
48. Orth RN, Kameoka J, Zipfel WR, Ilic B, Webb WW, Clark TG, et al. Creating biological membranes on the micron scale: Forming patterned lipid bilayers using a polymer lift-off technique. Biophysical Journal 2003;85(5):3066–3073.
49. Bergman AA, Buijs J, Herbig J, Mathes DT, Demarest JJ, Wilson CD, et al. Nanometer-scale arrangement of human serum albumin by adsorption on defect arrays created with a finely focused ion beam. Langmuir 1998;14:6785–6788.
50. Satriano C, Carnazza S, Licciardello A, Guglielmino S, Marletta G. Cell adhesion and spreading on polymer surfaces micropatterned by ion beams. Journal of Vacuum Science & Technology A 2003;21(4):1145–1151.
51. Lee K-B, Park S-J, Mirkin CA, Smith JC, Mrksich M. Protein nanoarrays generated by dip-pen nanolithography. Science 2002;295:1702–1705.
52. Kumar A, Whitesides GM. Features of gold having micrometer to centimeter dimensions can be formed through a combination of stamping with an elastomeric stamp and an alkanethiol "ink" followed by chemical etching. Applied Physics Letters 1993;63(14):2002–2004.
53. Singhvi R, Kumar A, Lopez GP, Stephanopoulos GN, Wang DIC, Whitesides GM, et al. Engineering cell-shape and function. Science 1994;264(5159):696–698.
54. McDevitt TC, Angello JC, Whitney ML, Reineck H, Hauschka SD, Murry PS, et al. In vitro generation of differentiated cardiac myofibers on micropatterned laminin surfaces. Journal of Biomedical Materials Research 2002;60(3):472–479.
55. Vogt AK, Stefani FD, Best A, Nelles G, Yasuda A, Knoll W, et al. Impact of micropatterned surfaces on neuronal polarity. Journal of Neuroscience Methods 2004;134(2):191–198.
56. Lee CJ, Huie P, Leng T, Peterman MC, Marmor MF, Blumenkranz MS, et al. Microcontact printing on human tissue for retinal cell transplantation. Archives of Ophthalmology 2002;120(12):1714–1718.
57. Renault J, Bernard A, Juncker D, Michel B, Bosshard H, Delamarche E. Fabricating microarrays of functional proteins using affinity contact printing. Angewandte Chemie International Edition 2002;41(13):2320–2323.
58. Renault J, Bernard A, Bietsch A, Michel B, Bosshard H, Kreiter M, et al. Fabricating arrays of single protein molecules on glass using microcontact printing. Journal of Physical Chemistry 2003;B23(107):703–711.
59. Lussi JW, Falconnet D, Hubbell JA, Textor M, Csucs G. Pattern stability under cell culture conditions – A comparative study of patterning methods based on PLL-g-PEG background passivation. Biomaterials 2006;27(12):2534–2541.
60. Keselowsky BG, Collard DM, Garcia AJ. Surface chemistry modulates fibronectin conformation and directs integrin binding and specificity to control cell adhesion. Journal of Biomedical Materials Research Part A 2003;66A(2):247–259.

61. Keselowsky BG, Collard DM, Garcia AJ. Integrin binding specificity regulates biomaterial surface chemistry effects on cell differentiation. PNAS 2005;102(17):5953–5957.
62. Lehnert D, Wehrle-Haller B, David C, Weiland U, Ballestrem C, Imhof BA, et al. Cell behaviour on micropatterned substrata: Limits of extracellular matrix geometry for spreading and adhesion. 2004;117(Pt 1):41–52.
63. Chen CS, Mrksich M, Huang S, Whitesides GM, Ingber DE. Micropatterned surfaces for control of cell shape, position, and function. Biotechnology Progress 1998;14(3):356–363.
64. Gallant ND, Capadona JR, Frazier AB, Collard DM, García AJ. Micropatterned surfaces for analyzing cell adhesion strengthening. Langmuir 2002;18:5579–5584.
65. Gallant ND, Michael KE, Garcia AJ. Cell adhesion strengthening: contributions of adhesive area, integrin binding, and focal adhesion assembly. Molecular Biology Cell 2005;16(9):4329–4340.
66. Endler EE, Nealey PF, Yin J. Fidelity of micropatterned cell cultures. Journal of Biomedical Materials Research Part A 2005;74A(1):92–103.
67. Thissen H, Johnson G, Hartley PG, Kingshott P, Griesser HJ. Two-dimensional patterning of thin coatings for the control of tissue outgrowth. Biomaterials 2006;27(1):35–43.
68. Revzin A, Tompkins RG, Toner M. Surface engineering with poly(ethlyne glycol) photolithography to create high-density cell arrays on glass. Langmuir 2003;19:9855–9862.
69. Dusseiller MR, Schlaepfer D, Koch MK, Kroschewski R, Textor M. An inverted microcontact printing method on topographically structured polystyrene chips for arrayed micro-3-D culturing of single cells. Biomaterials 2005;26:5917–5925.
70. Mrksich M, Chen CS, Xia Y, Dike LE, Ingber DE, Whitesides GM. Controlling cell attachment on contoured surfaces with self-assembled monolayers of alkanethiolates on gold. Proceedings of the National Academy of Science, USA 1996;93(20):10775–10778.
71. Britland S, Morgan H, Wojiak-Stodart B, Riehle M, Curtis A, Wilkinson C. Synergistic and hierarchical adhesive and topographic guidance of BHK cells. Experimental Cell Research 1996;228:313–325.
72. Britland S, Perridge C, Denyer M, Morgan H, Curtis A, Wilkinson C. Morphogenetic guidance cues can interact synergistically and hierarchically in steering nerve cell growth. Experimental Biology Online 1996;1(2):1–5.
73. Charest JL, Eliason MT, Garcia AJ, King WP. Combined microscale mechanical topography and chemical patterns on polymer cell culture substrates. Biomaterials 2006;27(11):2487:2487–2494.
74. Weiss P. Experinments on cell and axon orientation in vitro. Journal of Experimental Zoology 1945;100(3):353–386.
75. Walboomers XF, Croes HJE, Ginsel LA, Jansen JA. Growth behavior of fibroblasts on microgrooved polystyrene. Biomaterials 1998;19(20):1861–1868.
76. Johansson F, Kanje M, Eriksson C, Wallman L. Guidance of neurons on porous patterned silicon: is pore size important? Physics status solidi 2005;2(9):3258–3262.
77. Dalby MJ, McCloy D, Robertson M, Wilkinson CDW, Oreffo ROC. Osteoprogenitor response to defined topographies with nanoscale depths. Biomaterials 2006;27(8): 1306–1315.
78. Hamilton D, Wong K, Brunette D. Microfabricated discontinuous-edge surface topographies influence osteoblast adhesion, migration, cytoskeletal organization, and proliferation and enhance matrix and mineral deposition in vitro. Calcified Tissue International 2006;78(5):314–325.
79. Clark P, Connolly P, Curtis AS, Dow JA, Wilkinson CD. Topographical control of cell behaviour. I. Simple step cues. Development 1987;99(3):439–448.
80. Oakley C, Brunette DM. The sequence of alignment of microtubules, focal contacts and actin-filaments in fibroblasts spreading on smooth and grooved titanium substrata. Journal of Cell Science 1993;106:343–354.
81. Oakley C, Jaeger NAF, Brunette DM. Sensitivity of fibroblasts and their cytoskeletons to substratum topographies: Topographic guidance and topographic compensation by micromachined grooves of different dimensions. Experimental Cell Research 1997;234(2):413–424.

82. Clark P, Connolly P, Curtis ASG, Dow JAT, Wilkinson CDW. Topographical control of cell behavior. 2. Multiple grooved substrata. Development 1990;108(4):635–644.
83. Uttayarat P, Toworfe GK, Dietrich F, Lelkes PI, Composto RJ. Topographic guidance of endothelial cells on silicone surfaces with micro- to nanogrooves: Orientation of actin filaments and focal adhesions. Journal of Biomedical Materials Research Part A 2005;75A(3):668–680.
84. Teixeira AI, Nealey PF, Murphy CJ. Responses of human keratocytes to micro- and nanostructured substrates. Journal of Biomedical Materials Research 2004;71A:369–376.
85. Rosa AL, Beloti MM, van Noort R. Osteoblastic differentiation of cultured rat bone marrow cells on hydroxyapatite with different surface topography. Dental Materials 2003;19(8): 768–772.
86. Lee SJ, Choi JS, Park KS, Khang G, Lee YM, Lee HBHB. Response of MG63 osteoblast-like cells onto polycarbonate membrane surfaces with different micropore sizes. Biomaterials 2004;25(19):4699–4707.
87. Lossdorfer S, Schwartz Z, Wang L, Lohmann CH, Turner JD, Wieland M, et al. Microrough implant surface topographies increase osteogenesis by reducing osteoclast formation and activity. Journal of Biomedical Materials Research 2004;70A:361–369.
88. Rea SM, Brooks RA, Best SM, Kokubo T, Bonfield W. Proliferation and differentiation of osteoblast-like cells on apatite-wollastonite/polyethylene composites. Biomaterials 2004;25:4503–4512.
89. Perizzolo D, Lacefield WR, Brunette DM. Interaction between topography and coating in the formation of bone nodules in culture for hydroxyapatite- and titanium-coated micromachined surfaces. Journal of Biomedical Materials Research 2001;56(4):494–503.
90. Chehroudi B, McDonnell D, Brunette DM. The effects of micromachined surfaces on formation of bonelike tissue on subcutaneous implants as assessed by radiography and computer image processing. Journal of Biomedical Materials Research 1997;34(3):279–290.
91. Foley JD, Grunwald EW, Nealey PF, Murphy CJ. Cooperative modulation of neuritogenesis by PC12 cells by topography and nerve growth factor. Biomaterials 2005;26(17):3639–3644.
92. Recknor JB, Sakaguchi DS, Mallapragada SK. Directed growth and selective differentiation of neural progenitor cells on micropatterned polymer substrates. Biomaterials 2006;27(22):4098–4108.
93. Matsuzaka K, Yoshinari M, Shimono M, Inoue T. Effects of multigrooved surfaces on osteoblast-like cells in vitro: Scanning electron microscopic observation and mRNA expression of osteopontin and osteocalcin. Journal of Biomedical Materials Research Part A 2004;68A(2):227–234.
94. Wang JHC, Grood ES, Florer J, Wenstrup R. Alignment and proliferation of MC3T3-E1 osteoblasts in microgrooved silicone substrata subjected to cyclic stretching. Journal of Biomechanics 2000;33:729–735.
95. Jiang X, Bruzewicz DA, Wong AP, Piel M, Whitesides GM. Directing cell migration with asymmetric micropatterns. PNAS 2005;102(4):975–978.
96. Brock A, Chang E, Ho CC, LeDuc P, Jiang X, Whitesides GM, et al. Geometric determinants of directional cell motility revealed using microcontact printing. Langmuir 2003;19(5):161–1617.
97. Orth RN, Wu M, Holowka DA, Craighead HG, Baird BA. Mast cell activation on patterned lipid bilayers of subcellular dimensions. Langmuir 2003;19(5):159–1605.
98. Schmalenberg KE, Uhrich KE. Micropatterned polymer substrates control slignment of proliferating Schwann cells to direct neuronal regeneration. Biomaterials 2005;26:1423–1430.
99. Yamato M, Konno C, Utsumi M, Kikuchi A, Okano T. Thermally responsive polymer-grafted surfaces facilitate patterned cell seeding and co-culture. Biomaterials 2002;23(2): 561–567.
100. Lee CJ, Blumenkranz MS, Fishman HA, Bent SF. Controlling cell adhesion on human tissue by soft lithography. Langmuir 2004;20:415–4161.
101. Parker KK, Brock AL, Brangwynne C, Mannix RJ, Wang N, Ostuni E, et al. Directional control of lamellipodia extension by constraining cell shape and orienting cell tractional forces. FASEB Journal. 2002;16(10):1195–1204.

102. Nelson CM, Pirone DM, Tan JL, Chen CS. Vascular endothelial-cadherin regulates cytoskeletal tension, cell spreading, and focal adhesions by stimulating RhoA. Molecular Biology Cell 2004;15(6):2943–2953.
103. Vogt AK, Wrobel G, Meyer W, Knoll W, Offenhausser A. Synaptic plasticity in micropatterned neuronal networks. Biomaterials 2005;26(15):2549–2557.
104. Chen CS, Mrksich M, Huang S, Whitesides G, Ingber DE. Geometric control of cell life and death 1997;276:1425–1428.
105. Dike LE, Chen CS, Mrksich M, Tien J, Whitesides GM, Ingber DE. Geormetric control of switching between growth, apoptosis, and differentiation during angiogenesis using micropatterned substrates. In Vitro Cell Developmental Biology – Animal 1999;35:441–448.
106. Thomas CH, Collier JH, Sfeir CS, Healy KE. Engineering gene expression and protein synthesis by modulation of nuclear shape. Proceedings of the National Academy of Sciences of the United States of America 2002;99(4):1972–1977.
107. McBeath R, Pirone DM, Nelson CM, Bhadriraju K, Chen CS. Cell shape, cytoskeletal tension, and RhoA regulate stem cell lineage commitment. Developmental Cell 2004;6(4):483–495.

Chapter 11
Biosensors Micro and Nano Integration

Ravi Doraiswami

Abstract This chapter addresses micro and nanocomposite biocompatible integration techniques and processes for biosensors. It outlines the design parameter requirements for sensor systems interconnects. It describes in detail materials choice and properties, influence on interconnect parameters, fine pitch nano/micro interconnects and technologies and integration techniques. The chapter concludes by providing examples of miniaturized sensors and their fabrication.

Keywords: Microelectronics · Nanoelectronics · Biosensors · Nano-interconnects · Bumps · Intermetalics · MEMS

11.1 Introduction

A rapidly growing trend in analytical science in recent years has been the development of chemical and biosensor technologies. The driving force behind this trend is a desire to render analytical chemical measurements more timely and relevant in order to provide immediate feedback with respect to detection results. Lab on a chip gives the ability to do thousands of experiments at the same time. This is very much apparent in the biomedical field, it is even more a requirement in fields like environmental analysis (industrial, occupational, military, etc.) and process control. The Sensor field is interdisciplinary in nature, which attracts the expertise of chemists, biochemists, mechanical engineers, electronics engineers, and physists.

MEMS for biological or medical applications or involving biological component(s), so-called biomicroelectromechanical systems (bio-MEMS) [1] are becoming more and more popular. Depending on their applications, this is justified by the inherent benifits of miniaturization in bio-MEMS such as small size, low weight, potential low unit costs per device, efficient transduction processes, high reaction rate, low reagent consumption, multiple sensors per chip and the potential to

---

R. Doraiswami
G. W. Woodruff School of Mechanical Engineering, Georgia Institute of Technology

P. J. Hesketh (ed.), *BioNanoFluidic MEMS.*

manufacture minimally invasive devices and systems. One of the factors which contribute to miniaturization of the sensor while maintaining its electrical and mechanical performance is the choice of interconnects and interface materials and the process of integration. Nanotechnology is frequently discussed these days as an emerging frontier for the development of various future devices. Micro-and/or nano-electromechanical systems (MEMS/NEMS) are seen as the basis of future nanotechnologies, because they combine to miniature sensors and actuators with electronics.

This chapter addresses Micro and Nanocomposite bio compatible integration techniques and processes for biosensors. It will outline the design parameter requirements for sensor systems interconnect. It will describe in detail materials choice and properties, influence on interconnect parameters, fine pitch Nano/Micro interconnects and technologies, Integration techniques. The chapter will conclude by providing sited examples of miniaturized sensors and their performance.

11.2 Micro and Nanocomposite Bio Compatible Interconnect

Integration of the entire sensor system on a single chip is possible, for very low cost per die. Design criteria for single chip process depend on several considerations such as sensor target, environmental conditions, bio or non-bio compatibility, etc. The design and process parameters can generally be formulated to achieve, low cost, high trough put and excellent electrical/mechanical performance. Conventional chip packaging, is described in some detail below by way of reference. With much further miniaturization, the conventional approach is useful for high density global interconnects linking the sensor with the external world. The nano scale is closer to the sensor head and nano-scale techniques for transmitting data over electrical lines have not yet developed beyond dropping nano samples on nano wires and hope for the right contact to form in accordance to the laws of statistics. The most common parameters for high speed electrical are the capacitance, inductance and resistance of the interconnects. Process parameters depend on die size, interconnect pitch, passivation via (μm), Under Bump Metallurgy (UBM) diameter (μm), Bump Height (μm), Bump Diameter (μm), number of interconnects/sq cm and Encapsulation/under fill process. Figure 11.1 shows a schematic of two commonly used interconnect technologies for sensors a) Flip Chip Process b) Wire Bond Process.

11.2.1 Flip Chip Process

Flip chip is defined by how the functional surface of the chip is interconnected to the substrate. In this case the chips functional surface faces the substrate and is interconnected through solder bumps as shown in Fig. 11.1. Flip Chip process starts with the wafer level packaging (WLP). Wafer level packaging is one in which the die and "package" are fabricated and tested on the wafer prior to singulation.

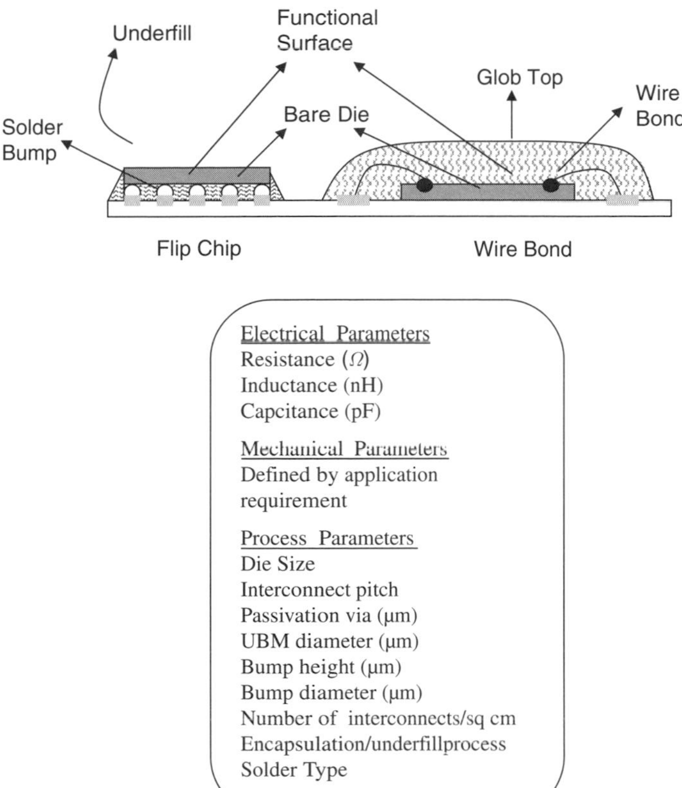

Fig. 11.1 Design criteria for two most commonly used interconnect techniques

This process eliminates many of the packaging process required using conventional packaging resulting in drastic reduction in manufacturing cost [2]. The benefits are

a. Small IC package size
b. Low cost of electrical testing
c. Lowest burn –in cost
d. Better electrical interconnects due to short interconnects
e. Cooling ease through back of the die

The flip chip design process starts by identifying the passivation layer material, Under Bump Metallurgy (UBM) composition, solder type and bumping strategy. The following are the selection criteria for these parameters:

a) Passivation material which the one which is coated on the surface of the functional chip to serve as protection to the strip line interconnects connecting the functional part of the chip and the bumps. The most widely used material is SiO_2. The thickness of this materials range from 1 micron to 3 microns depending on the bump height.

b) Under Bump Metallurgy (UBM), has a multilayered thin film composition. The materials which form the layers have specific tasks to perform. The first layer forms the barrier layer and will normally be Cr or Ni. This layer protects the strip line from exposure to solder. The next layer is a high strength barrier layer which protests the chip from electro migration and intermetalics from the solder bump. The last layer is the wetting layer which normally is a coating of gold.

c) Interconnect materials choice is based on the mechanical and electrical property of materials. Tin has emerged as the predominant material used in interconnects. NEMI has identified Sn3.5Ag and Sn3.5Ag0.7Cu as the lead free solders with the best of electrical and mechanical properties. Both these solders have tin as a predominant metal ranging up to 96% of the composition.

d) lead free solder electroplating has proven over the years to produce the highest yield per wafer.

Wafer bumping by electroplating however, has the largest potential for realizing highest I/O densities with a pitch range from 20 μm to 25 mm. It is particularly suited for high volume production of bumped wafers at a high-quality standard. As the value of wafers increases, the relatively high processing costs are less and less perceptible. The realization of bumps using electroplating can be divided into fundamental process steps, which are sputtering of the plating base, photoresist patterning, electroplating, differential etching, and, if required, a final solder reflow as shown in Fig. 11.2.

Today, a number of commercial electroplating baths for different metals are available. The bath chemistry must be compatible to the photoresist system and should be insensitive to out-bleeding. Exceptionally cleaned makeup ingredients ensure deposits with a high purity and consequently well-defined electrical and mechanical properties. For solder bumping, a low co-deposition rate of organics is required to guarantee a low volume of out gassing during the reflow and bonding process. For this reason, relatively slow depositing electrolytes with a small amount of organic additives are preferred. Furthermore, the deposits have to show a well-defined bump shape and solder-alloy composition, low internal

Table 11.1 Shows typical parameters used for flip chip bump process for 100 micron pitch [3]

Connection Metallurgy	100 Micron Pitch SnAg Lead Free Solder
DC Resistance	5 m Ω
Inductance	32 pH
Capacitance	17 fF
Bump Height	50 μm
Pitch	100 μm
IO/cm^2	2553

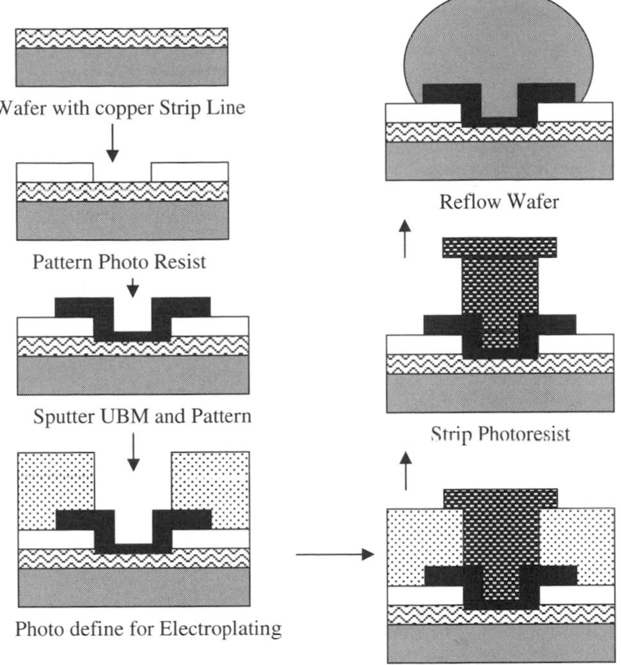

Fig. 11.2 Typical flip chip bumping process using electroplating technique

stress, non-porosity, as well as a negligible number of defects caused by dendrite formation, pittings or particle encapsulation. These attributes as a whole have to be adjusted by a suitable combination of all relevant deposition parameters like concentration of electrolyte components, current density and agitation strength. In particular cases, the optimization depends on the specific pattern layout as well [3].

Lead free solder 100 micron pitch assembly process is conducted by fabricated on a substrate have high Tg, high CTE and low cost. FR4 (Fire retardant 4) substrates are ideal for this purpose. They have a Tg of 125–135°C and CTE of 14 to 18 ppm. Figure 11.3 shows the process flow for flip chip assembly process and Fig. 11.4 shows a cross section of a 100 μm pitch flip chip assembly.

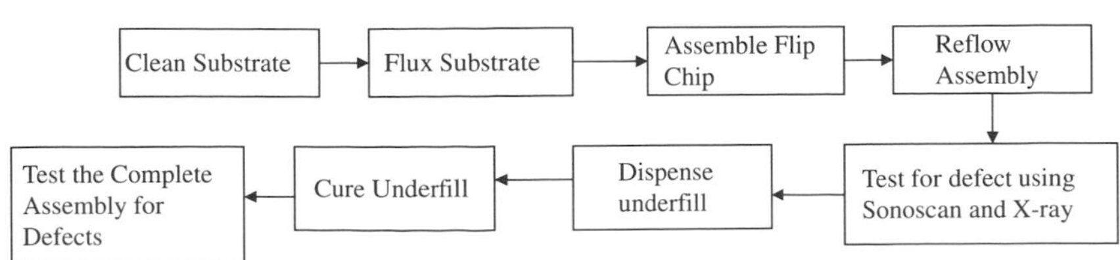

Fig. 11.3 Flip chip assembly process flow

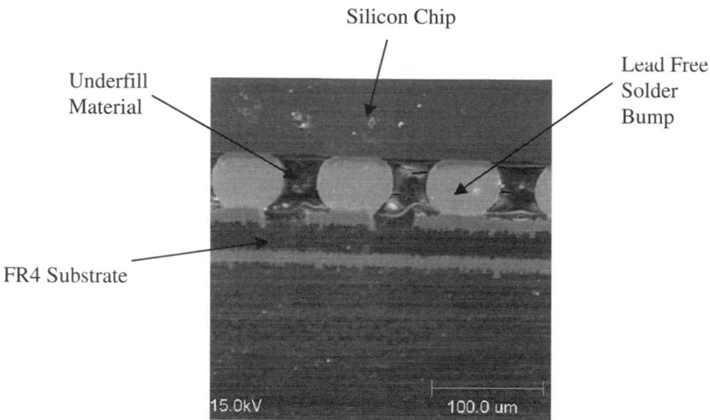

Fig. 11.4 Cross section of 100 micron pitch flip chip assembly [3]

11.2.2 MEMS Packaging

Packaging of Bio-MEMS functional systems depend on choice of appropriate materials and technologies. The selection of compatible processes and the planning of a fabrication workflow which would produce the least interference between biological component and the integration process. Bio-MEMS consist of several subunits for sample acquisition, sample preparation, sensing, data analysis, etc. The degree of integration (monolithic, hybrid, discrete subsystems), disposable and durable subsystems become very important in achieving low cost and high functionality. The use of biocompatible packaging can be achieved by encapsulating conventionally packaged parts of the bio-MEMS with a biocompatible material.

11.2.3 Material Trends in MEMS Packaging

Biocompatibility, flexibility and the ability to operate over a wide temperature range are all characteristics of micro sensors that are determined by the substrate material. PDMS, PMMA, parylene and polyethelene are some of the biocompactable polymers commonly used in MEMS for biomedical applications [4–6]

11.2.4 Challenge in Integration Technologies

Particular challenges consist in joining substrates made of different 1) materials (hybrid bio-MEMS) and 2) topographies like micromachined structures (e.g., microchannels). Also, the integration of various functionalities within a device (e.g., interfacing a microfluidic part to an optical detection system while maintaining optimal performance of both functions), and the compatibility of processes with metallic electrical connections and standard integrated circuit fabrication pose significant challenges. Assembly joints need to provide specific properties such as mechanical strength, biochemical resistance, water tightness but not necessarily hermeticity (e.g., gas permeable), biochemical compatibility, resistance to various chemicals, chemical inertness, and the absence of additional chemical substances such as outgassing or reaction products. Sealing without denaturing temperature-sensitive materials, especially biochemical surface modifications/coatings is a particularly important requirement for bio-MEMS as well as the ability of bonding at selected locations versus full wafer bonding. The preservation of structural dimensions during joining (no clogging, shrinkage, built-in stress resulting in distortions) and precise alignment strategies to provide high precision joining are also issues. Finally, automation of the process as well as its suitability to up-scaling and mass-production needs to be addressed.

When ultrathin chips are not an option, the use of standard thickness chips can still enable extremely small devices by the use of flip-chip technology. Hearing aids are an excellent example which demonstrates the evolution from large boxes worn around the neck, to those now small enough to incorporate high-performance digital signal processor (DSP), microcontrollers, microphones, and batteries in one tiny in-the-ear device.

In order to increase the functionality per unit volume and simultaneously keep a given process flow (such as surface-mount technology (SMT)) the mounting, of bare die chips in an SMD-compatible form factor is possible. Sometimes termed chip scale packages (CSP), these can be mounted together with other SMD components in a high-volume capable process [6] without the need for special high-precision bonding equipment and/or ultra-fine-line substrate technology. Examples here are pacemaker or defibrillator units, which employ digital and analog high-performance chips, which in a packaged or wire-bonded format would require significantly more space and weight.

The packaging of bio-MEMS demonstrate that the choice of materials and processes for packaging are much more stringent for bio-MEMS than for pure technical MEMS. The same applies in some cases to the required density of packaging. An example are tiny medical implants. Here, the costs are less important than the size with the consequence that the most sophisticated packaging techniques can be applied [7].

11.3 Temperature Dependents of Integration for Bio-MEMS Process

Interconnect assembly process for conventional flip chips require higher process temperatures than what bio compactable materials can with stand. Table 11.2 shows a list of lead free interconnect materials and their liquidus temperatures.

Table 11.2 Lead free solder process temperatures

Lead Free Solder Material	Liquidus Temperature
High Temperature Lead Free Interconnect Materials	
Sn-37Pb	183 °C
Sn-55bi	138 °C
Sn	260 °C
Sn-0.7Cu	227 °C
Sn-3.5Ag	221 °C
Sn-3.4Ag-0.9Cu	217–219 °C
Sn–Ag–Cu–Sb	213–218 °C
Sn3.4Ag-4.8Bi	210–217 °C
Sn-9Zn	199 °C
Low temperature interconnect Materials	
Conductive polymers	150 °C

Flip chip process is relatively new for MEMS based applications. Wire bond technique has played a predominant role in the integration. With the resent insertion of flip chip into MEMS integration, assembly materials and appropriate process selection also changed. In a flip chip assembly process the type of interconnect determines the choice of process materials and parameters. Process temperatures could affect low temperature bio materials.

There are three types of current and future interconnect technologies used for Bio-MEMS process: Flip Chip Stud Bump process, Pressure Bonding Process, Lead free solder process and wire bond.

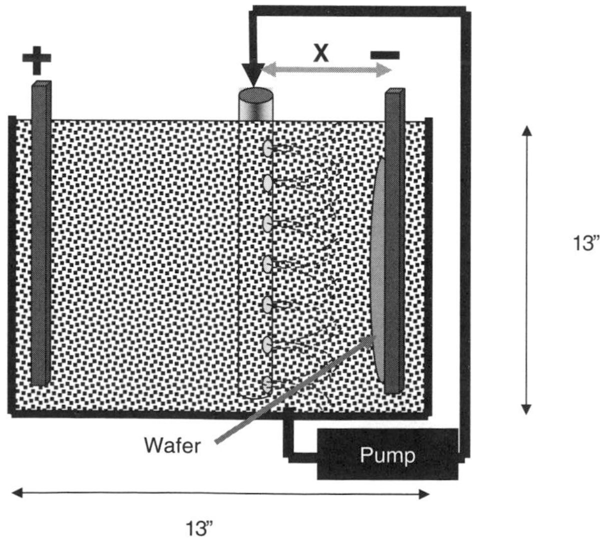

Fig. 11.5 Lead free nano composite electroplating technique (X=10 cm)

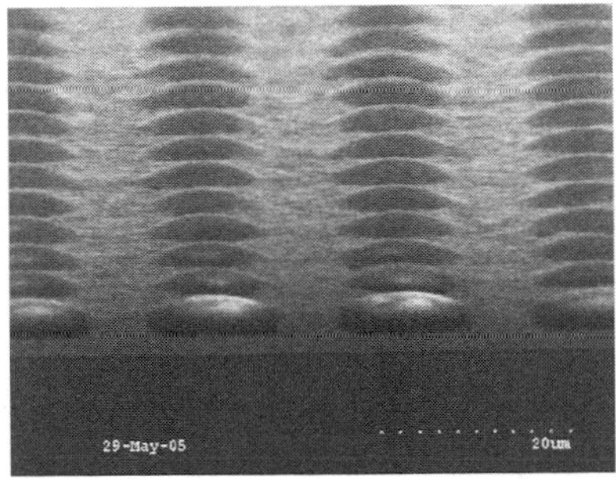

Fig. 11.6 Lead free nano composite 20 μm pitch interconnect [9]

11.4 Flip Chip Stud Bump Assembly

The bumping process starts by using a wire bonder to introduce stud bumps on flip chips. This a not a parallel process and requires the wire bonder to execute in quick steps the process of placing a studs on the bond pad. The technique is a peripheral array process and the pitch will depend on the stud bump diameter (~2x times the wire diameter). Gold stud bumps are most commonly used [7]. The wire bonder presses the sphere of gold ball on the bond pad by applying mechanical force (5–25 N), heat (125–150°C) and ultrasonic energy to create a metallic connection. The bond wire is broken just above the ball by a sharp cutter. The step is repeated on all the bond pads there by creating the stud bumps. Figure 11.7 shows the flow process of stud bumping and assembly of the flip chip.

11.4.1 Pressure Bonding Technique

In this technique solder (AuSn 80/20%) thin film material is deposited on the bond pads of the chips and the substrate. The chip is aligned and assembled on the substrate by apply pressure and temperature (278°C). Bond is created after the assembly cools down [8].

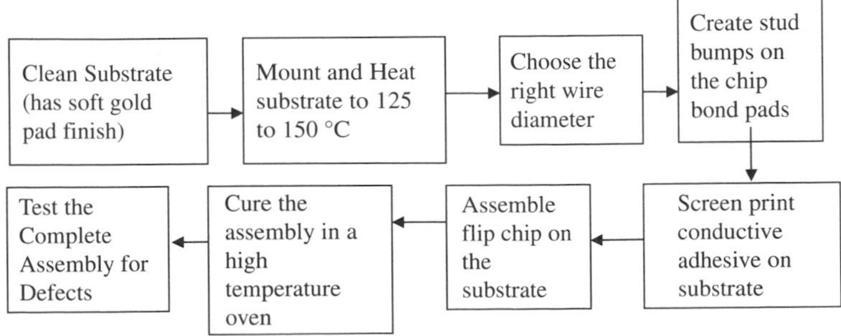

Fig. 11.7 Stud bump assembly process

11.5 Next Generation Nano Composite Interconnect Technique for Bio MEMS Systems

Electroplating requires an optimized plating rate and uniformity of thickness with fine grain size after plating. Experiments are conducted to evaluate the requirement for a seed layer during tin plating. Sputtering technique is used to deposit Ti/Sn, Ti/Cu and Ti/Au as seed layers on 4" wafers. Shipley's sulfric acid based tin electrolyte mixture is used. Pure tin is used as the anode. Care is taken to circulate the electrolyte for 4 hours so that the temperature and the electrolyte mixture stabilizes at 80 degrees centigrade. A slurry is made of the nickel nano particles, and applied on the wafer before electroplating. The slurry is left to dry in an oven at 80°C The plated nano interconnect nickel is reflowed using a profile in a 5 chamber reflow oven at 110, 125, 155, 220, 260°C. Care is taken to optimize the use of nitrogen in the reflow chamber. The nitrogen flow affects the shape of the nano interconnect. The electroplating setup used in this experiment is shown below Fig. 11.5. The distance "x" between the electrolyte spray system and the wafer is optimized to arrive uniform bump heights.

The electroplated wafer is diced and polished. Figure 11.6 shows the cross-section of Nano composite interconnect for 20 micron pitch and 15 micron pitch achieved through nano particle electroplating.

11.6 Examples Bio-Medical Packaging Applications

A prominent medical application for bio-MEMS packaging is the field of medical implants such as pacemakers, hearing aids and drug-eluting implants to name a few. It is obvious that miniaturization is a key desirable requirement for implantable devices. This is especially true for implants in very small organs or those which are inserted using minimally invasive surgical procedures where the maximum allowable size is restricted by the diameter of the working channel in an endoscope. Medical implants are often rather complex systems and can consist of many components such as power sources, transducers, control units, modules for wireless communication, etc., potentially resulting in rather bulky devices if inappropriate packaging technologies are used. Example packaging technologies suitable for miniaturized and high-density bio-MEMS includes bare die assembly techniques like flip-chip technology. These, provide thin, small, and lightweight features and can be implemented on a multitude of substrates such as ceramic, laminate, molded interconnect devices (MID) as well as on flexible substrates. Very often, these processes use materials which are not biocompatible. In this case, additional encapsulation steps are necessary to avoid a direct contact between nonbiocompatible materials and body fluids or tissue.

Another bio-MEMS category comprises biosensors or complete lab-on-a-chip systems for analytical tasks [10]. Biochips and biosensors are regarded as key elements for the development of multianalyte detecting instruments, especially of hand-held instruments for point-of-care or point-of-use testing. Compared to conventional microelectronic chips, chemosensing biochips additionally require the controlled mass transfer, i.e., the access of analytes and reagents to the biosensing surfaces of the chip. On the other hand, interferences by chemical compounds on the detection mechanism and the transducing elements of the sensor [11].

Many MEMS applications require miniature packages, for example, pressure sensor attached to the tip of the catheters used in heart-related diagnosis and surgeries, to achieve the miniaturization of the finished package, we can either decrease the die size or the volume of the packaging material itself. However, decreasing the die size even though may boost the throughout per wafer, may create difficulties in dicing and handling. Decreasing the volume of the package itself can require new materials and new processes . careful evaluations are needed and certain generic research and development on these two issues can also begin done by industry.

References

1. R. Basher, "BioMEMS: State-of-the-art in detection, opportunities and prospects," *Advanced Drug Delivery Review*, 2004: 56, 1565–1586.
2. A. A.O. Tay, M. K. Iyer, R. Tummala, V. Kripesh, E. H.Wong, M. Swaminathan, C. P.Wong, M. D. Rotaru, R. Doraiswami, S. S. Ang and E. T. Kang, "Next generation of 100 μm – pitch wafer level packaging and assembly for systems-on-package", 2004: 27, (2), 413–425.
3. M. J. Wolf, G. Engelmann, L. Dietrich and H. Reichl, "Flip chip bumping technology—Status and update", *Nucl. Instrum. Methods Phys. Res. A*, 2006: 565, 290–295.
4. S. Guillaudeu, X. Zhu and D.M. Aslam, "Fabrication of 2-μm-wide poly-crystalline diamond channels using silicon molds for micro-fluidic applications," *Diamond Rel. Mater.*, 2003: 12, 65–69.
5. H. Becker and L. Locascio, "Polymer microfluidic devices," *Talanta*, 2002: 56, 267–287.
6. J. J. Cefai and D. A. Barrow, "Integrated chemical analysis microsystems for life sciences research in space," *J. Micromech. Microeng.*, 1994: 4, 172–185.
7. I. Clausen and O. Sveen, "Die separation and packaging of a surface micromachined piezoresistive pressure sensor", *Sens. Actuators A: Phys.*, 2007: 133, (2), 457–466.
8. K-M. Chu, W-K. Choi, Y-C. Ko, J-H. Lee, H-H. Park and D. Y. Jeon, "Flip-chip bonding of MEMS scanner for laser display using electroplated AuSn solder bump", *IEEE Trans. Adv. Packaging*, 2007: 30, (1), 27–33.
9. R. Doraiswami, "Nano Composite Lead-Free Interconnect and Reliability", *Proceedings. 56th Electronic Components & Technology Conference (IEEE Cat. No. 06CH37766C)*, 2006, p. 3.
10. A. Manz et al., "Miniaturized total chemical analysis systems. A novel concept for chemical sensing," *Sens. Actuators B, Chem.*, 1990: 1, 249–255.
11. T. Velten, H. H. Ruf, D. Barrow, N. Aspragathos, P. Lazarou, E. Jung, C. K. Malek, M. Richter, J. Kruckow and M. Wackerle, "Packaging of Bio-MEMS: Strategies, technologies, and application", *IEEE Trans. Adv. Packaging*, 2005: 28, (4), 533–546.

About the Cover

The author gratefully acknowledges the cover photograph by Felice Frankel, Artist in Residence at the Massachusetts Institute of Technology and coauthor of *On the Surface of Things: Images of the Extraordinary in Science*.

This particular image, taken with Nomarski optics, presents a wafer-bonded piezoresistive pressure sensor. It is fabricated in the sealed-cavity process developed by Professor Martin Schmidt of the Massachusetts Institute of Technology with his graduate students, Lalitha Parameswaran and Charles Hsu. The piezoresistors are clearly visible, and the slight contrast across the central diaphragm region shows that the diaphragm is actually slightly bent by the pressure difference between the ambient and the sealed cavity beneath.

Index

Printed in the United States of America.